Astrolabes and Angels, Epigrams and Enigmas

BOETHIUS

Texte und Abhandlungen zur Geschichte der Mathematik und der Naturwissenschaften

Begründet von Joseph Ehrenfried Hofmann, Friedrich Klemm und Bernhard Sticker

Herausgegeben von Menso Folkerts

Band 56

Astrolabes and Angels, Epigrams and Enigmas

From Regiomontanus' Acrostic for Cardinal Bessarion
to Piero della Francesca's Flagellation of Christ

An essay by David A. King

inspired by two remarkable discoveries
by Berthold Holzschuh

Franz Steiner Verlag 2007

Bibliografische Information der Deutschen National-
bibliothek
Die Deutsche Nationalbibliothek verzeichnet diese Publikation in der Deutschen Nationalbibliografie; detaillierte bibliografische Daten sind im Internet über <http://dnb.d-nb.de> abrufbar.

ISBN 978-3-515-09061-2

ISO 9706

Gedruckt auf säurefreiem, alterungsbeständigem Papier.
Druck: Printservice Decker & Bokor, München
Printed in Germany

With gratitude to
Asger Aaboe, Frans Bruin, Bernard Goldstein,
Willy Hartner, Ted Kennedy, Otto Neugebauer,
Olaf Pedersen, David Pingree, Derek Price,
Carl Schoy, Gerard Turner and Ernst Zinner,
who, directly or indirectly, taught the author
how to look at historical astronomical tables and instruments,
if not how to look at Renaissance paintings.

Notes to the reader:

This text was produced in camera-ready copy on the author's Apple Mac OS X using MSWord 2004 for Mac. Even without the illustrations, the text with its sheer size, complex formatting and numerous cross-references was too much for the above combination – hardware, software and author – and several problems occurred. The culprit may be the Arabic transliteration font that I was using for a previous publication, from which I borrowed some items of the bibliography.

The numbering of the footnotes is automatic and hence supposedly consecutive, but in this text there are some gaps. This happened when sections of text with footnotes were moved from one place to another: ghost footnotes remained in the old positions at the same time as the "real" footnotes moved to the new positions. Cross-references to the original footnotes, however, remained as they were before, so that they became references to ghost footnotes. Where there are gaps in the footnotes, this is indicated by * for one missing footnote or ** for more than one. Cross-references to footnotes are automatic and a check has been made that there are no remaining cross-reference to ghost footnotes. Several dozen of the latter had to be corrected by hand.

Also, a few footnotes jumped forward a page or two, and large gaps appeared in various parts of the text, some of which alas remain.

The author and the publisher agreed that it seemed preferable not to reformat the text entirely to try to remove all of these inconsistencies.

Greek letters are in SYMBOL font, and are thus devoid of accents. Also a regular *sigma* is used for final *sigma*. Distinguishing features of certain Polish and Turkish letters have been suppressed and the transliteration of Arabic observes only the long vowels.

Summary:~

In Rome in 1462, the young German Regiomontanus, the leading astronomer in Europe, presented to his new patron, the ageing Greek Cardinal Bessarion, a beautiful little astrolabe, that is, an astronomical instrument representing the three-dimensional universe in two dimensions. This astrolabe bears a complex Latin inscription in verse, geometrically arranged, and in the form of an acrostic, with hidden meanings in eight vertical axes. The epigram is intimately linked to the image of an angel, representing the angelic patron saint of the Cardinal. Now that these hidden messages have been interpreted for the first time, we find that the epigram refers to two different astrolabes and two different Bessarions; features two different people named Ioannes; combines the names of two different people, one alive and one long dead, in a single word; and conceals four different dates in a single group of letters and numbers. Bessarion must have been very pleased with this ingenious gift containing hidden messages that only he would have understood. This was Regiomontanus' agenda: nothing more, nothing less.

However, when Bessarion saw the epigram and the angel, and perhaps with a little help from his friends, he saw things that Regiomontanus could never have intended: out of these was developed a scenario for an extraordinary painting featuring Bessarion's own personal and political agendas.

The epigram reveals that each and every one of the eight persons to be represented in the painting, as well as the classical god on top of the column behind Christ, might have had double or multiple identities, at least in the minds of those who conceived the painting. Their names can be derived from groups of letters, something like Byzantine Greek and Christian Latin monograms, in the appropriate places in Regiomontanus' epigram. One image, recalling the angel on the 1462 astrolabe, is typical of Piero's angels: this is Regiomontanus, but he embodies three other young men close to Bessarion, all deceased. There are now four persons named Ioannes (Latin), Ioanne/is (Greek) or Giovanni (Italian), and altogether over 20 persons are represented in the painting, as well as four classical gods.

The painting is "The Flagellation of Christ" by the mathematician-artist Piero della Francesca, datable about the same time as the astrolabe. It is the most enigmatic and controversial painting of the Quattrocento. The painting, whose basic geometry is closely related to that of the Regiomontanus' epigram, represents three-dimensional space in two dimensions by means of perspective. Further, it depicts several different scenes in two main groups of images. All previous interpretations have accepted that two of the eight persons represented have double identities: the figure of Christ somehow also represents the Church and the now lost Byzantium, and the person sitting on the throne of Pontius Pilate watching the symbolic flagellation is the Byzantine Emperor Ioannis VIII.

Now interpreted for the first time using a contemporaneous document, Piero's symbolic Flagellation of Christ, as well as of the Church and Byzantium, is intended to portray important aspects of the life of Bessarion, the main figure, including his feeling that the failed Union of the Churches had led to the fall of Byzantium at the hands of the Turks, his opinion that the Greek heritage was being misinterpreted by his major opponent, his mourning the recent loss of his brilliant godson, and his hope for the future personified in his new *protégé*, Regiomontanus.

The epigram with its hidden messages and the painting with its concealed personalities are two of the most remarkable intellectual achievements of the Renaissance. Both are monuments in the history of mathematics as well as in the history of art. The challenge is to explain how the former might have inspired the latter. We cannot claim that the epigram was interpreted some 450 years ago precisely as we have interpreted it here. However, we hope to have established at least that Regiomontanus' epigram is the key to understanding Piero's painting. All interpretations of the former will be hypothetical, as have been all interpetations of the latter until now. What I present is perhaps the 41st hypothesis regarding the painting, but it is the first to be based on a contemporaneous document. My findings for the three "moderns" confirm what some of the leading scholars have already suggested, and they show how those scholars could have come to mutually contradictory findings. The only new major player in the painting is Regiomontanus, and it was he who provided the document on which my hypothetical identifications of all of the players is based. He fits so well into the painting that his presence there could have been hypothesized even without the epigram.

Table of contents:~

PRELUDE

"Die ungelösten Probleme halten einen Geist lebendig und nicht die gelösten."* Erwin Guido Kolbenheyer (1878-1962).

The astronomical instruments of the Renaissance in Europe, like those of the earlier Islamic Renaissance[1] and the European Middle Ages, are scientific works of art. One of the most remarkable of these is a beautiful little astrolabe that was presented in Rome in the year 1462 by the young German mathematician-astronomer Regiomontanus to his ageing patron, the Greek Cardinal Bessarion: see **Figs. I-1** and **I-2**. The dedication on this is an almost perfect elegiac couplet.

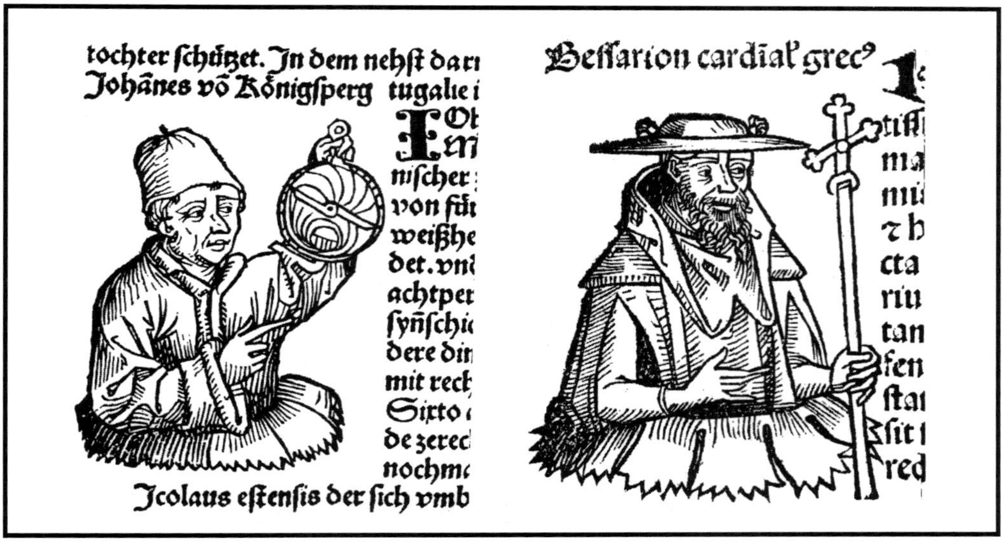

---

* "It is the unsolved problems that keep the mind alert, not those that have been solved already."

1 It is customary to use the term Renaissance only in relation to "our" reappropriation of the Greek and Latin heritage. When "others" appropriate things like Indian trigonometry and Iranian astrological doctrines – even when these are subsequently adopted by "us" and we do not know that they were originally Indian or Iranian – we are less interested. On astronomy during the Islamic Renaissance see further App. H.

I-1:~ Johannes Müller of Königsberg (b. 1436), known as Ioannes de Monte Regio or de Regio Monte, and, after his death, also as Regiomontanus, as depicted by the Nuremberg doctor Hartmann Schedel (1440-1514) in his *Liber chronicarum*, published in Nuremberg in 1493. The astronomer was in Nuremberg from 1471 to 1475 and was surely well known to Schedel, and so this is possibly what Regiomontanus looked like in his late 30s. He died in Rome in 1476, aged but 40 (see the end of Part IV below). [From *Venice 1994 Exhibition Catalogue*, p. 104.]

Fig. I-2:~ Cardinal Bessarion as he is shown in the Latin edition of the same work, also published in Nuremberg in 1493. He is faithfully represented, as we can judge from several other portraits, but very much showing signs of age. As we shall show, he was born in 1407, and it is well established that he died in 1472, so that he reached age 65. [From *Venice 1994 Exhibition Catalogue*, p. 292.]

Now why would Regiomontanus, renowned as a poet, not have composed a *perfect* elegiac couplet to serve as his dedication? This, as we shall see, has caused some problems in the recent past. However, in February 2005, Berthold Holzschuh realized that Regiomontanus' epigram is a brilliant acrostic with vertical axes containing "hidden meanings" and that it also has a sophisticated geometrical structure. This discovery immediately solved a problem that had been nagging this author for years, for I had published the instrument several times but never properly dealt with the inscription, which makes enough good sense in Latin to accept it at face value. But Kolbenheyer's *dictum* is universally valid, and I was not at ease with the Latin epigram, which Holzschuh had discussed some years previously in the same seminar. This was the reason I asked Holzschuh to make *yet another* short presentation on the epigram in February 2005. What he discovered – that the epigram is an acrostic geometrically arranged – was material for a significant paper in the history of Renaissance instruments.

In May 2005, however, Berthold Holzschuh found a similar geometrical structure in the most enigmatic and most controversial painting of the Quattrocento, "The Flagellation of Christ" by the mathematician-artist Piero della Francesca.[2] In this, the scene of the flagellation, which looks very contrived, is a background to three "moderns", a bearded foreigner, an angelic

2 The two discoveries were announced by Berthold Holzschuh at the conference "Mandarini Bizantini – Il mondo intellettuale tra Oriente e Occidente (XIV-XV secoli)" held at the Fondazione Giorgio Cini in Venice in June, 2005. Alas, no text was prepared for the *Proceedings*.

youth, and a potential sponsor. If there was indeed a connection between the epigram and the painting, the consequences were monumental. First of all, as Holzschuh immediately hypothesized, the bearded man could indeed be Bessarion and the young man in red his new *protégé*, Regiomontanus!

This author was confronted with a challenge, namely, to try to present the connection between the epigram and the painting and its consequences in such a way that might be convincing to a sceptical reader. This challenge was all the the greater as I realized that the connection and the consequences were far more complicated than Holzschuh or I first thought, and the more I wrote, the more complicated they became. The task is easier, and the results somewhat more convincing, with the appropriate graphics.

I shall attempt to show that the epigram with its eight vertical axes is intimately related to the painting with its eight persons depicted. I shall also show how the former can be used to identify all of the persons featured in the latter. Now identifying people in historical paintings is a very dangerous game, and I would not venture to begin to do this without a credible contemporaneous document. All previous attempts to identify the persons in the painting have assumed that, essentially, there is only one person to be associated with each image. Two exceptions are the figure of the Byzantine Emperor, Ioannis VIII, who apparently represents Pontius Pilate, and the figure of Christ, who surely also represents the Church, or at least the Eastern Church. Otherwise, scholars have tended to concentrate on identifying the three principal figures on the right. The epigram is not so much the key to understanding the painting, as it is the key to understanding the program that led to the painting.

I maintain first that the epigram on the astrolabe provided the ideas for the composition and the basic geometric structure of the painting, and second that each of the eight images across the painting represents at least two persons whose identities are deliberately partially concealed. In addition, the image of the classical god on top of the column behind Christ represents three gods, not one. Each of the names of these multiple personalities can be derived from Regiomontanus' epigram. The implication of all this is that Bessarion and Regiomontanus between them looked at the epigram in a certain way, not completely different from the way that I shall do, and, between them, or maybe with the help of some friends, not least Piero himself, deduced the scenario for the painting. There is actually no need to assume that Piero saw the epigram or was privy to all that it revealed: he painted what he was commissioned to paint. However, we should not underestimate Piero, for, if he had seen the epigram, he could have surely contributed to the "game" of interpreting some of its letters as names or background features for his painting.

This monograph remains a limited undertaking. I am not going to rewrite any part of history except the story of an angel and an epigram on an astrolabe and the story of a painting with a sort of angel in it. If I have interpreted these correctly, this work achieves little more than that. I hypothesize that the painting, with its angel, is related to the angel and epigram on the astrolabe. Interested readers will be guided to the more detailed studies available on the lives and works of the three principal persons involved – a German astronomer, a Greek cardinal and an Italian painter. My introductions to these individuals will necessarily focus on the encounters or exchanges, documented or undocumented, that took place, or which I think took place, between them. I do not intend to discuss how the astrolabe was made, and neither do I intend to discuss how the painting was achieved in practice. I shall, however, discuss how I think that the epigram on the astrolabe was conceived, and how I think that the programme for the painting was conceived. This is all highly speculative, and others might have come up with a completely different scenario starting from the epigram, some concluding that there is no connection whatsoever between the epigram and the painting. But much nonsense has been said and written about the 1462 astrolabe, and most of what has been written about "The Flagellation" has been highly speculative, much of it off the wall, and none of it based on any contemporaneous document. Now, for the first time, we have such a "document". What I have done in the interpretation of the document still does not prove that the epigram and the painting are related, and my general thesis remains nothing more than a hypothesis. I happily leave it to others to establish that what I claim to have found in the epigram is the result of wishful thinking on my part, or to demonstrate that the epigram and the painting are not really related at all.

I will apologize at the outset only to my colleagues in the history of mathematics, for there is much more that could have been written in this study about various Euclidean and Archimedean connections. However, no historian of mathematics has ever written on the geometry of the painting beyond the perspective, which is a monument in the history of mathematics. My only excuse is that I have been preoccupied with the astrolabe and the painting and the connection between them. My hope is that the recent identification of a manuscript of the works of Archimedes as being in the hand of Piero[3] and these new insights on his painting may arouse the interest of colleagues in the history of mathematics.

To delve more deeply into the background of the contemporaries of Bessarion represented in the painting would be far beyond my competence and I leave the task to others more familiar with the Quattrocento. However, readers will soon find that what I present goes beyond the realms of what is known about the Renaissance, for we are dealing with three of its most brilliant minds. Two of

---

3 See the text to n. 177 below.

these men were gifted in Latin and Greek, one more gifted in astronomy and mathematics, the other more in philosophy and theology, both gifted in poetry, one coming to terms with the heritage of the German-speaking world and the Humanism of Vienna, the other steeped in the Byzantine Greek heritage and the scene of the Lateran College in Rome and the courts of Italy, and both – as we shall see – fond of letters, numbers and games. In their games the two also had a great deal of luck, for much of what they achieved with letters and numbers was the result of pure coincidence. Out of these very serious but still rather playful games with letters, numbers and images, I maintain, stem the two most remarkable scientific works of art of the Renaissance: an astrolabe and a painting. The latter could only have been achieved by the third of them, a familiar of popes, cardinals and princes, who was as much a mathematician steeped in Euclid and Archimedes as he was the leading artist of the Quattrocento. It is small wonder that both the astrolabe and the painting have caused scholars so much trouble.

Regiomontanus' epigram was surely one of the most brilliant and most complex epigrams or acrostics ever conceived in the Renaissance. There appears to be nothing with which one can compare it.[4] Further, I am confident that no scholar of Renaissance epigrams has ever written anything of consequence on this one.[5] It is not an exaggeration to assert that only in the past 25 years scholars have started to realize that acrostics and configured letters are an important part of our

---

4 See the collection of studies of classical and neo-classical epigrams reprinted in Pfohl, ed., *Das Epigramm – Zur Geschichte einer inschriftlichen und literarischen Gattung*, or, say, Bradner, "The Neo-Latin Epigram in Italy in the 15th Century", and Hausmann, "Untersuchungen zum neulateinischen Epigramm Italiens im Quattrocento". Acrostics are not mentioned in Hutton, *The Greek Anthology in Italy*, although an early-17th-century Italian book on the epigram as a literary genre has a chapter on acrostics as well as an acrostic on the Pope (p. 68). IJsewijn & Sacré, *Companion to Neo-Latin Studies*, II, pp. 129-131, on figured poems and acrostics, offers nothing to compare with Regiomontanus' epigram.

The serious study of neo-Latin texts is a relatively young discipline. See IJsewijn, *Companion to Neo-Latin Studies*, with a second volume published jointly with D. Sacré on "literary, linguististic, philological and editorial questions", and the journals *Humanistica Lovaniensia – Journal of Neo-Latin Studies* and *Neulateinisches Jahrbuch – Journal of Neo-Latin Language and Literature*, and, for the German world, the collections of studies in Keßler & Kuhn, eds., *Germania latina – Latinitas teutonica*, and Ludwig, *Miscella Neolatina*. For a corpus of texts see Kühlmann, "CAMENA – Lateinische Texte der Frühen Neuzeit" on the Internet. A conference "The Neo-Latin Epigram, 15th-17th Century: Towards the definition of a genre", was held at the Royal Dutch Institute, Rome, during April 12-15, 2006.

5 See n. 527 below.

cultural heritage.[6] Particularly the works of the 9th-century Benedictine monk and later Bishop of Mainz, Hrabanus Maurus, and especially his *Liber de laudibus sanctae crucis* with its 28 brilliant and spectacular configured poems based on the Cross, have now attracted the attention they deserve.[7] It is not

---

6 I suspect that few neo-Latin specialists will like Regiomontanus' acrostic. This situation is hinted at in the article "Acrostics" in the *Dictionary of the Middle Ages*, where it is stated:

> "Man's fascination with the physical and semantic properties of words can be illustrated in all ages, … … . More difficult to parallel is the technical complexity sometimes achieved in medieval Latin examples (of acrostics and wordplay) and the apparent seriousness with which literate society treated works frequently dismissed by modern commentators as frivolous excess."

This attitude problem amongst specialists is well exemplified by a remark by Edward Courtney ("Greek and Latin Acrostichs", p. 5, my italics), when referring to the man who influenced Alcuin and Hrbanus Maurus (see n. 7 below), namely Porphyrius (Constantinople, *ca*. 260-335):

> "The surrealist patterns of Publilius Optatianus Porphyrius *have lost all contact with reality and do not merit discussion*."

W. Levitan ("Optatian Porfyry", p. 246, his italics) is far more generous because he understands:

> "Frankly, (Optatian's) poems make entirely unremarkable, even banal reading – competent verse for the most part but repetitive and very tired. But it is not *reading*, as the word is commonly understood, that the poems invite; rather *wonder*, to say the least, at the appalling genius responsible for them, a genius that won Optatian a fair measure of renown, if we can judge from the testimonia, and allowed his work to survive in the Middle Ages in some two dozen copies. Optatian is not a good poet; he is not even a bad poet. His poems are prodigies, monsters in the literal sense. They are extreme and elaborate examples of the ancient technopaegnion, a type of literary *grande pirouette* that first appears in the West in the third century BCE and has had a very long history not only in Europe but in literate cultures throughout the world. And in this history, the figure of Optatian looms large. In fact, of all similar writers of every age and race, Optatian is supreme: this, at least, in the opinion of one of his editors, Lucian Mueller. We may be inclined to humor an editor's partiality, but I, for one, would be very much surprised if in this judgment Mueller were mistaken."

For another author cognisant of the historical importance of the poems see Barnes, "Publilius Optatianus Porfyrius".

7 For medievalists, at least, the 1980 exhibition in Mainz on the works of Hrabanus Maurus was an eye-opener: see *Mainz 1980 Exhibition Catalogue*. This was followed by an exhibition on historical configured letters in Wolfenbüttel in 1987-88: see *Wolfenbüttel HAB 1987-1988 Exhibition Catalogue*. In February, 2005, the Frankfurt Seminar was privileged to see a second exhibition in Mainz on Hrabanus Maurus featuring the Vatican manuscript of his *Liber de laudibus sanctae crucis*: see *Mainz 2005 Exhibition Catalogue*. On the author see the articles "Hrabanus Maurus" in

unreasonable to suppose that they were known at the monastery of Klosterneuburg outside Vienna, where Regiomontanus might have made his astrolabe. Also, Bessarion came from a milieu in which acrostics were part of daily life.[8]

Furthermore, whilst I am fully aware of the dangers of focussing on a single painting of a major artist, I accept the challenge presented by this one, for it is unlike any other. I am confident that others will be better able to construct a more carefully documented historical context for the painting once the new documentary evidence relating to the painting known as "The Flagellation of Christ" – a modern title – is assessed.

I have felt bound to present the discoveries in a form in which at least Holzschuh's conclusions and my subsequent conclusions are clearly recorded.

The two landmark astrolabes that I shall discuss – one with Greek inscriptions that Bessarion brought to Italy from Constantinople and the other with Latin inscriptions that Regiomontanus made for Bessarion – have been either much abused in the scholarly literature or deliberately omitted from it, in each case as the result of various misunderstandings that I shall attempt to correct here. I alas have to be prepared that colleagues with knowledge neither of 15th-century Viennese instruments – there are some 45 known from 15th-century Vienna – nor of fake astrolabes – I assert that there are none from early 19th-century Italy – will claim once again that the astrolabe supposedly made by Regiomontanus is a fake, as was maintained in England (but not elsewhere) in 1990 (and disproved shortly thereafter). I find myself in the unenviable position of having to defend the astrolabe against its erstwhile detractors so that no misapprehensions may cloud its connection to Piero's painting.

Furthermore, no less detailed account would do justice to my interpretation, inevitably subjective and hypothetical, of this new evidence relating to Piero's painting. I have to be prepared that art historians will claim that I have no knowledge of art history and so cannot contribute to the subject. What nonsense to claim that a painting owes its initial inspiration to an epigram on an astrolabe! However, an epigram on an astrolabe is out of the domain of art historians, even though this one inspired the composition of a painting. So I have decided to try to put together a credible interpretation of the painting in the light of what I can piece together out of the debris of the literature that has been written on it, much

---

Lexikon des Mittelalters and *Dictionary of the Middle Ages*, also Haarländer, *Rabanus Maurus*, and North, *The Ambassadors' Secret*, pp. 375-380. See also n. 254 below.

8 See Krumbacher, "Die Akrostichis in der griechischen Kirchenpoesie", and Weyh, "Die Akrostichis in der byzantinischen Kanonesdichtung".

10 * Walker, "Esoteric Symbolism", pp. 218-220, quoted in the text of Interlude I between Parts 1 and 2 below (see n. 169).

of which, fortunately for me, need not even be reviewed. I shall, however, present an up-to-date list of previous hypotheses and look at some more closely.

This study is an essay in deciphering for the first time two cryptic enciphered "documents" – uncovering the hidden meanings of a Latin epigram and identifying the hidden persons in a painting – and the use of geometry in both a configured text and a very carefully planned painting. Two kinds of coded scripts were used in this period.[10] In one variety, ciphers of one sort or another are used so that the text is accessible only to those who know, or can crack, the code.[11] In the other, which, fortunately for us, is the kind conceived by Regiomontanus, the text reads like a normal text but has hidden meanings for those like Bessarion with eyes to see them. As we shall see, if we define a disordered matrix of letters, something akin to Byzantine Greek or Christian Latin monograms but with less order, as an acceptable kind of code, Regiomontanus' epigram, once deciphered, turns out, *mainly quite by chance*, to contain materials for a quite different scenario from the one that he originally intended.

Such activities are very much in the spirit of the *illuminati* of the Quattrocento, the century of Leon Battista Alberti (1404-1472), the father of modern cryptography.[12] Both Regiomontanus and Bessarion had been child prodigies,

---

11 For some visually spectacular examples, conceived by Daniel Schwenter (1585-1636), professor of Mathematics and Oriental Languages in Nuremberg, in his book *Steganologia et steganographia* (1620), and by Giovanni Battista Porta of Naples (*ca.* 1535-1615) in his *De furtivis literarum notis vulgo de ziferis* (1558) and *De occultis literarum notis* (1593), see King, *The Ciphers of the Monks*, pp. 228-234.

12 See Meister, *Geschichte der italienischen Kryptographie des XV. Jahrhunderts*, and *Die Geheimschrift im Dienste der päpstlichen Kurie*, for an overview of Italian ciphers and coded scripts. On medieval coded scripts see Bischoff, "Übersicht über die nichtdiplomatischen Geheimschriften des Mittelalters", and the article "Geheimschriften" in *Lexikon des Mittelalters*. Because these works are in German and not readily accessible outside the German-speaking world, overviews of these subjects are missing from popular works in English such as Kahn, *The Code Breakers*.

The medieval numerical and alphanumerical ciphers discussed in Bischoff, "Die sogenannten 'griechischen' und 'chaldäischen' Zahlzeichen des abendländischen Mittelalters", had a particularly colourful history from Antiquity to the 20th century that is now documented: see King, *The Ciphers of the Monks*.

Until recently, there was no proper survey of Islamic cryptography. As Kahn, *op. cit.*, p. 93, wrote:

> "Cryptology was born among the Arabs. They were the first to discover and write down the methods of cryptanalysis".

The important work listed as Mrayati & Mir 'Alam & Tayyan, *Arab Cryptology*, containing translations of three of the most significant Arabic treatises, is now published

both were brilliant linguists and poets,[13] and both were intellectually versatile: even Regiomontanus could lecture on classical Latin poetry,[14] and whilst Regiomontanus apparently computed an entire ephemeris – solar, lunar and planetary positions for each day of a complete year – when he was 11 years old,[15] even Bessarion could compute planetary positions.[16] It is therefore hardly surprising that this double project of ours should involve many complex procedures and concepts. Similar complexities have been encountered previously when the attention of historians of science has been drawn to a Renaissance artefact or painting that has previously defied interpretation by art historians.[17] Also, I would argue, the methodology of the history of the exact sciences – based on texts and linguistic insights, and access to the appropriate primary and secondary sources – can help to unravel complicated historical phenomena in

---

but alas is not easily accessible. See the other references to Arabic cryptography in King, *The Ciphers of the Monks*, pp. 72-81.

13 On Bessarion and his Greek poetry see Ronchey, "Bessarione poeta".

14 Zinner, *Regiomontanus*, pp. 78, also 40, 73, 130, 276, 309. See also the text to nn. 44 and 152 below.

15 *Ibid.*, p. 13.

16 Tihon & Mercier, *Pléthon – Manuel d'astronomie*, pp. 118-127 and 243-249.

17 I list five examples:

Hartner, "The Mercury Horoscope of Marcantonio Michiel of Venice" (1955); and *idem*, "The Vaso Vescovali in the British Museum – A Study on Islamic Astrological Iconography" (1975).

Neugebauer, "From Assyriology to Renaissance Art" (1989), traces a single parameter, the mean length of the synodic month, from cuneiform tablets to a Greek papyrus fragment of a Babylonian ephemeris, to the Jewish calendar, and on to an early-15th-century *Book of Hours*. More information on the last is given in *idem*, "Astronomy in the Très Riches Heures".

Pingree, "A New Look at *Melencolia I*" (1980), identifies the astrological significance of various features in a print by Albrecht Dürer. On p. 257 he wrote:

> "Upon recently reading William Heckscher's fascinating study of Camerarius's description of Dürer's *Melencolia I*, I was led to consider some interpretations of this enigmatic print which I believe have not been noticed hitherto. I present them in this note for others more expert than I both in iconography and in Dürer's art to appreciate or to deprecate. Certainly I do not pretend to have any evidence that these meanings were intended by the artist. The basis of the new interpretations is a consideration of the composition as a whole, whereas previously scholars have tended to concentrate on individual elements. ... ... "

Bennett, *The Measurers – A Flemish Image of Mathematics in the Sixteenth Century* (1995).

North, *The Ambassadors' Secret – Holbein and the World of the Renaissance* (2002/2004). See also North's new interpretation of the geometry underlying Piero's *The Baptism of Christ*, *ibid.*, pp. 95-102 and 361-370.

completely different fields, such as religious iconography.[18] And at least we are used to the situation where a single discovery can completely change what we know about a particular topic.[19] Furthermore, we are used to the fact that the very existence, let alone the commission or execution, of artefacts of major historical significance went mostly unrecorded in any textual sources, contemporaneous or later. This is the case of hundreds of instruments made

---

18 See, for an example, Panaino, *I Magi evangelici*.

19 One example from Antiquity is the complex geared mechanism discovered off the island of Antikythera in 1900 by Greek sponge-divers. This remarkable object, which dates from *ca.* 80 B.C., completely changed what we knew about ancient Greek technology. See Price, *Gears from the Greeks*, and a new study to appear in *Nature*.

Another example from Antiquity but with relevance to the entire Islamic Middle Ages into the modern period, is the discovery that the rectangular base of the Ka'ba in Mecca is astronomically aligned. This was known in the Middle Ages, and it accounts for the often surprising orientations of many medieval mosques. Its discovery led to the realization that these mosques were indeed oriented towards Mecca, but by methods that were quite distinct from those of the mathematicians and astronomers of medieval Islam, who calculated the direction of Mecca using sophisticated formulae and geographical coordinates. See Hawkins & King, "The Orientation of the Ka'ba"; King, "Astronomical Alignments in Medieval Islamic Religious Architecture"; and *idem*, *In Synchrony with the Heavens*, VIIa.

Another example, this time from the Islamic Renaissance, is represented by a group of world-maps centred on Mecca, engraved on brass and fitted with a sophisticated grid preserving direction and distance to the centre. Since the three known world-maps of this kind, located between 1989 and 2001, date from the 17th century, it was necessary to hypothesize an earlier date for their inspiration. Meanwhile, texts have been found from 10th-century Baghdad and 11th-century Isfahan in which the underlying mathematics is documented. (I had been looking in texts on the mathematical determination of the *qibla* or sacred direction toward Mecca; the relevant texts, dealing primarily with conic sections, were located by Prof. Jan Hogendijk of Utrecht.) These maps completely change our understanding of Islamic mathematical geography. See King, *Mecca-Centred World-Maps*, and *idem*, *In Synchrony with the Heavens*, VIIc.

An example from the European Middle Ages is a geared astrolabe from 14th-century Hainault, which came to my attention only in April 2006. This is extremely carefully constructed and will force a complete reappraisal of astronomical and technical knowledge in this region. One similar object from N. France, far less carefully made, has been known for decades, and the only other known astrolabe from the region, with inscriptions in the Picard dialect of Old French and numbers engraved in monastic ciphers, has similar technical defects (which do not lessen the historical imortance of the instrument). See further the text to n. 431 below on the geared astrolabe, and n. 433 on the other two pieces.

An example from the European Renaissance is the corpus of illustrated texts on mechanical devices from Siena that predate the diagrams of Leonardo da Vinci, and whose discovery forced a reassessment of Renaissance Italian engineering. See Galluzzi, *Les ingénieurs de la Renaissance de Brunelleschi à Léonard de Vinci*.

between 750 and 1500, including, inevitably, the 1462 astrolabe of Regiomontanus.[20] And this is the case of the majority of Renaissance paintings too.

By May 2006, 60 million copies of Dan Brown's novel *The Da Vinci Code* had been sold worldwide, and a film had been released. This was very much a case of the audience deserving the book (as one commentator put it, "them" writing for "them"). Now Brown's novel is about the deciphering of codes, those particular codes being conveniently written in English and fairly transparent at that. The present venture is somewhat more complicated, for we are dealing with a coded text in Latin that even when decoded could still reveal a lot more. It will help the reader to have secondary-school Latin; to understand what an astrolabe is; to be aware of the significance of Cardinal Bessarion; to recognize the historical and religious significance of Christ's Passion; and to be aware of the importance of Piero della Francesca in the history of art. The bibliographical notes at the end of this study are intended to guide the reader to works on the lives and significance of the main persons in our drama; the lives of saints that nobody has ever heard of; detailed discussions of astrolabes; the "golden section" or *proportio divina*; acrostics and configured letters; the standard abbreviations in Latin that were the spice of life for the *litterati* of the Renaissance; as well as to summaries of the diverse earlier hypotheses about the iconography, environment and dating of Piero's painting. Appendixes A-G contain more general information on the astrolabe and more details about those mentioned in this study.

We should try to imagine the excitement that Regiomontanus must have experienced when Bessarion showed him an old Byzantine astrolabe, and the satisfaction that Bessarion must have enjoyed when Regiomontanus presented him with a brand new astrolabe. We should try to imagine Bessarion's delight as he unravelled all that Regiomontanus had intended with its coded message. However, between the two of them, they found more in the epigram than Regiomontanus ever intended, enough, eventually, to inspire a complex painting. The persons who conceived the scenario for the painting must have had a lot of fun together.

This is a study of people and their emotions and their intellectual games as much as it is a study of three inanimate objects. The objects, however, are waiting to speak to us.

---

20 Only one scientific instrument from the period 750-1500 known to me is mentioned in a text, in this case, contemporaneous: see the text to n. 406 below. Slightly later instruments that are mentioned in texts are viewed with suspicion: see the text to n. 434 below.

22 * On St Bessarion see the text to nn. 101-106.

As I shall attempt to show, it is necessary to understand what Regiomontanus achieved in his astrolabe epigram before one can begin to understand Piero's painting. And in order to understand what Regiomontanus was doing, we have to start with an old Byzantine astrolabe.

PART I:~ CARDINAL BESSARION AND HIS TWO ASTROLABES

> "The large astrolabe (dated 1062) in the Museo dell'Età Cristiana at Brescia is remarkable not only because its Byzantine origin makes it an object of great rarity, but also on account of the inscriptions which it bears. One of these, in five iambic verses, declares the general uses of the instrument, ending with the name and nationality of the person who had it constructed." O. M. Dalton, "Byzantine Astrolabe" (1926), p. 133.

> "The antique astrolabe (dated 1462) here described is of unique historical and sentimental interest. ... ... It is a wonderful thing that this most significant historical relic should have been preserved ... ." Derek J. de Solla Price "The First Scientific Instrument of the Renaissance" (1959), pp. 26 and 30.

From Eastern monk to Roman cardinal, from Trebizond to Vienna

In Constantinople in 1423, when the gifted young Basileios from Trebizond became a monk, he followed the custom of the Order of St Basil (the English form of Basileios) in adopting the name of a saint whose name started with the same letter as his own. He chose the name Bessarion, after a 5th-century Egyptian desert-father.[22] He surely noticed at some stage that all but one of the letters (the "L") in the Latin form of the name BASILEIOS are contained in the name BESSARION. This distinctive name would serve him well during his later Italian period, and he sometimes used the name Ioannes (Latin) or Ioannis (Greek).[23] The reason for this appears to be that:

---

23 This has led to some confusion in the 20th-century literature, which explains, for example, why "John Bessarion" is to be found on the Internet. The topic is discussed in Bianca, *Bessarione*, pp. 141-149, but, inevitably, without the author realizing that Bessarion actually used two first names, let alone why.

(1) the standard medieval Latin abbreviation of that name, IO,[24] is embedded in both the names BasileIOs and BessarIOn, and indeed

(2) each of the letters I-O-A-N-E-S is embedded in the name BESSARION, as each of the letters I–Ω–A–N–H–Σ (where H, capital η, *eta*, corresponds to Latin E or I) is embedded in the Greek ΒΗΣΣΑΡΙΩΝ.

I barely need point out that Ioannes was a very popular name in both Byzantine Greek and Christian Latin traditions, as were its equivalents in the Jewish and Muslim traditions. Also, as is well known, Bessarion's teacher at Mystra during the early 1430s had been the celebrated philosopher Giorgios Gemistos (*ca.* 1355-1452), who in 1439 had also adopted another "name", Plethon (Πλεθων), apparently with the arguments that the sounds of the name of the great Plato (Πλατων) were embedded in it yet avoiding the presumptuosity of adopting the great philosopher's actual name.[25]

As we shall see, all this was a good start for some games with letters when Bessarion met the gifted young German astronomer Johannes from Koenigsperg (modern Königsberg, a small village in Franconia) whose name means "king's mountain". Born in 1436, he had adopted the Latin name Ioannes de Regio Monte. He is known to posterity by the "humanist" name Regiomontanus,[26] but he apparently did not use that name himself, for it is documented only after his death.[27] This is surprising not least because this name *also* has I-O-A-N-N-E-S embedded within it, as Bessarion would surely have taken pleasure in pointing out to him. When the two men met in Vienna in May 1460, Bessarion doubtless relished the fact that the young German also had the letters IO in his name

---

24 The standard work on Latin abbreviations is Cappelli, *Lexicon abbreviaturarum*. See in particular his introductory remarks on abbreviations by truncation, contraction and use of special signs on pp. xi-lvi. See also Bischoff, *Latin Palaeography*, pp. 150-168.

On IO for Ioannes see Cappelli, *op. cit.*, pp. 184-185.

We shall have occasion to mention the abbreviation *rctus* for *rectus* (n. 122), *VI* for Vienna and *VE(n)* for Venice (n. 139); *MA* for Maria, *MA DI* for *Mater Dei*, and *VRGO* for *Virgo* (n. 150); *9ire* for *convenire* (n. 156); and *Ia* and *Iac* for *Iacobus* (n. 559). On the use of a symbol resembling a "9" as an abbreviation see Cappelli, *op. cit.*, pp. 68-85, and also n. 156 below. On the use of abbreviations by Piero della Francesca see n. 179 below.

25 Anastos, "Pletho's Calendar and Liturgy", p. 188, quoting Schultze, *Georgios Gemistos Plethon* (not seen), pp. 72-74, and C. Alexandre's edition of Pletho's *Nomoi* (not seen).

26 There is a charming introduction to such names in Ashley, "Classical Pseudonyms".

27 Zinner, *Regiomontanus*, p. 7.

IOannes de RegIO Monte,[28] and the further coincidence that the name *Basileios*, "imperial or royal" in Greek is derived from *basileus*, "emperor" or "king", equivalent to the Latin *rex*, "king", from which is derived the word *regius*, "royal", whence the ablative *regio*. So there is a sense in which they both had the same name, in two different languages. Given the standard abbreviations for Byzantine Greek and medieval and Renaissance Latin, and the passion of intellectuals for using them, one could go further and say that each of the two men could be identified by the other, and by others in the know, by the letters B IO for B(asileios) B(essar)IO(n) and IO RGO for IO(annes de) R(e)G(i)O (monte).[29] The letters B CARD NICENI are enough to identify Bessarion, Cardinal of Nicaea,[30] Regiomontanus signed some of his letters to colleagues "IO. Germanus".[31] Alas, we do not have any letters from Bessarion to Regiomontanus. But, as we shall see, we do have a "document" composed by Regiomontanus for Bessarion, in which the former in one place accentuated the B and the IO in BessarIOn and elsewhere intended the letter B or the

---

28 Also the form Iohannes was considered correct Latin. See the signature OPVS IOHANNIS on the Giovanni di Paolo's Virgin and Child (Siena, *ca.* 1440-45): see *Frankfurt 2006 Exhibition Catalogue*, pp. 187-188 and 301 (no. 43).

29 On abbreviations in Greek see Gardthausen, *Griechische Paläographie*, II, pp. 319-352; also, for example, Guarducci, *Epigrafia greca*, I, pp. 398-406, and the article "Abbreviations" in *The Oxford Dictionary of Byzantium*. There is a website introducing such abbreviations: see Halsall, "Greek Abbreviations".

30 Labowsky, "Cardinal Bessarion's Precious Manuscript of Euclid", p. 130.

31 Published in Curtze, "Regiomontan's Briefwechsel". The extant correspondence includes (some of) his exchanges with the astronomer Giovanni Bianchini in Ferrara (1463-64), Jacob von Speyer, court astronomer in Urbino (1465), and Christian Roder in Erfurt (1471). An unpublished German translation of the non-technical parts is in Zinsmeister, "Der Briefwechsel Regiomontans", kindly made available to Berthold Holzschuh by the author. Some, but not all of these historically interesting passages are in Mett, *Regiomontanus*, pp. 69-84. For a summary of the exchange with Roder see Rose, *The Italian Renasissance of Mathematics*, pp. 102-104.

On Giovanni Bianchini see the article in *Dizionario biografico degli Italiani*, X, pp. 194-196, and Rose, *The Italian Renaissance of Mathematics*, pp. 92, 98-99, *etc.*

On Jakob von Speyer see also the text to n. 570 below and the other references there cited.

On Roder's activities within the context of mathematics and astronomy at Erfurt in the 15th century see Kleineidam, *Universitas Studii Erffordensis,* II, pp. 63-73, especially pp. 68-70. On an astrolabe possibly made in Erfurt in the preceding century (no other instruments survive that can be surely associated with Erfurt) see item III-1 in Appendix D.

For this signature see Curtze, *op. cit.*, p. 209.

combination BA or BAS for Basileios.[32] And we can be sure that both men liked the abbreviation IO bcause of its physical resemblance to the number 10, favoured by the Pythagoreans as the sum of the *tetraktys* 1 + 2 + 3 + 4, represented by an equilateral triangle of four levels of dots with apex 1 dot and base 4 dots.[33]

At the Council of Ferrara and Florence in 1438-39, Basileios Bessarion, then in his early 30s as Metropolitan of Nicaea in the Eastern Church, supported the union of the Eastern and Western Churches. The Council was a spectacle the likes of which Italy had not seen for a long time.[34] The Byzantine Emperor, Ioannis Palaiologos VIII, and the Patriarch, together with a retinue of some 700 clerics and laymen, stayed in Italy for 18 months, attending a series of splendid official happenings, including encounters with the Pope and the cardinals. The Council moved to Florence on account of an outbreak of the Plague in Ferrara. The Union was officially proclaimed in Florence in July, 1439, but, once back in Constantinople in 1440, Ioannis VIII was confronted with considerable local opposition against it. The orthodox patriarchs had a lot to lose, and they were realistically suspicious and extremely wary of those "Christians" who had ransacked Constantinople in 1204 and committed countless atrocities against the Greek population, of being subject to the Pope, and of losing out on various theological and doctrinal issues on which their opinions were unacceptable to the Western Church. The Union was publicly repudiated by the Orthodox clergy in 1441, and two crusades sent by the King of Hungary were halted at Varna in 1446 and Kosovo in 1448, respectively.[35] Ioannis VIII died in 1448 without having implemented the Union. His successor Constantine XI was a staunch supporter of the Union, and at the urging of Isodore of Kiev had it proclaimed in 1452 from the pulpit of Hagia Sophia. But this was too late: without the much sought after military assistance from Europe, Constantinople fell to the Ottoman Turks in 1453. Bessarion may well have thought thereafter that Byzantium and

---

32 For a Greek a "B" will suffice to call to mind Basileus or Basileios or derived forms thereof. See Gardthausen, *Griechische Paläographie*, II, p. 344; and also n. 451 below, item 4, for β as an abbreviation for βασιλικω in a Byzantine inscription. Likewise in Latin an "R" will suffice to denote *rex* or *regina*: when I was a child, the letter-boxes in Britain were marked "GR".

33 See Hopper, *Medieval Number Symbolism*, pp. 42-43; Grattan-Guinness, "Numerology and Gematria", p. 1587; also the article "Pythagoras" by Kurt von Fritz in *Dictionary of Scientific Biography*, especially pp. 220b-221a. See also the text to n. 147 below.

34 The splendour of the event is captured in Gouma-Peterson, "Piero's Flagellation", p. 220, where references to the secondary literature are also given.

35 On the former see "Varna, Crusade of" in *The Oxford Dictionary of Byzantium*, and the new study Imber, *Crusade of Varna*. On the latter see the corresponding *Dictionary* article "Kosovo Polje".

the Church had been "betrayed" by the Byzantine Emperor and the Ottoman Sultan, the one by his ineffectiveness and the other by force, who had come together in one, *convenerunt in unum*, almost as Christ had been betrayed by Judas, and Herod and Pontius Pilate had "come together in one" against God and His Messiah. The phrase *convenerunt in unum*, "they came together in one", was familiar from the *Psalms* and the *Acts of the Apostles*, and it was also used in the Good Friday mass.[36] In spite of these events, Bessarion remained in favour with Pope Eugenius IV, who sent a mission to Constantinople in 1439 to invite him to Rome as cardinal in the Western Church, for which purpose he arrived in Italy in 1440. Perhaps the emissary was the jurist Giovanni Bacci, clerk to the Camera Apostolica, to whom an epithet equivalent to "emissary to Caesar" was applied in a 17th-century source; however, there is no other documentary evidence that Bacci was involved with Bessarion in this way.[37]

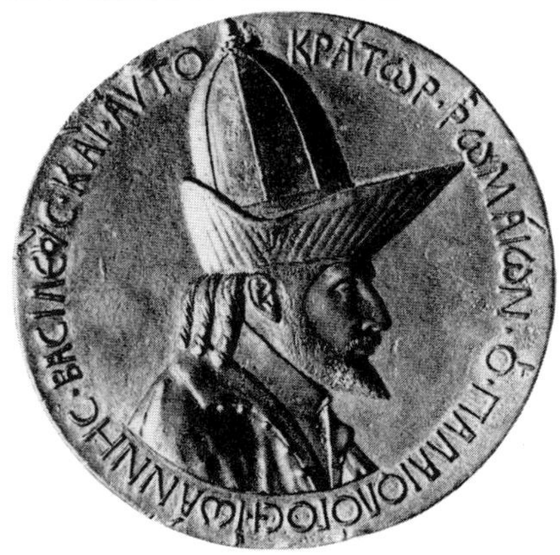

Fig. I-3a:~ A medallion by Pisanello featuring Ioannis VIII in 1438/39. [From Fiaccadori, *Bessarione e l'Umanesimo*, p. 66.]

36 The English renditions here are from the King James' edition. *Psalms* II:2: "*Adstiterunt reges terrae, et prinicipes convenerunt in unum adversus Dominum et adversus Christum eius,*" "The kings of the earth set themselves, and the rulers take counsel together, against the Lord, and against his anointed", and *Acts*, IV:26-27: "*Convenerunt in unum adversus Dominum et adversus Christum eius. Convenerunt enim ... Herodes, et Pontius Pilatus, cum gentibus et populis Israel,*" "The kings of the earth stood up, and the rulers were gathered together against the Lord, and against his Christ. ... both Herod, and Pontius Pilate, with the Gentiles, and the people of Israel, were gathered together." See further, for example, Ginzburg, *The Enigma of Piero*, pp. 49-51, 53-54 and 65-66.

37 See the text to n. 297 below.

For reasons that will become clear below, we now jump 20-odd years in the life of Bessarion. In 1460, he was still grieving the loss of his young godson Buonconte, the son of Federigo da Montefeltro, Count of Urbino. Buonconte had been a linguist gifted in both Latin and Greek, whose abilities had much impressed Bessarion, but, still in his late teens, he died of the Plague in the summer of 1458. He had been on a mission to the King of Naples with another talented youth in his teens, Bernardino, the son of Ottaviano Ubaldini dalla Carda, Federigo's right-hand man. (He was so named because his family seat was at La Carda, a hill near Urbino.) On their way back to Urbino they both succumbed to the plague: Buonconte died in Aversa, and Bernardino died in Federigo's summer residence at Casteldurante (now Urbania).[38] The effect on the Court and the city of Urbino was devastating, and Bessarion was heartbroken. We shall mention later the death of a third young man, Vangelista, son of Ludovigo Gonzaga, Marquis of Mantua, a close friend of Ottaviano and Bessarion.[39]

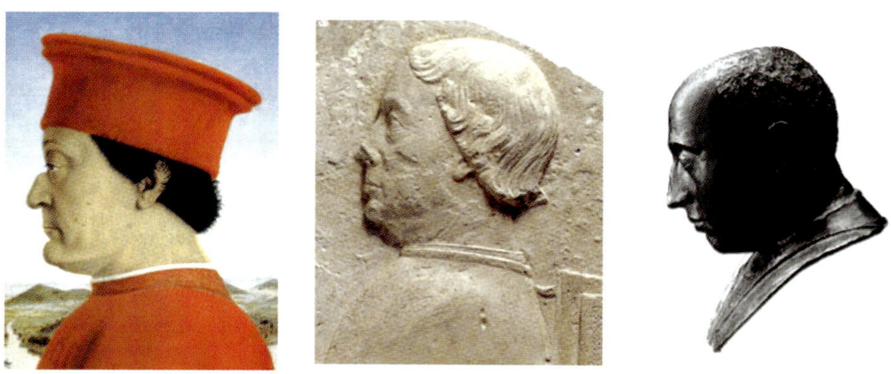

Fig. I-3b: The three bereaved fathers. From the left, Federigo da Montefeltro, whose natural son was Buonconte; Ottaviano Ubaldini dalla Carda, father of Bernardino; and Ludovigo Gonzaga, father of Vangelista. Bessarion, who was godfather to Buonconte, had close ties to each of these men. [Courtesy of the Abumax Archive, Frankfurt.]

---

38 On Buonconte see Franceschini, "La morte di Buonconte da Montefeltro"; Clough, "Cardinal Bessarion and Greek at the Court of Urbino", pp. 161-163; Monfasani, "An Epilogue to Bessarion's Relationship with the Court of Urbino", pp. 95-98, and Ginzburg, *The Enigma of Piero*, pp. 82-85, n. 89 on pp. 98-99, and pp. 117-118.

39 See the text to n. 200 below.

On February 28, 1460, Bessarion, with an entourage of 15-20 persons, arrived in Nuremberg, and on May 4 in Vienna. His principal mission was to try to persuade the Holy Roman Emperor Frederick III to launch a campaign against the Turks.[40] However, he had another purpose in going to the imperial capital, namely, to encourage the celebrated Humanist and astronomer Georg Peuerbach and his younger colleague and former student, Ioannes de Regio Monte, to complete the Latin epitome of the *Almagest* of Ptolemy (*ca.* 125), the ultimate work of ancient Greek astronomy.[41] (As a result of such ventures, the door was essentially closed not only to the many improvements to Ptolemy's astronomy that had been made by Muslim scholars, but also to the many advances in observational and computational astronomy that they had made.[42]) Vienna was the leading centre of astronomy in Europe in the mid-15th century, as well as the foremost centre for the production of astronomical instruments. Some 45 instruments from late-15th-century Vienna survive to this day.[43] As Peuerbach was on his deathbed in April, 1461, he made the 24-year-old Regiomontanus promise to complete the *Epitome*.

Bessarion probably had little difficulty persuading the young German astronomer to accompany him to Rome in order to participate in his project to render the most important Greek scientific works into Latin. Regiomontanus was forced to cancel his series of lectures on the *Georgics* of Vergil announced for the Winter Semester of 1461 at the University of Vienna.[44] The Vienna workshop would be left in the capable hands of the Benedictine monk, Hans Dorn, who had been a student of both Peuerbach and Regiomontanus.[45]

---

40 On this journey, see, in addition to the standard sources, Meuthen, "Zum Itinerar der deutschen Legation Bessarions", and Enepekides, "Die Wiener Legation des Kardinals Bessarion".

41 On Ptolemy see the article in the *Dictionary of Scientific Biography* by Gerald Toomer. On the *Almagest* see Toomer, *Ptolemy's Almagest*, and Pedersen, *A Survey of the Almagest*. A useful introduction to Ptolemaic astronomy is J. Evans, *The History and Practice of Ancient Astronomy*.

42 These remained mainly unknown in Europe until the 19th and 20th centuries, when they were investigated by orientalists. See further Appendix H.

43 See King, "Astronomical Instruments between East and West", pp. 183-188, for a preliminary list. See also the text to nn. 484 and 535 below. A new updated list is presented in Appendix D.

44 See the references cited in n. 14 above, and also the text to nn. 152-153 below.

45 Zinner, *Regiomontanus*, pp. 46 and 245, and *idem*, *Astrononomische Instrumente*, pp. 292-297. Dorn's instruments, of which the dated pieces are from 1480-86, show several of the innovations developed by Regiomontanus. See further Appendix D, especially Section VIII.

Table 1

Selective chronology of Bessarion's and Regiomontanus' activities mainly from 1460 to 1464

(For more details see Zorzi, "Cenni sulla vita di Bessarione", and Rigo, "Bessarione e Regiomontano".)

1407

Jan. 2 — Basileios born in Trebizond; studies in Constantinople before 1423.

1423

Basileios joins the order of St Basileios in Constantinople, adopting the name Bessarion and writing a panegyric of St Bessarion.

~1432

B studies with Plethon at Mystra.

1436

June 6 — Johannes Müller (= R) born in Königsberg, Franconia.

1437

B nominated Metropolitan of Nicaea.

1438-39

An imposing Greek delegation, including Ioannis VIII and B, attends the Council of Ferrara, resumed (because of Plague) at Florence. The Union of the Western and Eastern Churches was proclaimed in Florence in July 1439 but never properly implemented in Constantinople thereafter.

1439

Papal mission (Giovanni Bacci??) goes to B in Constantinople inviting him to become a cardinal, an offer that he accepts (Dec. 18).

1440

B arrives in Italy as cardinal.

1444

July 22 — Murder of Oddantonio in Urbino. His brother Fedeigo assumes control.

1448

Death of Ioannis VIII in Constantinople.

1453

	Fall of Constantinople.
	~1448-1450
	JM (= R) attends university at Leipzig.
	1450
April	JM moves to Vienna and studies with Peuerbach (b. 1423).
	1452
Jan. 16	R (= JM) awarded Bachelor's degree at age 15.
	1457
Nov. 11	R with Master's degree granted permission to teach.
	1458
Summer	Death from the Plague on the way from a diplomatic mission in Naples back to Urbino of B's godson, Buonconte, son of Count Federigo da Montefeltro, and Bernardino, son of Ottaviano Ubaldini dalla Carda, Federigo's right hand man in Urbino.
	1459
	B drafts *In calumniatorem Platonis* against George of Trebizond (printed 1469).
	B attends the Council of Mantua, with Pope Pius II present, hosted by the Marquis, Ludovigo Gonzaga.
	~1459
	Vangelista, teenage son of Ludovigo Gonzaga, dies of meningitis.
	1460
whole year	R in Vienna since 1450.
Jan. 29	B arrives in Venice.
Feb. 28	B arrives in Nuremberg, with Perotti and 15-20 attendants (also visits Worms).
May 4	B comes to Vienna as papal legate to the court of Friedrich III, stays until Sept. 1461.
	1461
April 8	Death of George Peuerbach, aged 37.
late summer	Fall of Trebizond; B receives news only at the beginning of 1462 (?)
ca. Sept. 10	B and R depart from Vienna.
Sept. 18-20	stay in Leoben.
Oct. 23	arrival in Bologna.

Oct. 31	arrival in Ravenna; B suffering from kidney stones.
Nov. 20	arrival in Rome.
	1462
date uncertain	R presents B with his new astrolabe in Rome.
Jan. 2	B's birthday.
April 26	B delivers a speech in Rome on the occasion of the presentation of the head of St Andrew (brought to Italy by Thomas Palaiologos) to Pope Pius II.
May	B taking a cure in Viterbo, visited by R.
June 6	R's birthday.
	1463
date uncertain	R presents B with a completed copy of the *Epitome* of the *Almagest*.
Jan. 2	B's birthday.
April	B nominated Bishop of Negroponte.
May	B in Viterbo, visited by R.
	1463
date uncertain	R presents B with the completed *Epitome* of the *Almagest*, completed in 1462, printed in 1496,
April 6	B writes from Rome to Cardinal Nicholas of Cusa.
April 28	B appointed Patriach of Constantinople.
July 5	B and R leave Rome for Venice, where B has been appointed papal legate *in latere*, avoiding plague-ridden Ferrara, where B had wanted to visit Bianchini on the way.
July 22	B and R arrive in Venice, staying at the monastery on the island of S. Giorgio.
July 27	R writes from Venice to Bianchini in Ferrara.
Aug. 29	B presents the reliquary of the True Cross to the Scuola della Carità in Venice.
late '63 or Jan. '64	R writes to Bianchini again, mentioning that he needs to go to Milan for an unspecified purpose (*pro certo negotio*); it is not known whether he ever went.
	1464
date uncertain	R presents B with his new work "On Triangles".
late Feb. or March	R sends a third letter to Bianchini, presumably from Venice, mentioning his encounter with Leon Battista Alberti.
April ~	R at Padua for a course of lectures on al-Farghânî, probably also for June and July.

late July	B sails from Venice to Ancona, probably with R, to greet Pope Pius II, who arrived there very sick and died there on Aug. 13; planned Crusade aborted.
late Aug.	B returns to Rome, probably with R.
Aug. 29	Pope Paul II elected by Cardinals. Regiomontanus at some later time made or had made for him a Vienna-type pocket-sundial (preserved in Nuremberg).
Aug. 29/30	R in Rome writes a treatise against the planetary theory of Gherardo da Sabbioneta.
Oct. 6	R observes an eclipse at Rome.
	1465
date uncertain	R presents his new book on an armillary instrument to B.
Feb. 15	R writes from Rome a letter to Jakob of Speyer in Urbino.
late April, May	B in Viterbo, visited by R.
June 19	R observes a solar eclipse (in Viterbo?)
August	B returns to Rome.
	1466
whole year	R presumably in Rome with B.
	1467
summer	R in Gran, Hungary, at the invitation of Archbishop Johannes Vitéz. He stayed in Hungary until 1471, when he moved to Nuremberg.
	1472
Nov. 17/18	Death of Bessarion in Ravenna (buried in Rome on Dec. 3).
	1475
July 28	R's last observation in Nuremberg.
	R comes from Nuremberg to Rome to assist in the reformation of the calendar.
	1476
ca. July 8	Death of Regiomontanus (by murder? or of the Plague?) in Rome aged 40.

Another thing that would have united the two men was their piety. The young German used as a mark of possession on some 17 surviving manuscripts of his a symbolic Cross on a hill.[46] In Rome in 1462-63 he made his own copy of the Greek New Testament.[47] Bessarion would have told him already in Vienna that he had brought to Italy fragments of the True Cross – a reliquary that he presented to a religious society in Venice in August 1463, with Regiomontanus surely present.[48]

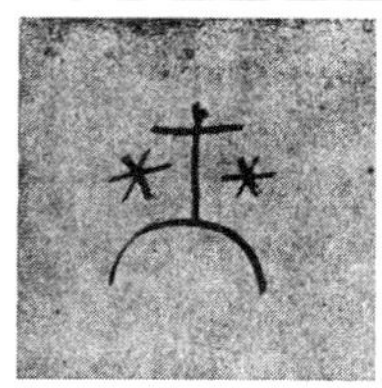

Fig. I-4:~ Regiomontanus' mark of ownership used on his manuscripts. [From Zinner, *Regiomontanus*, p. 295.]

46 Zinner, *Regiomontanus*, pp. 275 and 295.

47 *Ibid.*, pp. 130, 275 and 331.

48 It seems that he had brought it to Italy himself; I have not found any references to its earlier history. It consists of two pieces of wood supposedly from the Cross and two pieces of cloth supposedly from Christ's cloak. These were encased in an elaborate gilded silver frame with a hinged cover. (See Polacco, "La stauroteca del cardinal Bessarione", especially p. 372 and fig. 182 on p. 373 for the inscription, and also in the same volume *Venice 1994 Exhibition Catalogue*, pp. 148, 279, 280 and 520, for images of Bessarion carrying the reliquary, and pp. 160-161 for an 18th-century lithograph of it, further the article "Bessarion's Reliquary" in *The Oxford Dictionary of Byzantium*.)

The Crucifixion scene on the cover corresponds to the items featured inside, so that, for example, the three Roman soldiers with the cloak on the front cover the two little pieces of cloth. There are two angels above the Cross, not yet poised to catch the blood from the pierced hands of Christ. The two images are each surrounded by seven Stations of the Cross. To the left of the Cross, underneath the three Marys and St Anne, we find St Helena, who recovered the Cross from Jerusalem in the 4th century, and under the mourning apostles we find St Constantine, the Emperor, son of Helena, who was designated the 13th apostle in the East. A Latin dedication that Bessarion surely devised himself is engraved on the back of the reliquary.

The other relic with which Bessarion was involved was the head of St Andrew. See further the text to n. 271.

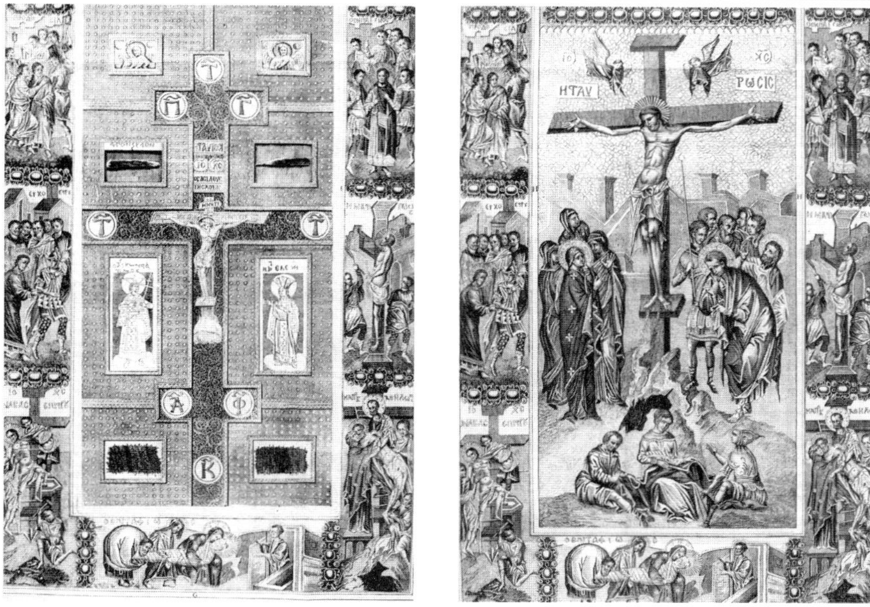

Fig. I-5:~ The reliquary of the True Cross brought to Venice by Bessarion. The silver box was thought to contain fragments from the True Cross and from Christ's robe. It was presented by Bessarion to a religious foundation in Venice in 1463. [From a 1767 lithograph, taken from *Venice 1994 Exhibition Catalogue*, pp. 160-161.]

Further, he would have told him that an acquaintance of his in Italy, a mathematician intensely interested in Euclid and Archimedes, was currently painting fantastic frescoes depicting "The Legend of the True Cross" in a church in Arezzo. See further Parts III and IV.

We should also mention here the earlier translation of and commentary on the *Almagest* by George of Trebizond (b. Crete, 1395, d. Rome, 1472 or 1473), completed *ca.* 1450. See **Fig. I-6** for a 16th-century image of him. Bessarion had a singular contempt for this former friend of his, who had adopted the name of the city of his family, not only because of his poor scholarship and his scorn for Plato and Theon, but also because of his rapprochement with the Ottoman Sultan, Fâtih Mehmet II.[49] This was a Greek who had pronounced that the

[49] See Monfasani, *George of Trebizond* and *Collectanea Trapezuntiana*, also Bádenas, "Byzantine Intellectuals at the Ottoman Court", pp. 4-5, and Bisaha, *Creating East and West*, pp. 153-156. On his scientific work see Rose, *The Italian Renaissance of Mathematics*, pp. 41-42 and 101-102, and now Shank, "Regiomontanus against George

Prophet Muhammad was a second Plato, more sensible than the first. And he was perhaps referring to Bessarion when he mentioned a fourth Plato who would spread the creed of *voluptas*.[50] Bessarion in 1464 referred to George as a calumniator of Plato in the title of one of his books, and Regiomontanus *ca.* 1475 announced his intention to publish a book to expose the weaknesses of his "worthless" commentary on Ptolemy's *Almagest*. This was not to be, for Regiomontanus died in 1476.[51] The venomous controversy between Bessarion and Regiomontanus on the one hand and George on the other was one of the most vehement amongst academics in the Quattrocento.

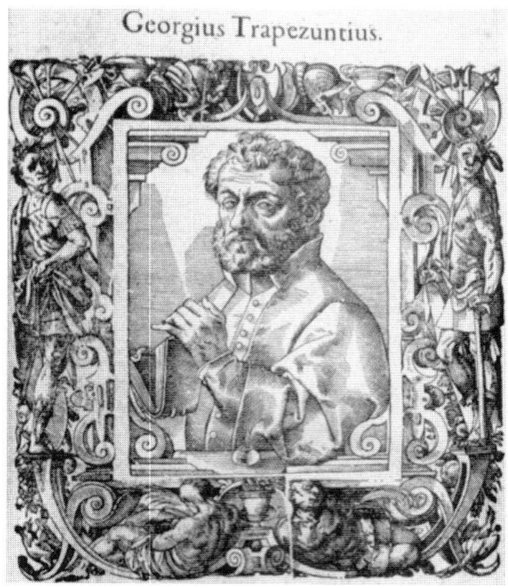

Fig. I-6:~ A woodcut of George of Trebizond, published by P. Giovio in 1577. [From Fiaccadori, ed., *Bessarione e l'Umanesimo*, p. 85.]

Another contemporary of Bessarion, and his correspondent (he wrote to Bessarion to console him on the fall of Trebizond), was George Amirutzes (b.

of Trebizond". See also the text to nn. 284 and 360 below. On the problems associated with images of George of Trebizond see Monfasani, *George of Trebizond*, pp. 234-235, n. 21.

50 Monfasani, *George of Trebizond*, pp. 158-159.

51 See the text to n. 361 on Regiomontanus' death.

Trebizond, date?, d. Istanbul, 1475).[52] He was regarded by some as a "traitor" and "renegade" because of his close ties to the Ottoman Sultan, and his support for the latter's cultural policy, which included proposed new translations of the *Almagest* into Arabic and Turkish (that were never undertaken).[53]

The Byzantine astrolabe of 1062

Now Bessarion brought with him to Italy hundreds of Greek manuscripts, including numerous scientific ones. His library of some 750-odd volumes, of which over 480 were in Greek, eventually formed the basis of the Biblioteca Marciana in Venice.[54] He surely also brought with him a splendid Byzantine astrolabe with Greek inscriptions commissioned by the consul Sergios "the Persian" in Constantinople in 1062:~ see **Figs. I-7** and **8** and Appendix C. Sergios did not actually make the instrument himself; his input seems to have been a "decree and command" for somebody else to make it for him, with an inscription that he himself conceived. The piece is well known because it is preserved in what was until recently the Museo dell'Età Cristiana in Brescia (now the Santa Giulia Museo della Città). It was published by the British Classicist Ormonde Maddock Dalton in 1926 and has been displayed in several exhibitions on Byzantine art.[55] It bears the number #2 in the International Instrument Checklist.[56]

---

52 See Bádenas, "Byzantine Intellectuals at the Ottoman Court", pp. 3-4, also the curious poem by Amirutzes in praise of Mehmet II in Reinsch, "Byzantinisches Herrscherlob für den türkischen Sultan". Amirutzes' Arabic translation of Ptolemy's *Geography* for the Ottoman ruler was completed with the help of his son in 1465. See Ihsanoglu *et al.*, *Ottoman Geographical Literature* (in Turkish), I, pp. 12-13, and the facsimile edition of the remarkable unique Aya Sofya manuscript in Sezgin *et al.*, eds., *Klaudios Ptolemaios – Geography – Arabic Translation (1465 A.D.).*

53 As noted above (n. 42), competent *Arabic* translations of the *Almagest* had been made in the 9th century. These were not available in Europe.

54 See Labowsky, *Bessarion's Library*, also Rose, *The Italian Renaissance of Mathematics*, pp. 44-46, and Zorzi, *La Libreria di San Marco*, pp. 23-85.

For an English version of the documents relating to the donation and the contents of the chests of books see Englander *et al.*, *Anthology of Medieval Sources on Culture and Belief*, pp. 146-156. See also the quote at the beginning of the bibliography at the end of this study.

55 See Dalton, "Byzantine Astrolabe at Brescia". For more information see Appendix C below.

56 These are essentially the numbers used in Gunther, *Astrolabes* (see n. 407) up to 336, those assigned in Price *et al.*, *Astrolabe Checklist* (see n. 62), up to 4000, and those used in the unpublished Frankfurt catalogue of medieval instruments (see n. 410), up to 9999. See further King, *In Synchrony with the Heavens*, XIIIa-3.

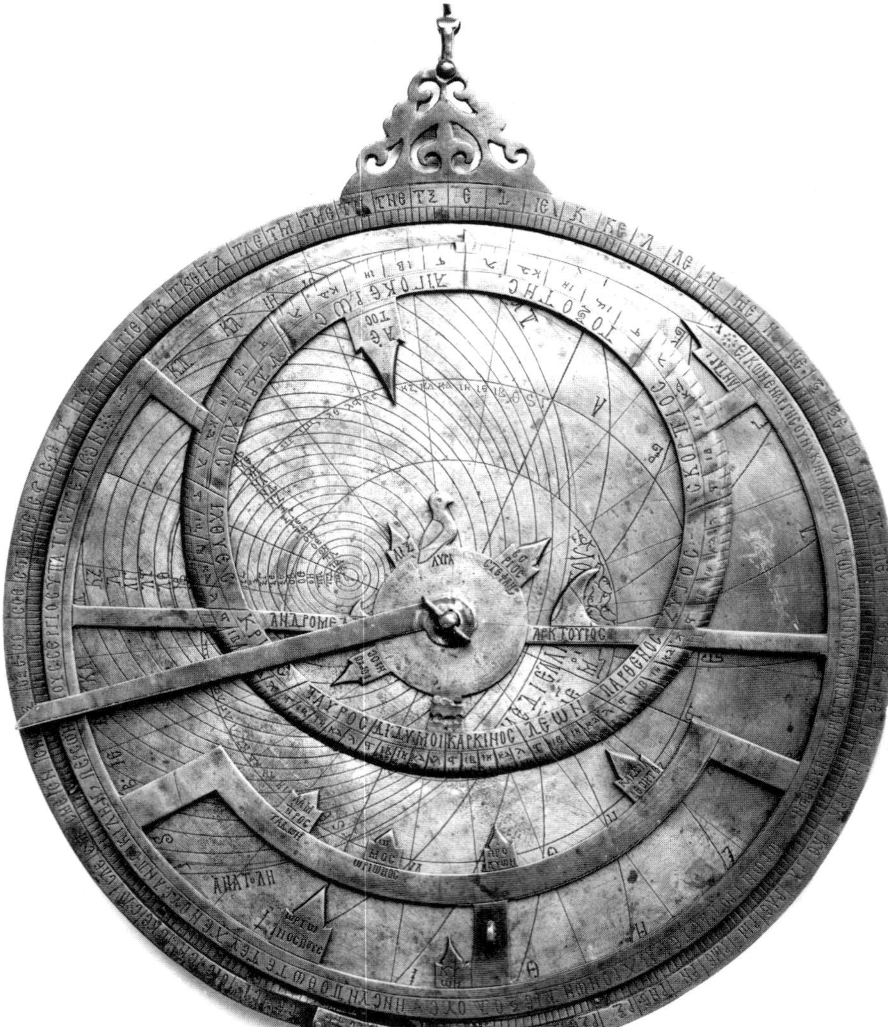

Fig. I-7:~ The sole surviving Byzantine astrolabe (#2), made for the consul Sergios in Constantinople in the year 1062, is an imposing 37.5 cm in diameter. The front, shown here, shows not one *iota* of visible Islamic influence, contrary to what is generally maintained. The plates for different latitudes inside the astrolabe serve Constantinople, Hellespont, and Rhodes. [Courtesy of the former Museo dell'Età Cristiana, Brescia (now the Santa Giulia Museo della Città).]

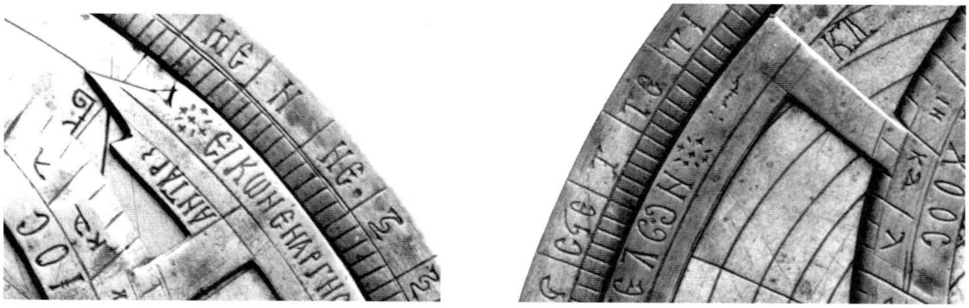

Figs. I-8a-b:~ The beginning and end of the inscription on the front of the astrolabe of Sergios. Note the colon and tittle (" :~ ") at the end, indicating that the inscription is continued on the back of the astrolabe.

This astrolabe would have represented for Bessarion a Byzantine model of the Ptolemaic universe. It is a standard astrolabe, displaying the three-dimensional celestial sphere in two dimensions, achieved by a projection, known as stereographic, from the south celestial pole. The celestial part, called the rete, bears pointers for certain prominent stars and a ring representing the ecliptic, that is, the apparent path of the sun against the background of fixed stars. The terrestrial part, in the form of a circular plate, shows the horizon and meridian of a particular latitude and the altitude circles up to the zenith. The passage of time within 24 hours is displayed by the rotation of the rete with respect to the latitude-dependent plate. Time within a year is displayed by the movement of the sun on the ecliptic ring on the rete. Such astrolabes were the favourite instruments of Byzantine, Muslim and European astronomers.[57] The astrolabe is a veritable mirror of the universe, as even the terminology used for it confirms.[58]

For practical purposes, if not for didactic ones, the Byzantine astrolabe was useless by the 15th century because the star-positions were 400 years out of date. Nevertheless, an inscription in iambic metre on the front describes in the most

57 The best introductions to the theory and use of the astrolabe in English remain North, "The Astrolabe", and *Greenwich Astrolabe Booklet*. For other introductions see n. 383 below. For an introduction to the use of astrolabes, Islamic and European, as historical sources, see King, *In Synchrony with the Heavens*, XIIIa. See also Appendixes A-B below.

58 On notions of the origin of the astrolabe in the Middle Ages and the Renaissance see *ibid.*, XIIIe. On the notion of the astrolabe as a mirror of the universe, apparently first mentioned by the great 11th-century scientist al-Bîrûnî, see *ibid.*, section 7, and on the idea of the throne as being analogous to the handle of a mirror – first suggested to me by Derek Price around 1970 – see *ibid.*, p. 584.

eloquent and timeless terms the essence of what an astrolabe is, namely, an icon of the rotating universe, and much more besides. The inscription begins:~

⁕ ΕΙΚΩΝ ΕΝΑΡΓΗϹ ΟΥΡΑΝΟΥ ΚΙΝΗΜΑΤΩΝ
ΣΑΦΩϹ ΤΡΑΝΟΥΣΑ ΤΟΝ ΔΡΟΜΟΝ ΤΩΝ ΑΣΤΕΡΩΝ … … … :~

which Dalton translated elegantly as:~

(This is) an image manifest of the heaven's movements,
making distinct and clear the course of the stars, … … :~ ,"

although actually it is the astrolabe, an icon, that speaks. The colon and tittle (or tilde) indicate that the inscription is continued on the back of the instrument.[59]

As I shall show, Bessarion probably took this instrument with him to Vienna. He would not have been the first cardinal in the 15th century to travel around with astronomical instruments in his baggage.[60] Regiomontanus had surely seen this instrument and its remarkable inscription, probably in Vienna in 1460-61. He surely showed Bessarion some of the instruments made by the Vienna workshop. Bessarion surely expounded to him the intricacies of the Greek poem on the Byzantine astrolabe, as he would have discussed the failure of the Union of the two Churches and the consequent loss of Constantinople.

---

59 On punctuation marks in Greek, including the asterisk, colon and horizontal line, see Gardthausen, *Griechische Paläographie*, II, pp. 394-415. See also the text to no. 123 below on these marks in Latin.

Du Cange, *Glossarium Graecum* (1688), cols. 3-5 of the second pagination, presents a list of punctuation marks in Greek based on Isidore (Seville, *ca.* 560-636). Here we find an αστερισκοσ indicating that text seems to be missing, and a straight horizontal line called οβελοσ for what we now call a comma (*virgula*) on its side (*iacens*) indicating places where the text seems to be unnecessarily repeated or where a reading is incorrectly noted (?). A horizontal line with a dot above and below to the right of centre is called a *limniscus* or *virgula inter geminos punctos iacens* and indicates places where interpreters of extracts from scripture with the same sense or meaning transfer them from different sermons (?). An asterisk followed by a vertical bar (αστερισκοσ *cum obelo*) was used by Aristarchos (the literary critic and grammarian of Samothrace, d. 145 B.C.E.) in his poetry to indicate verses that were misplaced (?).

60 On the instruments purchased in Nuremberg in 1444 by Cardinal Nicholas of Kues see Hartmann, "Die astronomischen Instrumente des Kardinals Nikolaus Cusanus", especially p. 8 on the purchase. See also items II-1-3 in Appendix D. Also one could argue that Cardinal Niccolò Albergati had brought a little Italian astrolabe with him to Bruges in 1431 or to the Congress of Arras in 1435, for it is depicted with him, much enlarged, in the painting *St Jerome in his Study* associated with Jan van Eyck and probably executed after an original by him, now lost (see nn. 401 and 542 below). On the problems of the attribution of the painting see van Cleempoel, "Representations of Astrolabes in Western Art", p. 105.

It is 50 : 50 whether it was Regiomontanus or Bessarion who first noticed that the name REGIO is embedded in the name SERGIOS. The young Regiomontanus had a clever and rather playful idea.

Regiomontanus' 1462 astrolabe

In Vienna in April, 1461, Peuerbach died. In September, Bessarion had to borrow money to finance his return to Italy. In the late summer of that same year, Bessarion's beloved hometown of Trebizond, on which he had written a eulogy, fell – the last bastion of the Byzantine Empire – to the Turks; although he was surely aware of the possibility of this, he learned of it only in Rome early in 1462.[61] Surely as part of an effort to help Bessarion improve his spirits, Regiomontanus made a new astrolabe for him, a beautiful work of art and technical expertise and a modest 11.6 cm in diameter:~ see **Figs. I-9** and **10** and Appendix E.[62] This astrolabe has been allocated the International Instrument Checklist number #640.[63] It is possible that Regiomontanus completed the astrolabe without the inscription in the Vienna workshop already in 1461, and that the inscription was added in Rome in 1462. It is also possible, and indeed probable, that he did not "make" the piece with his own hands. At least in Vienna he would have had access to a group of craftsmen who would have been making instruments under his direction. Maybe he only engraved it? We simply do not know. Bessarion and Regiomontanus travelled together from Vienna to Rome between the beginning of September and late November, 1461. It is known that they passed through Bologna and also Ravenna, where Bessarion became so sick that he almost died. It is certainly difficult to imagine Regiomontanus "making" the astrolabe in Bessarion's official residence or even in his summer villa on the Via San Sebastiano, near San Cesareo, in the early

---

61 See Mohler, *Kardinal Bessarion*, I, pp. 304-306, on the letter about the disaster from George Amirutzes (see n. 52 above), that Bessarion received in Rome only in early 1462. Clearly, it would have taken several months for the letter to reach him. See the articles "Tarabzun" in the *Encyclopaedia of Islam*, and "Trebizond" in *The Oxford Encyclopaedia of Byzantium*. On Bessarion's treatise, published by O. Lampsides in 1984, see Stormon, "Bessarion's Early Writings", pp. 140-142.

62 First published in Price, "The First Scientific Instrument of the Renaissance" (1959). The instrument is described in more detail and in the light of ten related astrolabes in King & Turner, "The Astrolabe Presented by Regiomontanus to Bessarion" (1994). See further Appendix E.

In the late 1960s and early 70s Price and two doctoral students at the Yale History of Science Department, Sharon Gibbs and Janice Henderson, compiled a handlist of Islamic and European astrolabes with one-line entries on each piece: see Price *et al.*, *Astrolabe Checklist*. It was with the help of this valuable reference tool that this author, in the early 1990s, was able to locate 10 astrolabes closely related to the 1462 piece.

63 See n. 56 above.

months of 1462. Between March and July of that year he oscillated between Rome and Viterbo, where Bessarion was taking a cure.

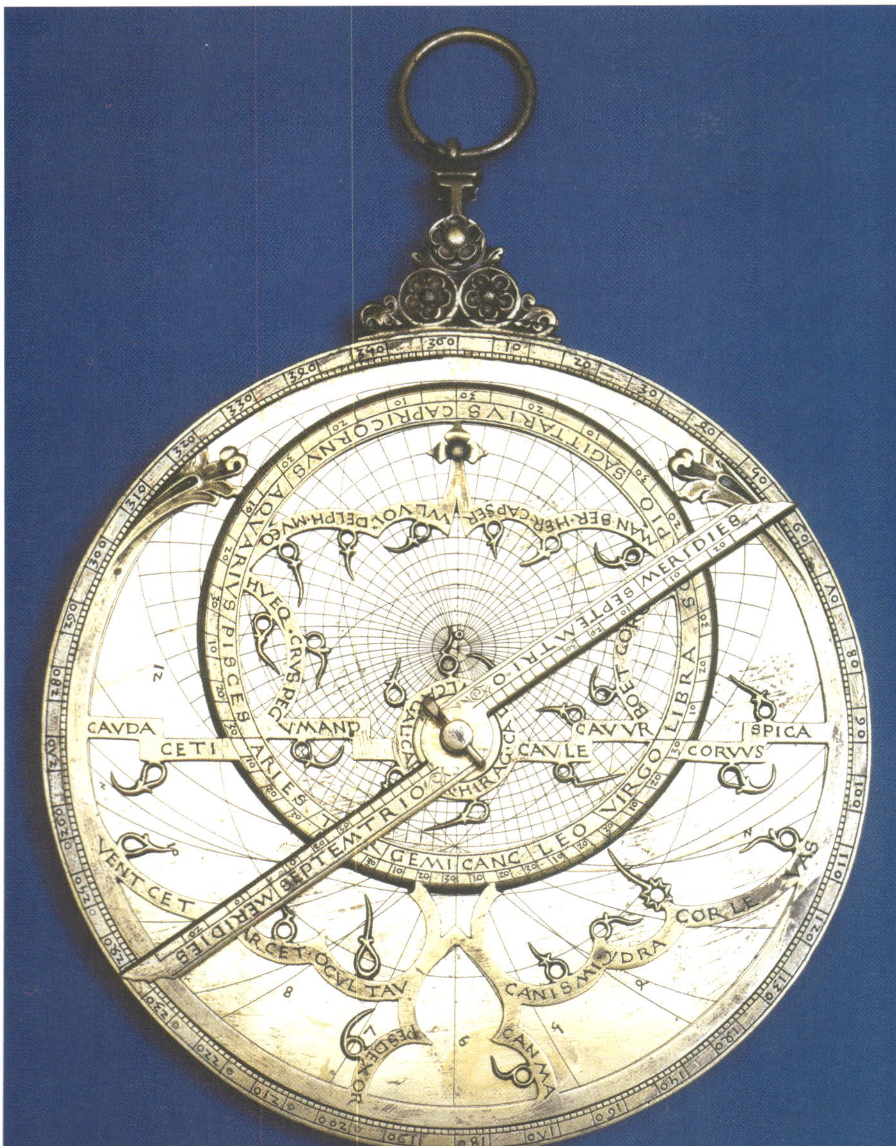

Fig. I-9.

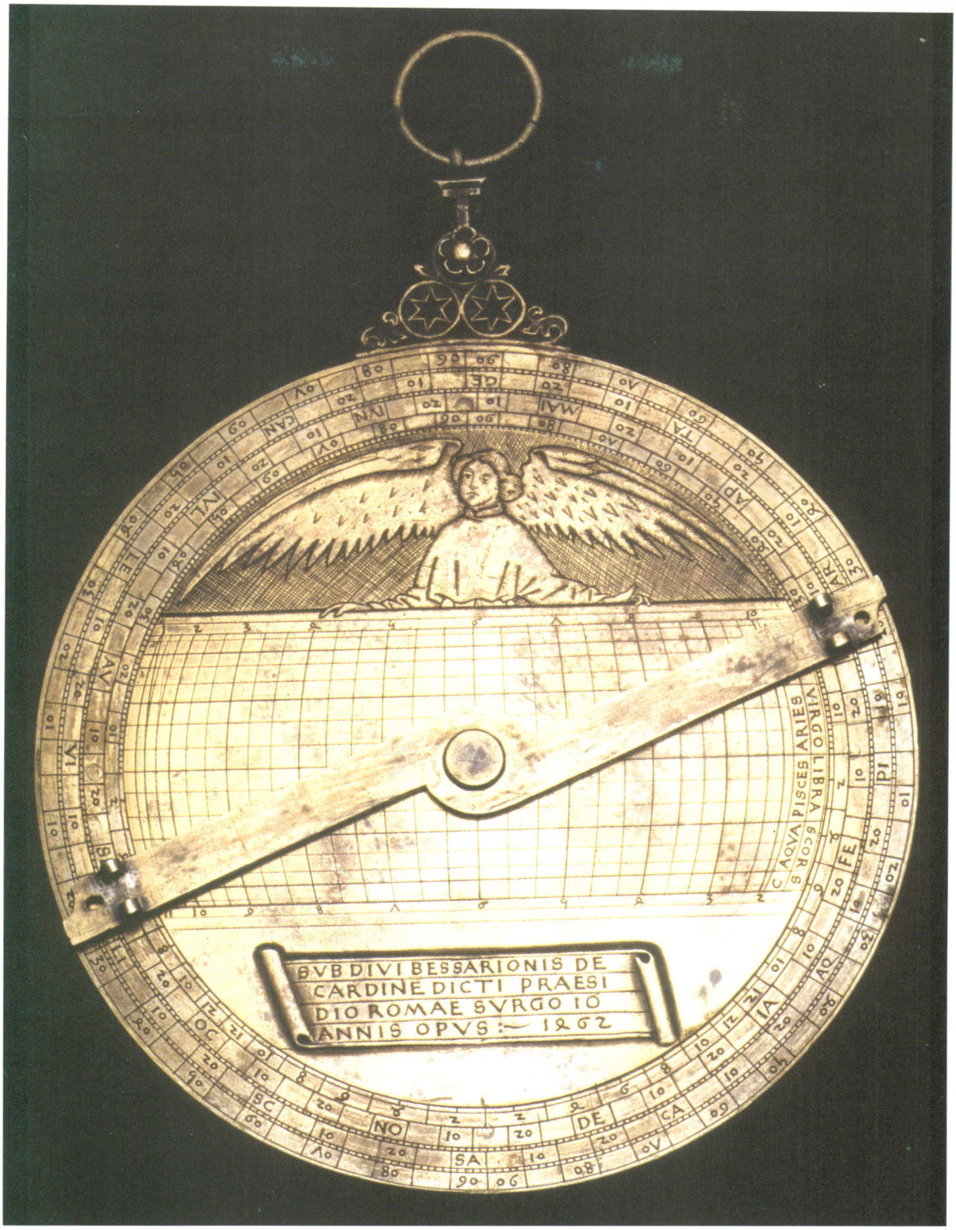
SVB DIVI BESSARIONIS DE
CARDINE DICTI PRAESI
DIO ROMAE SVRGO IO
ANNIS OPVS:— 1462

Fig. I-10.

Fig. I-9:~ The front of the astrolabe of Regiomontanus (#640). The design is the same as that on ten other astrolabes from 15th-century Vienna, and is rich with Marian symbolism, in the form of a lily with three petals and rosettes with five petals. Identical engraving is found on one other astrolabe (#452) from this workshop. One of the various plates inside the instrument is specifically marked for the latitude of Rome. [Courtesy of Christie's, London.]

Fig. I-10:~ The back of Regiomontanus' astrolabe showing the the *organum Ptolemaei*, and the same new astronomical markings are found on one other piece (#256) from the same workshop. In this study we shall be concentrating on the angel and the dedicatory epigram, as well as the idiosyncratic labelling of the solar / calendrical scales.

This new astrolabe was a standard astrolabe, based on stereographic projection. However, on the back, Regiomontanus included a universal timekeeping device associated with Ptolemy and called *organum Ptolemaei*, and on which he had penned a treatise *ca.* 1456: the markings result from a projection from infinity of the part of the celestial sphere bounded by the solstitial circles onto the plane of the solstitial colure.[64] There would have been two reasons why he included such a device. First, it was a very satisfactory substitute for the universal horary quadrants that were used on earlier medieval European astrolabes and which functioned only approximately and even unsatisfactorily in European latitudes.[65] The markings of the *organum* provide an accurate solution to the problem of timekeeping. Second, these markings are beautifully symmetrical, and provided Regiomontanus with the opportunity to include hour scales from left to right above, and from right to left below, as well as solar scales running up and down on each side. These scales are a kind of prelude to his epigram.

The angel

There was space for an image above these markings on the back and for an inscription below. Regiomontanus engraved an angel on the top, without any identifying attributes. In appearance it was virtually identical with the angels decorating an astronomical calendar in a triptych made in Vienna in 1461, which

---

64 Zinner, *Astronomische Instrumente*, pp. 131-134, especially p. 132. On this device see now King, *In Synchrony with the Heavens*, XIIb-15. Its early history demands a detailed investigation. See further the text to n. 513 below.

65 See the text to nn. 390 and 394 below.

survives in Nuremberg:~ see **Figs. I-11a-b**.[66] However, this typically German angel[67] was engraved on the astrolabe for a specific purpose, and this was not to represent Gabriel or Michael or any other archangel, for it is intimately linked to the inscription. It was certainly not a Greek angel, such as would have been known to Bessarion, but whilst he certainly brought his old astrolabe to Vienna, there is no need to suppose that he travelled with a Byzantine imperial banner featuring the Archangel Michael: see **Fig. I-12**.[68] See further Part II.

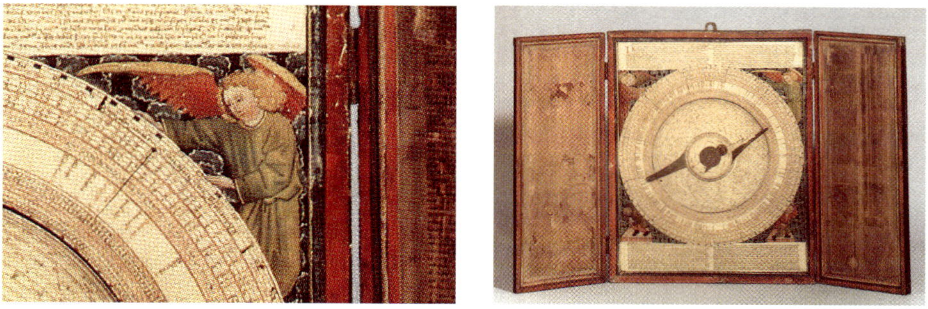

Figs. I-11a-b:~ An ephemeris computed in Vienna in 1461, and a detail of one of the angels near one of the hinges of the triptych that supports it. Note the pose and the way the fingers overlap the outer border of the circular rings of tables. [Courtesy of the Germanisches Nationalmuseum, Nuremberg.]

---

66 For a description by Johannes Willers see *Nuremberg 1992-93 Exhibition Catalogue*, II, pp. 512-514, where Nuremberg is to be replaced by Vienna. It is a pleasure to acknowledge some fruitful telephone conversations with Dr. Johannes Willers.

67 The literature on 15th-century German and Austrian religious art is voluminous. For the ultimate group of angels in Nuremberg, the Engelsgruß in the St. Lorenzkirche by Veit Stoß (1517-1518), see Stolz *et al.*, *Der Englische Gruss des Veit Stoss*, and Bauer & Stolz, *Engelsgruß in St. Lorenz*.

68 This banner (76 × 76 cm) is preserved in the Palazzo Ducale in Urbino (inv. no. 1990 DE 216). See Dal Poggetto, *La Galleria Nazionale delle Marche ... nel Palazzo Ducale di Urbino*, pp. 69-70 and fig. 54, also *Venice 1994 Exhibition Catalogue*, p. 166.

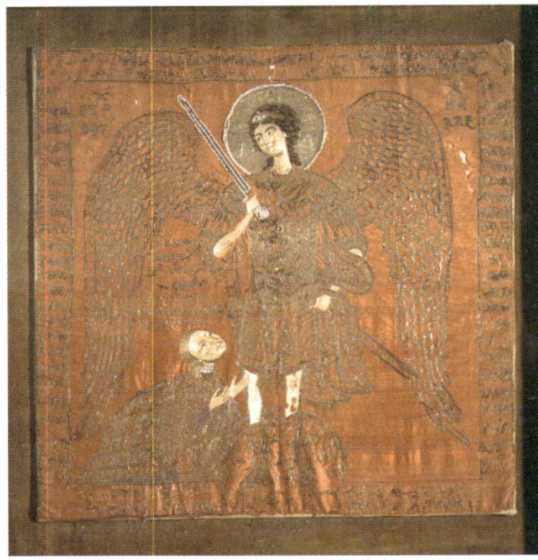

Fig. I-12:~ A very Greek angel – the Archangel Michael – on the banner of Manuel Palaiologos, datable before 1411. [Courtesy of the Soprintendenza per il Patrimonio Storico, Artistico ed Etnoantropologico delle Marche, Urbino.]

The fingers of the angel

The angel is trying to tell us something. His finger on the left points to the 4th hour on the upper scale of the astronomical markings, which correspond to the 8th hour on the lower scale, which is divided into intervals of 20 minutes. His fingers on the right also seem to point to the 8th hour, which corresponds to the 4th hour on the lower scale. Surely every time Regiomontanus thought of the numbers 4 and 8 he thought of the two Archimedean solids and the number of their vertices: four for the regular pyramid and eight for the cube.[78] As we shall

---

78 ** See the reference cited in n. 96 below.

see, the identification of "4" and "8" on either side is significant, for we shall encounter hidden features in eight basic parts of the four-line inscription.

The epigram

Regiomontanus inserted a Latin inscription that is almost a perfect elegaic couplet (see **Fig. I-13**):~[79]

SVB DIVI BESSARIONIS DE CARDINE DICTI
PRAESIDIO ROMAE SVRGO IOANNIS OPVS:~ 1462

This can be rendered:~

"Under the protection of the divine Bessarion, said to be from the *cardo*,
I arise in Rome as the work of Ioannes in 1462."

The astrolabe, like the Byzantine one, is speaking to us, a feature not without parallel on medieval and Renaissance objects and in books; in fact, two other astrolabes come to mind that "speak".[80] The verb *surgo*, "I arise" or "I spring up", used in relation to an astrolabe, is very surprising.[81] The noun *cardo* is used as a pun, but not only in relation to Bessarion's status as cardinal[82] – see further below. Whatever connotations *cardo* had for the pair of scholars, we shall find in Part 4 another play on the words *carduus*, Latin for "thistle" and the origin of the associated notions of carding cloth, *etc.*, and *cardo*, Latin for "hinge, pole, axis, *etc.*", but Italian for "thistle".

---

79 The problem lies in the scanning of the word DIVI. I leave the details to the specialists.

80 For such an inscription identifying the maker of an English astrolabe (#292) dated 1342 (... *me fecit* ...) see Gunther, *Astrolabes of the World*, II, pp. 468-469. For another identifying a change of hands in 1522 between two Humanist scholars (... *me dono dedit* ...) of a 14th-century Picard astrolabe (#202) with numbers represented in monastic ciphers see Gunther, *op. cit.*, II, p. 349, and further King, *The Ciphers of the Monks*, pp. 141-142, also 422-423. On this latter piece see also n. 157 below.

81 Lewis & Short, *Latin Dictionary*, p. 1817.

82 *Ibid.*, p. 291 (*cardo* and *carduus*); Niermeyer, *A Medieval Latin Dictionary*, p. 140 (with an early medieval attestation of *in cardine* and *de cardine* as relating to clerics who assisted the Pope at Mass); and Prinz, ed., *Mittellateinisches Wörterbuch*, cols. 277-278. See also the articles "Cardo" in Pauly-Wissowa, *Realenzyklopädie*, and "Cardo, Kardo" in *Der neue Pauly*.

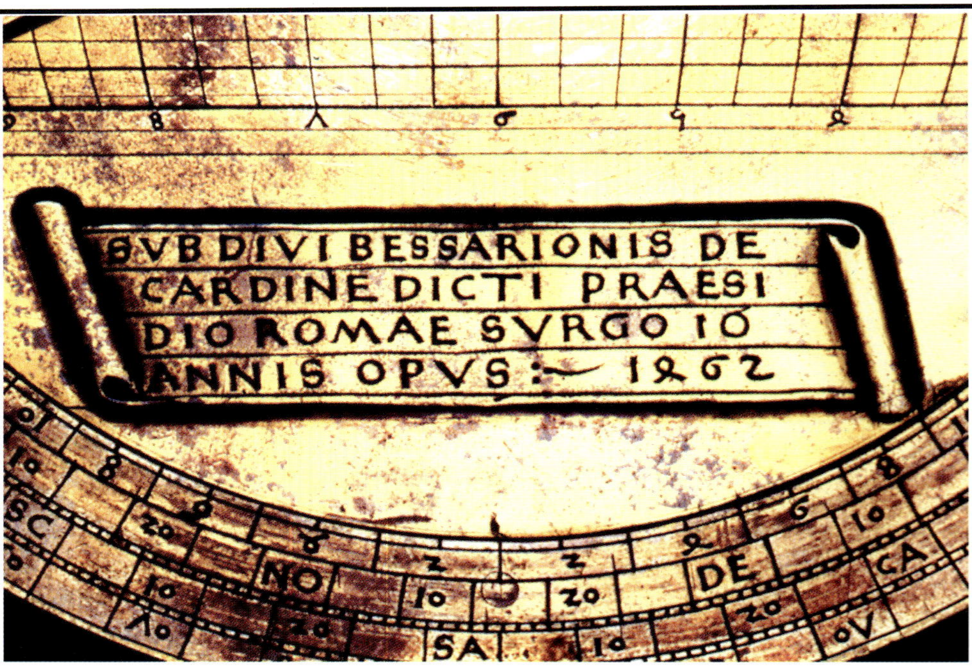

Fig. I-13:~ The inscription on the Regiomontanus astrolabe. The text reads as if it were a straight neo-Latin elegaic couplet. However, the metre is not perfect, for the word DIVI does not scan properly. Also, the epigram is arranged over four lines, with some letters squashed together and others pulled apart. Before 2005, nobody had ever asked: why? The letter "B" and the "I" of IO for IO(annes) in the name B-essar-IO-n define two principal vertical axes, roughly corresponding to the divine cuts of the main body of the inscription (without the initial "S"). The text is an acrostic with eight vertical axes. One can read downwards and even backwards. The akrostych, or left-hand axis, defines a time scale with the words SVB CD ANNIS marking the 400 years since the Byzantine astrolabe was made. Two mesostychs (at about 3/8 and 5/8 of the main text) identify the configured letters BE-NE-DICTI ROMAE, and the (much contested) year of birth of Bessarion, "I I V", followed by " :~ ", indicating "keep on reading" as on the Byzantine astrolabe, and then "14"(00). "I I V" is to be read as "I + I + V", so that the date is 1407. (Alternative readings would yield 1397 or 1403, but these do not accord with certain statements made about Bessarion or by him.) The medial axis at the R in the middle of the word BESSARIONIS serves to identify the north-south axis or *cardo* that leads to the angelic St Bessarion above. The telestych, or right-hand axis, contains a cryptic chronogram that yields the dates 1462 and 1062, a boustrophedon in which we can read the age of IOannes de RegIO monte as 26, and his date of birth as 14"26" + 10 = 1436. Also in the telestych are the configured words DEI and DEO, to be read "around" the central "S".

The words OPVS and DIVI and the phrase DE CARDINE DICTI can be combined in the phrase OPVS DICTI DE CARDINE DIVI, "a work explaining the axis or pole of the sky, *i.e.*, the rotation of the heavens", which recalls the

87 * A universal history of acrostics does not seem to exist. The standard collection of classical Latin is Buecheler & Lommatzsch, *Carmina Latina Epigraphica*, and various supplements by other authors are readily accessible on the Internet. A universal history of configured letters, with numerous acrostics, is Ernst, *Carmen figuratum*. Yet other examples in *Wolfenbüttel HAB 1987-1988 Exhibition Catalogue*. See also various encyclopedia articles such as "Akrostichon" and "Figurengedicht" in *Lexikon des Mittelalters*, and the articles Ernst, "Zahl und Mass in den Figurengedichten der Antike und des Frühmittelalters", and "The Figured Poem". Older works such as Dornseiff, *Das Alphabet in Mystik und Magie*, pp. 146-151, are still useful.

beginning of the Greek inscription of the 1062 astrolabe: see further below. The expression OPVS DICTI is a *genitivus objectivus*, the noun *dictum* indicating the purpose of the *opus* (compare *liber explicationis*, "a book of explanation"). Now DIVI relates to *divum*, "sky", rather than to *divus*, "divine". There are certainly other, simpler ways of rendering "rotation of the heavens" in Latin, but, as we shall see, Regiomontanus needed DE CARDINE and DICTI for other purposes. Also, the expression DIVI DE CARDINE DICTI could be taken to mean "those divines said to be from the *cardo*", and could conceivably be construed as a reference to the cardinals. The use of the word *Cardinalis* would not have been appropriate in Regiomontanus' rather playful inscription for his patron, or, to put it another way, in colloquial English, it would have been a "dead giveaway". In any case, Berthold Holzschuh has suggested that Regiomontanus might have conceived the words OPVS DICTI DE CARDINE DIVI in a different epigram *before* he deliberately concealed them by rearranging them in the final epigram, with some letters and spaces stretched out and others squashed together. See further below.

Regiomontanus certainly had a good deal of luck that his resulting epigram was almost a perfect elegiac couplet, and that it lent itself to concealing much more than appears at first sight.

The acrostic

The inscription *as it is engraved on the astrolabe* has further "hidden messages", for it is in the form of an acrostic, with vertical axes, configured letters, cryptograms and numerograms.[87] Letters are said to be configured if they are deliberately situated in a text in a certain way so that they can be read in special ways other than simply horizontally from left to right. If the text is in manuscript, and is then printed without due regard to the configuration, the significance will be lost immediately. We must look at Regiomontanus' epigram as he engraved it to see the configured letters.

A	**M**	**T**
K	**E**	**E**
R	**S**	**L**
O	**O**	**E**
S	**S**	**S**
T	**T**	**T**
Y	**Y**	**Y**
C	**C**	**C**
H	**H**	**H**

The simplest form of acrostic has a single axis at the front, the akrostych. Any rendition in prose is likely to hide the significance, and translation into another language will lose it altogether. The most obvious examples are certain *Psalms*, in which the verses begin with words starting with the successive letters of the

Hebrew alphabet.[88] The standard form of the rectangular acrostic has three main vertical axes: the *akrostych* at the left-hand side, the *mesostych* at the centre, and the *telestych* at the right-hand side. We can imagine that Bessarion told Regiomontanus about acrostics in the Byzantine liturgy.[89] Regiomontanus' acrostic is more refined, with several vertical axes. Acrostics with five vertical axes are known from Angilbert (Abbeville, *ca.* 800) and one late 16th-century German example.[90] Possibly other examples survive in which the axes have been lost in printing the epigram.

In Regiomontanus' "octacrostic", there are two mesostychs at about three-eighths and five-eighths of the width of the inscription, and one in the middle. In addition, there are other vertical axes at the beginning and end of the epigram. Again, we must look at the epigram as he engraved it on the astrolabe.

The eight vertical axes

As we have already noted, the fingers of the angel point to "4" on the left and "8" on the right of the upper hour scale of *organum Ptolemaei*. But the scale can also be read backwards, indeed, on the lower hour-scale it is marked backwards from right to left. The "4" on the upper scale corresponds to "8" on the lower scale. The "8" on the upper scale corresponds to "4" in the lower scale. Regiomontanus is perhaps telling us to look for eight things. He has deliberately arranged his epigram in *four* lines so that *eight* different groups of letters are aligned in clearly-identifiable vertical axes. I label the letters defining the axes in bold face, keeping in mind that the 4th and 6th of these seem to be intended to have particular significance:~

---

88 See, for example, Minkoff, "As Simple as ABC: What Acrostics in the Bible Demonstrate".

89 See n. 8 above.

90 Ernst, *Carmen figuratum*, pp. 202-209. Here there is no attempt to conceal the axes. In one example, all axes read DEVS MISERERE downwards. In another, the vertical text continues from one axis to the next. For the late 16th-century German example of a "pentakrostichon" in which the five vertical axes in a epigram are accentuated by inscribing them on columns, so that there is no attempt to conceal them, see *ibid.*, 549-550. Here, however, the letters in each of the columns all spell MICHAEL, the name of the person for whom the epigram was composed.

I	II	III	IV
SVB	**B**	**V**I	**B**E
C	**A**	CARD**I**NE	**NE** DICTI
D	IO	ROMA	RO**M**AE
ANNIS	ANNIS		

V	VI	VII	VIIIa	VIIIb
R	**I**	**(I)**	DE	
CT	**I**	**R**	ESI	
SV	**V** O O	GO	**O**	IO
	:~ 14	**I**		1462

Here I simply mention the axes; their significance will become clearer below.

I The words SVB CD ANNIS indicate "on the occasion of 400 years". As we shall see, this relates to the Byzantine astrolabe.

II The letters BA IOANNIS can be construed as a reference to the two first names of Bessarion, namely, Ba(sileios) and his adopted "ΙΩΑΝΝΗΣ" or "Ioannes".

III The letters VI indicating VI(enna) lead down the *cardo*, that is, the north-south axis, to ROMA, calling to mind the journey of the two men in late 1461.

IV The words BENEDICTI ROMAE can be construed as a reference to the Papacy, or perhaps, the Church. On the other hand, the V (or the cluster V+E+I) in axis III taken together with the BENEDICTI calls to mind the phrase "*Venite benedicti ...* " from the *New Testament* and refers to Christians in general.[91]

V The R is at the middle of the 11 letters of the word BESSARIONIS, and its vertical bar coincides with the axis of the astronomical markings above the inscription. The R is possibly to be associated with the letters CT and VS to read R(E)CTVS, a straight (line). See further below.

---

91 *Matthew* XXV:34: "Come, ye blessed of my Father."
The two words are engraved on a book in the hands of St Bonifatius on a 9th-century limestone stela; on the other side is a cross marked SANCTA CRVX SALVA NOS. See *Mainz 2005 Exhibition Catalogue*, pp. 98-101.

VI The letters I I V connected to the 14(00) by means of the colon and tilde are deliberately intended to represent a date – see below.

VII The letters RGO are arranged as in an equilateral triangle, to signify R(E)G(I)O.

VIIIa The letters DEO and DEI are deliberately arranged around the central S, which is one of the three possible letters for organizing such configured letters in the Middle Ages (the other two being O and M).[92] They are perhaps to be understood in the context of SVB PRAESIDIO DEI and DEO, "under the protection of God", and "for God".

VIIIb The numbers in the lower right corner can be read as 1462 and 1062. The latter again refers to the Byzantine astrolabe.

I shall not insist on the other sets of letters that one can detect and that are arranged more or less in vertical axes, for I see no purpose in them within the framework of Regiomontanus' original intention. These are:

- IDRI downwards from the 5th letter on the first line, not deliberately aligned, and with no apparent significance;
- IN downwards from the 7th letter. This, which could be conceived as the abbreviation for Ieusus Nazarenus, is probably coincidental, but Regiomontanus certainly noticed it later.

We can only marvel at Regiomontanus' playful genius in arranging his dedication in this way. The epigram is, somehow, magic.

---

92 See, for example, the article "Figurengedichte" in *Lexikon des Mittelalters*, especially col. 442.

TRIBUTE I:~ Berthold Holzschuh's interpretation of Regiomontanus' acrostic

In February 2005, Berthold Holzschuh realized that there was a lot more to Regiomontanus' epigram than had been prevously recognized. He noticed that the words OPVS DICTI DE CARDINE DIVI corresponded to the meaning of the beginning of the Greek poem on the 1062 astrolabe. He then recognized the cryptogram for 1062, and then discovered that the epigram was an acrostic. See after Part III for his other discoveries. His brilliant achievement in cracking part of the secret of the epigram is represented in the following scheme:~

AKROSTYCH MESOSTYCHES TELESTYCH

Akro	**Meso 1**	**M2**	**M3**	**M4**	**Tele**	***Carmen figuratum***
|	|	|	|	|	|	|

S V B D I V I B E S S A R I O N I S DE | **DE**

C	**T**	**I**	**R**		**ESI**
D		**V**	**GO**	**IO**	**O**
ANNIS	**:~14**			**62**	

-stych	hidden text	BH's interpretation
akro	SVB **C** **D** ANNIS	400 years → Byzantine astrolabe of 1062
tele	IO	1062

	62	→ Byzantine astrolabe of 1062
meso 1 and 3		left and right golden section
meso 1	**B**E N**E** DICTI RO**M**AE	Benedicti Romae → Bessarion's association with Church
meso 2	**R**	central axis of the epigram; middle of the word BESSARIONIS; the letters AR recall Archimedes; the letter R with the T below, with the vertical stems of the R and the T deliberately aligned, would – if rotated around the axis – generate a sphere, and a cone, and with some stretching of the imagination, a cylinder.
meso 3	I I V :~ 14	IIV = 3 → 1403, year of Bessarion's birth (contested)
meso 4	**I** **R** **GO**	RGO → Regiomontanus
tele	RGO IO 62	Regiomontanus' age in 1462 is 26 (*boustrophedon*)
carmen figuratum	**DE** **ESI** **O**	words DEI and DEO to be read around central **S**
Golden section	1062	hint of the golden section (ratio ~62:100)

In addition, the "Klartext" of the epigram, that is, the additional meaning that Regiomontanus wanted to convey to Bessarion – a play on the message in the poem on the Byzantine astrolabe – was:~

SVB BESSARIONIS PRAESIDIO
SVRGO
IOANNIS *OPVS DICTI DE CARDINE DIVI*
ROMAE 1462

"Under the protection of Bessarion, I arise in Rome in 1462
as a work of Ioannes explaining the axis of the sky,
i.e., the rotation of the heavens."

Holzschuh does not think that the epigram has any other hidden messages playing on the names of the two principal players, or that the angel is St Bessarion. He thinks that the B / A / IO / ANNIS axis is a deliberate reference to the Byzantine Emperor Basileus Ioannis VIII and the DE followed by CARDINE is a deliberate reference to (Ottaviano Ubaldini) dalla Carda. See further after Part III.

Remarks

I can agree wholeheartedly with the interpretations of the hidden messages in the akrostych, two of the three mesostychs, and the telestych, although I think that Regiomontanus intended a lot more, also elsewhere in the epigram. I would prefer to interpret the date in the second mesostych as I+I+V+1400 = 1407, if only in deference to the sources identified by John Monfasani.[93] As far as the "golden section" is concerned, I, as a mathematician, would ask: "golden section" of what? Holzschuh's insight into the letters AR in BessARion's name at axis M2 is not at all farfetched; in fact, it is brilliant, given what more we can find out about the interest of the two men in Archimedes. See further below.

---

[93] See n. 125 below.

INTERLUDE I:~ On symmetry and the divine proportion

> "The Golden Number literature abounds with claims wherein the author does not realize that approximate measurements of drawings do not suffice to prove the claim. A proof requires some sort of documentary evidence that the designer of the object in question had the Golden Number in mind as a theoretical basis." Roger Herz-Fischler, "The Golden Number", p. 1583.

We shall talk of the "space" of the epigram as the rectangular block of letters, without the "S" of "SVB".[94] And we should not forget that we are dealing with a space about 3.8 cm wide.

Now the boundaries of this space on the left and right are not clearly defined. More important, they are not defined at all from a mathematical point of view. The left "boundary", somewhere between the initial S and the first column of letters (V-C-D-A), and the right "boundary", within, say, a millimetre of the wobbly last column (E-I-O-2), are, from a mathematical point of view, undefined.

Consider the two complementary fractions 3/8 and 5/8. These are complementary because their sum is unity. Now I can happily assert that the **B** and the **I** in the name **B**ESSAR**I**ON are *roughly* at 3/8 and 5/8 of the width of the epigram. However, any more extravagant claims would be unfounded.

Now suppose that we *could* define a first axis for the B and that we take the vertical down the middle of the I as the second axis for the I. And suppose those axes were originally intended to be *exactly* at 3/8 and 5/8 of the width of the epigram, as yet undefined. Then the distance between the two axes, which we can measure, would be 2/8 of the width, which is therefore now defined. In other words, the width is four times the distance between the two axes, and the boundaries are at 3/8 of the width to the left of the first axis and the same distance to the right of the right axis.

Now if one takes 5/8 of 5/8 one gets 25/64, which is about 3/8. In other words, not only are the axes roughly symmetrical to each other with respect to the centre of the epigram, but the left axis divides the space between the left

94 See the text to n. 140.

boundary and the right axis in about the same ratio as the right axis divides the entire width.

Now let us start again. Let us suppose that the I-axis divides the epigram in a certain ratio $\phi : 1$, where ϕ is a known quantity (roughly equal to 5/8), and the B-axis divides it in the complementary ratio $(1 - \phi) : 1$ (that is, about 3:8). We can again measure the distance between the axes, and, since we know ϕ, we could calculate the width if we wanted to. But we do not.

Now suppose that number ϕ was such that if one multiplies it by itself, one gets $1 - \phi$. In other words, in the language of mathematics, $\phi^2 = 1 - \phi$. It follows that

$$\phi = {}^1\!/_2\,(\surd 5 - 1) \approx 0.618034\ldots .$$

The ratio $\phi : 1$ was known in ancient Greek mathematics as "*the* ratio". It is intimately related to the mathematics of the regular pentagon and decagon. It was known to Regiomontanus because he was familiar with the *Elements* of Euclid.[95] Furthermore, Regiomontanus was interested in the mathematics of this proportion, for he later wrote about it twice in his scholarly correspondence.[96]

The incommensurable number ϕ is approximated by 0.62 to two decimal places. The value one can derive from Regiomontanus' trigonometric tables is 0.6180340, which is accurate to 7 decimal places.[97] Because of its remarkable mathematical properties, it was called in the Quattrocento *divina proportione*, and as such it was discussed by Piero della Francesca, whom we shall encounter again. In modern scientific terms, it is well known that the divine proportion underlies many patterns in nature.

---

95 Zinner, *Regiomontanus*, p. 20, also Kaunzner, "Regiomontanus als Mathematiker", p. 127, n. 10, records that at the University of Vienna, a *baccalaureus* required knowledge of Book I of Euclid's *Elements*, and the permission to teach required knowledge of Books I-V. See also Folkerts, "Regiomontans Euklidhandschriften".

96 On Regiomontanus' correspondence see n. 31 above. The two references are in a letter dated 1465 from Rome to Jakob von Speyer, the astronomer-astrologer of Federigo da Montefeltro, and in another dated 1471 from Nuremberg to Christian Roder in Erfurt. See Curtze, "Regiomontan's Briefwechsel", p. 297, '14, and pp. 332-333, 't. The references are, alas, not to a rectangle with a vertical axis dividing the width in the divine proportion but to the Archimedean polyhedra. On these, see further Field, "Rediscovering the Archimedean Polyhedra". The surviving manuscripts from Regiomontanus' Vienna days do not contain any materials relating to the divine proportion: see Folkerts, "Die mathematischen Studien Regiomontans in seiner Wiener Zeit".

97 From Regiomontanus' trigonometric tables (calculating 2 sin 18° with the appropriate base) we can derive 6180340:10000000 as an approximation for the divine proportion, which is accurate to 7 digits. See Glowatzki & Göttsche, *Die (trigonometrischen) Tafeln des Regiomontanus*, pp. 14 and 130. In Herz-Fischler, *Mathematical History*, p. 158, the author states that this value was first derived by Michael Maestlin, teacher of Kepler.

Now we may assume that Regiomontanus had noted that the years 1062 and 1462 more or less divided the 11th and the 15th centuries in the divine proportion. Here is yet another reason for him to have *thought of using* the proportion in his epigram.[98]

I should not be surprised if it crossed Regiomontanus' mind to place these two axes so that they divide the width of the inscription, however we might define it, in the divine proportion, once from the left, and once from the right.[99] I cannot prove this, and no measurements I can make will prove or disprove it because neither the axes nor the boundaries of the epigram are defined. However, we are dealing with an acrostic conceived by the leading mathematician of the 15th century, himself a poet, who had been a child prodigy.[100] As we have seen, the purpose of the two mesostychs was to highlight the B and the I in the name **B**ESSAR**I**ON. Deliberately putting them at roughly three-eighths (0.375) and five-eighths (0.625) of the width of the inscription was Regiomontanus' way of giving the epigram an additional measure of symmetry. If the width were defined, putting them as near as possible to the two divine cuts – at ~0.382 and ~0.618 of the width – would have been theoretically possible for Regiomontanus. We should always keep in mind that the width of the inscription is a mere 3.8 cm, so that, at least on the astrolabe itself, we could not with the naked eye tell the difference between the two pairs of axes anyway.

Traditionally, the divine cut of a segment is on the right-hand side. But, there is no reason why one should not define a second one on the left-hand side:~ see the figure below. The left-hand divine cut can be thought of as the right divine cut of a rectangle viewed backwards. The pair of symmetrical axes provides an invitation to look at the inscription backwards. The epigram with its two mesostychs at or close to the two divine cuts is in a very real sense an acrostic to end all acrostics. For in the same way that the front part of an astrolabe, representing the heavens, can be turned around and still represent the same universe, the text can be read not only horizontally and vertically, and, if necessary, backwards. Also, one can read words inside other words. Further, the symmetry that has been achieved can only be fully appreciated by someone adept in mathematics.

---

98 See also the text to n. 318 below.

99 For a study of the divine proportion within the history of mathematics, see Herz-Fischler, *A Mathematical History of Division in Extreme and Mean Ratio*, and "The Golden Number" (summary). The best modern introductions to the geometry of the regular pentagon and decagon are Aaboe, *Episodes from the Early History of Mathematics*, pp. 54-72, and Lawlor, *Sacred Geometry*, pp. 36-37 and 48-64. For Euclid's treatment see Heath, *Euclid's Elements*, I, pp. 402-403 (II:11), and pp. 96-100 (IV:10). See further the text to nn. 239 and 240 below.

100 See n. 16 above.

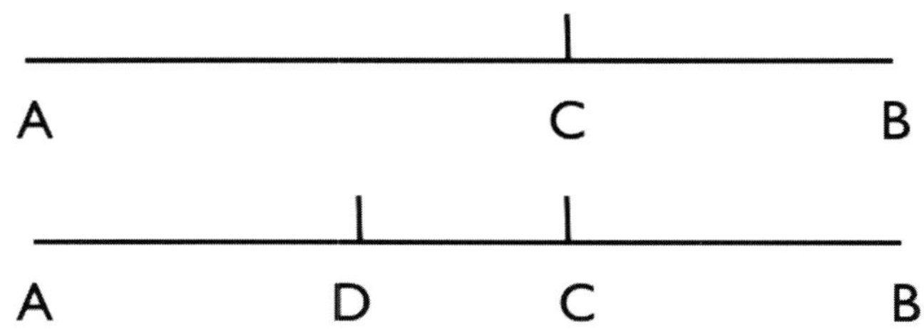

The point C divides the segment AB at the divine cut if AC:CB = AB:AC. If AB is 1 unit, then AC is approximately 0.62 units, actually, $\frac{1}{2}(\sqrt{5}-1) \approx 0.618034\ldots$. We label C the "right" divine cut, because we can also envision a "left" divine cut, here at D, where AD:DB = DB:AB. Because of the remarkable properties of the divine proportion, D is at the right divine cut of AC. Claiming that there are intentional divine cuts in inscriptions and paintings is an exercise fraught with danger.

What Regiomontanus was thinking we shall never know, and we cannot claim that the two mesostychs have anything to do with the divine proportion. What we can claim is that on an epigram of width 3.8 cm, it is not possible with the naked eye to distinguish axes at 3/8 and 5/8 of the width from those at 38/100 and 62/100 of the width. Investigating modern enlarged copies of the epigram will still not achieve the necessary definition of the axes or the boundaries. In any case, I shall refer to the two main mesostychs as **the principal axes** of the epigram.

I have no problem in identifying the right-hand principal axis as the vertical down the middle of the I in BESSARION. As we shall see, Regiomontanus may have thought of the left principal axis through the centre of the O in the word OPVS. If this was the case, we can see his signature in the I+O. On the other hand, we could imagine the left-hand axis as the vertical between the left-hand edge of the vertical segment of the B, which continues down the right-hand edge of the vertical segment of the E below, and is tangential to the O in the bottom line. Then if we then assert that these two axes divide the width of the epigram in the divine proportion, we can determine the width. Its left-hand limit is just before the initial S of the first line, but the right-hand limit is too far for comfort beyond the last letters on the right. This is all very tantalizing, especially as the letters for the words DIVINA and PROPORTIO are found in the neighbourhood of these two axes. (The reader will soon get used to jumping around the epigram, and using a given letter twice when necessary.) In any case, we are missing definitive evidence that Regiomontanus deliberately used the divine proportion in his epigram.

PART II:~ BESSARION'S NEW ASTROLABE DECODED or REGIOMONTANUS' AGENDA REVEALED

Regiomontanus clearly had a lot of fun preparing his gift for Bessarion. We now go beyond all previous renditions of the epigram on the 1462 astrolabe to uncover its hidden meanings. But it is not only the epigram and the angel that we shall consider, for Regiomontanus has even played around with the scales on the back of the astrolabe that surround the plate on which they are engraved.

The hidden identity of Regiomontanus' angel

We have already drawn attention to the angel, a typical 15th-century German angel but without distinguishing attributes. Which angel did Regiomontanus intend with this image?

Let me draw attention to a curious feature of Byzantine Christianity:[101]

> " ... the summit of the monastic achievement was to reach a completely immaterial state of being, like that of the angels. From the time of the early desert fathers, monks had been compared to the bodiless angels, the *asomata*. The sixth-century writer Cyril of Scythopolis said that Saint Euthymios, at the time of his death, possessed the form of an angel."

Now St Bessarion,[102] the patron saint of Cardinal Bessarion, had, like St Arsenios and St Pelagia, been venerated as an angel in the Byzantine liturgy at

---

101 For an introduction to this phenomenon see Maguire, *Icons of Saints*, pp. 24-25, where the example of St Euthymios is considered, and pp. 66-74 (the quote is from p. 67), on the "angelic" life of monks and ascetes, with several more examples. On monastic status as the "angelic life", see also the article "Angel" in *The Oxford Dictionary of Byzantium*. There is no mention of this phenomenon in Heiser, *Die Engel im Glauben der Orthodoxie*, especially pp. 187-192 ("Die Heiligen und die Engel").

For a remarkable depiction of a personification of the city of Constantinople as an angel on a gold goblet from the 7th or 8th century see Evans *et al.*, *The Arts of Byzantium*, p. 33, and also n. 263 below.

102 See the articles "Bessarione" by Joseph-Marie Sauget in *Bibliotheca sanctorum*, III (1963), cols. 140-141, and "Bessarion, Saint" by Lucien Regnault in *The Coptic Encyclopedia*, II (1991), p. 379. Various later saints were named after him: on Visarion,

least since *ca*. 800. In the *Triodion*, *Canon* of Theodoros Studites (759-826), for the morning of the Saturday before the 4th Sunday of Lent, we encounter:~[103]

"The great Bessarion, another Angel, living an angel-like life."

His feast-day in the Byzantine Church was June 6.[104] In the *Kondakion* or hymn for his feast-day, we find for Bessarion, the wonder-worker:~[105]

"Emulating the Heavenly Host, you lived your life in the way of the birds, consigning the present to oblivion. Drawn through constant yearning towards the heavenly beauty of Christ the King, until you reached His presence, o Bessarion; pray without ceasing for us all."

The young Basileios had adopted the name of his guardian angel-saint when he became a monk in 1423, and his panegyric on the saint survives in his own hand. In this, the monk Bessarion writes about the angel-like (αγγελικοσ) desert

---

bishop of Smolensk (17th century), Bessarion I and II, bishops of Larissa (14th-15th and 15th-16th century), and Visarion Saraj, a Bosnian confessor in Transylvania, see *Bibliotheca sanctorum orientalium*, I, cols. 428-432.

St Bessarion is alive and well on the Internet. A Google-assisted search for "Bessarion" together with "Wonderworker" yields over 500 sites, and changing "Bessarion" to the Slavic form "Vissarion" yields another 50 or so. In Toronto there is a subway stop named Bessarion.

103 Anonymous, "Die Engelstellen im Triodion der byzantinischen Liturgie", p. 4; also Kirchhoff, *Byzantinische Hymnen*, p. 157. This veneration of St Arsenios as an angel is not mentioned in a discussion of the veneration of the saint in Macrides, "Saints and Sainthood in the Early Palaiologan Period", pp. 73-79, perhaps indicating that by the 13th century this peculiar custom had been abandoned.

104 Mercenier, *La prière des églises de rite byzantin*, II:1 (*Grandes fêtes fixes*), p. 59; also von Maltzew, *Menologion der Orthodox-Katholischen Kirche des Morgenlandes*, II, p. 397.

105 Translated from the German in Bohác *et al*., eds., *Apostolos*, p. 650:

> Menologion – 6. Juni – Fest unserer heiligen Väter Bessarion des Wundertäters und Hilarion des Jüngeren, des Abtes vom Dalmatos-Kloster. ... ... Kondakion vom heiligen Bessarion, 2. Ton: Den Scharen in der Höhe nacheifernd, führtest du dein Leben nach Art der Vögel, das Zeitliche der Vergessenheit übergebend. Durch beständige Sehnsucht zu [*sic* for nach] der himmlischen Schönheit Christi des Königs hingezogen, bis du ihn erlangtest, o Bessarion; bete unaufhörlich für uns alle.

In an account of his life in Slovak in Michal Carny, *Zivot svatych*, Presov, Slovakia, 2003, pp. 302-303, it is stated that the saint could fast for 40 days, sleeping standing or sitting. Hence God gave him the ability to work miracles. On account of this severe asceticism, he compared himself to the Heavenly Host, for which reason he is depicted as an angel.

(I owe these two references to the kindness of Prof. Michael Kunzler of Paderborn.)

father Bessarion and the numerous miracles that he was reported to have performed, leaving out the ones for which he did not care.[106]

The angel on the astrolabe, I suggest, is intended to represent the angelic St Bessarion. A sceptic might protest that a German angel cannot represent a Greek angelic saint:~ he or she should try to picture Regiomontanus and Bessarion in Vienna, with the young German showing the Cardinal the very German angels he himself had drawn (or had had drawn) on his 1461 ephemerides, and the Cardinal unable to resist telling him about his patron saint, the αγγελικοσ St Bessarion, venerated as an angel in the Byzantine liturgy. One can hardly expect Regiomontanus to have engraved a Greek-type angel on his astrolabe, resembling, say, the Archangel Michael on the standard of Manuel Palaiologos that Bessarion had brought with him to Italy: see **Fig. I-12** above.

As we shall see, Bessarion is very much at the centre of the inscription: one could say that he is illustrated directly above it, just below the north celestial pole of the astronomical markings. When we encounter his name in the inscription, it will be as if he were coming from the pole (*cardo*) or down the celestial axis (*cardo*).

Yet there is a sense in which the face of the angel probably resembles that of the young Regiomontanus, as first suggested by Derek Price in 1959:~[107]

> "Diametrically opposite this inscription there is a portrait engraved, showing a young man with his hands spread and resting on the upper line of the *Organum Ptolomei* astrolabic projection which occupies this inner rotating disc on the dorsum of the instrument. He is garbed in a loose-fitting cloak, and has at the neck level a very beautiful pair of angel's wings. I believe that this is no random portrait, but the author himself, in the iconographic guise of learning spreading his wings over the world. Certainly the portrait cannot represent the patron, Bessarion, for he was an old and bearded man at the time, but Regiomontanus was just in his mid-twenties then."

This was criticized by V. Scholderer in 1960:~[108]

> "The bust has been explained as a portrait of Regiomontanus himself 'spreading wings of learning over the world', but this appears doubtful; not only is such an idea alien to fifteenth-century thought but it would surely

---

106 Edited with commentary in Ioannou, "Opuscule du cardinal Bessarion". See also Stormon, "Bessarion's Early Writings", pp. 131-132.

107 Price, "The First Scientific Instrument of the Renaissance", pp. 28-29. See further the text to nn. 135 and 518 below.

108 Scholderer, "A Note on the Astrolabe of Regiomontanus", p. 209.

have been the height of impertinence for a young man of humble origin to introduce so presumptuous (not to say sacrilegious) a representation of himself on a gift to a famous cardinal."

So I suggest that the angel represents Bessarion, but not the Cardinal. The angel might also represent Regiomontanus, so that perhaps Price was right after all. But it is the angelic St Bessarion who maybe looks a bit like the young German.

Hidden meanings in Regiomontanus' solar / calendrical scale

Regiomontanus has deliberately labelled in a special way the zodiacal signs and the months of his solar / calendrical scale: see **Fig. I-10** above. The names of the signs and months, all abbreviated to two or three letters, are engraved in the final 5-degree or final 5- or 6-day interval for the appropriate sign or month:~

|| – | 10 | – | 20 | – | MONTH || ,

except for February, where the last space is for only 3 days, so we find:~

|| – | 10 | – | 20 | FE | – || .

As a result of this, the labels are some 25 degrees counter-clockwise from the beginnings of the signs and months. This arrangement, which has caused some confusion in modern times,[109] is ingenious,[110] but until very recently it was not at all obvious why Regiomontanus adopted it. The motive behind Regiomontanus' idiosyncratic scale, together with the potential astrological implications, was proposed by Robert Dietz in the Frankfurt Seminar in January 2006[111]

The letters that are underneath the inscription are:~

months:		NO		DE	
signs:	SC		SA		CA

These letters correspond to pairs of letters in roughly the same order across the epigram. It is assumed that the plate is in its "rest position", with the celestial axis vertical.

---

109 See the text to nn. 508-509 and 522 below.

110 See one of my contributions to King & Turner, "Regiomontanus' Astrolabe for Bessarion", on p. 171, where the layout is explained as correct, but nevertheless idiosyncratic: "only to save space".

111 See also the text to n. 138 below.

113 * The basics of astrology for historical purposes are to be found in Ptolemy, *Tetrabiblos*. The writings of the Muslim astrologer al-Qabîsî mentioned by Regiomontanus in his lecture at Padua are now available in both Arabic and Latin, with an English translation of the Arabic, listed in the bibliography under al-Qabîsî, *Introduction to Astrology*. For some basic concepts of astrology represented on Islamic

Clues to finding the hidden vertical axes

It is clear that Regiomontanus chose this arrangement to identify some of the vertical axes in his epigram. The letters SC for Scorpio on the solar scale are precisely "opposite" the letters SC on the left hand side of the epigram, suggesting that one read down the akrostych (axis I): SVB **C D A**NNIS. The letters NO for November correspond to the letters in (CARDI)**N**(E) and (R)**O**(MAE) below it, suggesting the next axis (II). The letters SA for Sagittarius on the same scale are directly below the SA in the word BESS**A**RIONIS in the epigram, perhaps suggesting that if one follows SA one comes to R, which defines another axis (V). The letters DE for December invite one to look at the same letters in the telestych (axis VIIIb) to discover a *carmen figuratum*, in this case, the words DEO and DEI around a central S. The letters CA for Capricorn correspond to the CA that has jumped to the beginning of the second line of the epigram, and possibly would have suggested *cardo* to Bessarion, with all of its potential significations for the two men.

Hints of astrology

Now when one begins to mention months and zodiacal signs, one enters into the potential sphere of influence of astrology, an art as popular in the Renaissance as it was in Antiquity and the Islamic and European Middle Ages.[113] It is hardly surprising that Regiomontanus has arranged his solar / calendrical scale in this unusual way for yet other purposes.

We know Bessarion was born on January 2, when the solar longitude was *ca.* 290°, that is, the sun was in the sign of CApricorn. So the CA in the solar scale also points to the CArdinal and his birth-sign CApricorn.

---

astrolabes see Hartner, "The Astrolabe", A-B. For examples of horoscopes in Antiquity, the Islamic world, and medieval Europe see Neugebauer & van Hoesen, *Greek Horoscopes*; Kennedy & Pingree, *Astrological History*; and North, *Horoscopes and History*. See also North, *The Ambassadors' Secret*, pp. 218-243, on astrological aspects of a Renaissance painting. The most recent overview of the subject is Von Stuckrad, *Geschichte der Astrologie – Von den Anfängen bis zur Gegenwart* (2003). For some of the most recent scholarship on the history of astrology see Oestmann *et al.*, eds., *Horoscopes and Public Spheres – Essays on the History of Astrology*. For some of what Regiomontanus said and wrote about astrology see Swerdlow, *The Renaissance of Astronomy* (forthcoming). (My thanks to Noel Swerdlow for sending me the sections dealing with Regiomontanus.)

Further, Girolamo Cardano (1501-1576) states that Regiomontanus, born at *ca.* 4^h40^m p.m. in Urfinden, on June 6, 1436, when the sun was in Gemini 23°16′ (modern 24°9′), supposedly had Scorpio 15° as ascendant.[114] Thus, one could argue, SC points to Regiomontanus, whilst the sign of his birth is at GE, up by the *cardo* at the top of the solar scale.

But there is another connection to GEmini. The feast-day of St Bessarion in the Byzantine liturgy was also June 6.[115] This is one additional reason why Regiomontanus has moved the labels on the scales. Also, conveniently, we just happen to find the number "06" twice at the top of the scales, namely, as "90" upside down. I shall suggest below that Regiomontanus may have given the astrolabe to Bessarion on June 6, 1462, which was a day of significance to both of them, especially given the kind of games in which they both revelled.[116]

There is perhaps some further significance that the two men would have seen in the purely coincidental situation of NO and DE under the epigram. They arrived in Rome on November 20, 1461, but we have no information on their activities in December of that year. Alas, we may never know where and when Regiomontanus engraved the plate bearing the angel, the *organum*, and the epigram.

Hidden meanings in Regiomontanus' epigram

First, I should mention that the epigram exploits the standard abbreviations of medieval and Renaissance Latin and even Byzantine Greek.[117] Also, the numbers "1" and "0" are engraved identically to the letters "I" and "O", as elsewhere on the astrolabe.

The axes that I see in the epigram are as follows:

---

114 Swerdlow, *op. cit.*, contains an analysis of the horoscope of Regiomontanus presented by Cardano in his *Liber de exemplis centum geniturarum* (1547), where *geniture* means "nativity horoscope". On Cardano see the article by Mario Gliozzi in *Dictionary of Scientific Biography*, and for his adventures with monastic ciphers also King, *The Ciphers of the Monks*, pp. 206-210.

115 See the text to n. 104 above.

116 See the text to n. 163 below.

117 On Latin and Greek abbreviations see nn. 24 and 29 above, respectively.

S V B D I V I B E S S A R I O N I S D E

I II III IV V VI VII VIII

axis	hidden text		my interpretation
I	S**V**B **C** **D** **A**NNIS	 IO 62	400 years 1062 → Byzantine astrolabe
IIa	**B** **A** **I**O **A**NNIS		**Basileios** and **Ioannis** → the two names of Bessarion
IIb	**B** **A**R **I**O		Possibly a reference to the fact that B(essarion) was responsible for introducing AR(chimedes) to IO(annes)
III	**VI** CARD**I**NE ROMA		From Vienna to Rome by the north-south road (*cardo*) → **Bessarion and Regiomontanus travelling together**
IV	**B**E N**E** DICTI RO**M**AE		Church, Rome → Bessarion's affiliation with Rome
V	A**R** **T**		axis, centre → Bessarion is at the centre of the epigram; hidden reference to Archimedes
VI	**I** **I** **V** **:**~ 14		Bessarion-axis → **Bessarion born in 1407**
VIIa	**I** **R** GO		Regiomontanus-axis → **Regiomontanus hidden in IO RGO**

	I	
VIIa	**IS** **R** S RGO **I**	Sergios-axes → **Sergios hidden in IO RGO**
VIII	Boustrophedon	
	RGO IO 1462	→ Regiomontanus age 26 in 1462 and his year of birth 1426+10 = 1436
VIII		
	D E E **S** I O	*Carmen figuratum* around a central S → Dei / Deo meaning *sub praesidio Dei* and *Deo*

We now consider these hidden meanings in greater detail.

Hidden references to an out-dated astrolabe

Sergios and the Byzantine astrolabe of 1062

The chronogram in the lower right corner (axis VIIIb) reveals the date 1062. The akrostych (axis I) reveals the 400 years elapsed since 1062: SVB CD ANNIS, with SVB meaning "in, within, during, at, or by", so that the phrase can be taken to mean "on the occasion of the passing of 400 years". The number 400 just happens to have some significance in the Bible.[118] The name SeRGiOs is hidden in the word SVRGO reached by axis VII (see further below). By implication, Constantinople is embodied in these references to the old astrolabe. In addition, the words OPVS DICTI DE CARDINE DIVI, "a work explaining the pole or axis of the sky", that is, "a work explaining the rotation of the heavens", correspond to the first verse of the poem on the 1062 astrolabe (see below). Furthermore, Regiomontanus has exercised considerable ingenuity in incorporating into his epigram the colon and tilde found on the 1062 astrolabe:~ see below.

---

118 See Meyer & Suntrup, *Lexikon der mittelalterlichen Zahlenbedeutung*, cols. 833-834, citing *Genesis* XV.13, and various verses in *Kings* 1-2. The hundred suggests perfection and everlasting life, and the four, the evangelists, the cardinal virtues and stages of exegesis.

The words of the epigram rearranged to reveal a hidden meaning

One of the insights that led Berthold Holzschuh to recognize this connection with the 1062 astrolabe was his realization that the same words can be rearranged and rendered with a quite different meaning:~

SVB BESSARIONIS PRAESIDIO
SVRGO
IOANNIS *OPVS DICTI DE CARDINE DIVI*
ROMAE 1462

"Under the protection of Bessarion, I arise in Rome in 1462
as a work of Ioannes explaining the axis of the sky, *i.e.*, the rotation of the heavens."

Thus the "clear" or "Klartext" is revealed. It reproduces the sense of the beginning of the Greek verse on the Byzantine astrolabe. We can be confident that this was intended by Regiomontanus, and it was perhaps from this version that he conceived the final text and construed the epigram as it is engraved on the astrolabe.

However, there is a lot more hidden in the epigram and the image of the angel.

The angel and the Bessarion axes

The angel Bessarion and the Cardinal Bessarion conveniunt in unum

The name BESSARION in the dedication refers both to the image of the angel and to the Cardinal. Regiomontanus in his correspondence used the word *divus* to refer to his patron.[119] The word *divus* was sometimes used in the Renaissance in reference to exceptionally pious persons, revered religious authorities or highly respected rulers, or as *diva*, for women of exceptionable beauty.[120] The

---

[119] See the inscription DIVVS FE. VRBINAT. DVX on a bas-relief attributable to Francesco di Giorgio Martini or Paolo da Ragusa preserved in Museo civico di Pesaro, featured on the cover of Tommasoli, *La vita di Federico da Montefeltro*. This and another, bearing the inscription DIVI FEDERICI VRBINI DVCIS, are illustrated in Lauts & Herzner, *Federico da Montefeltro*, p. 388.

Bessarion, in a letter to Buonconte, even refers to "*divinus Plato*". See Clough, "Cardinal Bessarion and Greek at the Court of Urbino", p. 162.

Niermeyer, *Mediae latinitatis lexicon minus*, p. 455, translates it as "holy, sacred" and states that it was used "of deceased as well as living emperors and kings".

[120] On Regiomontanus' correspondence see n. 31 above. In a letter dated 15th February, 1465, from Rome to Jakob von Speyer, the court astronomer of Federigo da Montefeltro in Urbino, he wrote:

> *Cui quero plus favet divus ille Bessarion, dominus meus Reverendissimus, quam Umbrorum principibus Federico et Octaviano ?* To whom, I ask you, is *divus*

name Bessarion thus embodies the saint who was long dead and the cardinal who was, at least mentally, very much alive. The name BESSAR-IO-N, which has the same abbreviation IO for Ioannes, Greek (IΩ)αννησ, embedded in it, and even all the letters I-O-A-N-E-S,[121] has been given due prominence:~ see below. His two names, the baptismal BA(sileios) and the adopted IOANNIS come together in one in axis II.

Regiomontanus put the "R" in the middle of the 11-letter word BESSARIONIS exactly at the centre of the inscription (axis V) as it is engraved on the astrolabe, that is, at the vertical axis of the astronomical markings above the inscription, which passes through the celestial poles (with north at the top). This axis through the vertical bar of the "R" passes through the vertical bar of the "T" below it, which has a "C" before it, and then on to "SV" in the third line. We can conceive here RCTVS, an acceptable abbreviation (usually as *rct9*) for RECTVS, "straight".[122] One of the many meanings of the phrase DE CARDINE DICTI is that one could say the angel embodied in the BESSARION of the inscription was from the pole (*cardo*) or from the axis (*cardo*) of the astronomical markings above, which is where we find the angel.

Bessarion's names and rank

The hidden messages in axis II are that B / A / IO / ANNIS refer to Bessarion's baptismal name BAsileIOs and his adopted name IOANNES (Ioannis in Greek). The fact that there is another axis (VI) through the I of BESSARION again draws attention to the fact that IO for IOannes is embedded in his official name. The phrase DE CARDINE DICTI is a word-play on his rank of Cardinal. He is one of the DIVI DE CARDINE DICTI. The letters in axis IV indicate that he is one of the BENEDICTI ROMAE, which can be taken as a reference to the Pope and the Cardinals of the Lateran College.

Bessarion's year of birth

The combination of the colon and tilde " :~ ", also found on the Byzantine astrolabe, tells one to keep on reading (or to keep on looking!).[123] The year of

---

> Bessarion, my much respected master, more devoted than to the Dukes of the Umbrians, Federicus and Octavianus?

See Curtze, "Regiomontan's Briefwechsel", p. 292, and Zinsmeister, "Der Briefwechsel Regiomontans", p. 25: " … der große (göttliche) Bessarion … ". On Jakob von Speyer see also the text to n. 557 below.

121 See the text to n. 23.

122 Lewis & Short, *Latin Dictionary*, col. 1552b; Cappelli, *Lexicon abbreviaturarum*, p. 319 (*r°* for *recto*), 322 (*rcs* and *rcus* for *rectus*).

123 See already n. 59 on the colon and tilde in Byzantine Greek. Bischoff, *Latin Palaeography*, pp. 169, observes that the ancient grammarians up to Isidore (Seville, early 7th century) gave clear and simple instructions for the use of punctuation marks:

birth of Bessarion is revealed by an ingenious combination of letters and numbers in the right mesostych or axis VI: "I I V", necessarily, by the rules for acrostics, to be read downwards, connected by the colon and tilde to the "14(00)" of 1462. One does not even have to drop the "62", because 1400 is implied by the "14" together with the "0" above it, and with the other "0' directly above the "62". In any case, the result is I + I + V + 1400, or 1407.[124]

Now Bessarion was born in Trebizond on January 2, but his year of birth is nowhere stated directly in the known sources and is much contested. The date 1403 can be extracted from a remark by Capranica *ca.* 1525 to the effect that Bessarion died in December 1472 aged 69. (Capranica had delivered the funeral oration for Bessarion in 1472.) John Monfasani has argued convincingly for 1408, but the evidence he presents, although it excludes 1403, does not exclude 1407. In fact, a remark by Bessarion himself suggests 1407,[125] which confirms

---

low point (*comma*) for a short pause; medial point (*colon*) for the middle pause, and high point (*periodus*) for the end of a sentence. See his notes for collections of, and studies on Latin texts on punctuation before the Renaissance. On Renaissance Latin punctuation see also the interesting remarks in IJsewijn & Sacré, *Companion to Neo-Latin Studies*, II, pp. 473-474. However, a Neo-Latin epigram mimicking a colon and tilde found in an 11th-century Byzantine Greek inscription was (alas) not known to them.

124 The addition of numbers in this way is acceptable in chronograms: see White, "Chronograms". I dismiss the reading of "IIV" as 3, which would give the years 1403 or even 1397 because these are untenable: see the next note.

On mixed dates from the 15th and 16th centuries composed of both Roman and Hindu-Arabic numerals see King, *The Ciphers of the Monks*, pp. 284-286.

125 Monfasani, "Bessarion's Latin Eulogists and his Date of Birth", first published in 1986, reprinted in *idem*, *Byzantine Scholars in Renaissance Italy*, VI, especially pp. 112-123. (Earlier scholars had proposed 1403, and some (such as Krumbacher, *Geschichte der byzantinischen Litteratur*, I, p. 117) *ca.* 1395.)

Bessarion himself said he was "*not* born twenty-four years", that is, he was 23 years old, going on 24, when he was "honoured by Greek princes". According to Monfasani, this event occurred very late in 1430 or in early 1431, when Bessarion moved to Greece. If Bessarion had been born in 1407, he would have been "not born 24 years" in late 1430 and would have attained age 24 in January 1431. If he had been born in 1408, as proposed by John Monfasani, he would have been "not born *23* years" in late 1430 and would have attained age 23 in January 1431.

In the spring of 1438, Ambrogio Traversari wrote that Bessarion was younger than the other Greeks at Ferrara, in fact, that *est enim tricenarius*, "for he is indeed (only) 30 years old". If we interpret this terse phrase as "only in his early 30s (and not like the others in their 40s and 50s)", rather than "precisely 30 years old and not yet 31", 1407 would fit here as well.

John Monfasani considered these two pieces of evidence the other way round, taking the date 1408 from Traversari's statement of 1438 and seeking to corroborate it from Bessarion's own statement.

the date that Regiomontanus seems to have been trying to convey. As we shall see, Trebizond is also there, implicitly, and even explicitly (see below).

Hidden references to Regiomontanus

Regiomontanus' name

The letters RGO reached by the I in axis VII are adequate in medieval Latin to denote REGIO. However, around that axis one also finds the scattered letters IOANES, and in the next space (axis VIII) the letters IO which are not only the abbreviation of his name but also the beginning of that name, continued on the fourth line.

Regiomontanus' age in 1462 and his year of birth

By a playful combination of numerical manoeuvres, including reading a pair of numbers backwards, and again, *quite by chance*, a cryptogram in the lower right corner reveals the age Regiomontanus attained in 1462, namely, 26 (from "RGO IO 26" read as a boustrophedon) and also his year of birth, 1436 (= 14"26" + 10).

Other dual personalities in one

The angel Bessarion and Regiomontanus conveniunt in unum *(?)*

The angel is depicted as a beautiful youth. All young men, one could argue, have a certain beauty, not least for an ageing cardinal. Regiomontanus was probably 25 years old when he engraved the angel, which is a typical 15th-century Nuremberg or Vienna angel.[134] Furthermore, is not inconceivable that Regiomontanus looked better in 1462 than he did later when he was depicted by Schedel (see **Fig. I-1**).

We now know that Regiomontanus put the angel on the astrolabe to please Bessarion, for his epigram indicates that the image is of the angelic St

---

The date 1408 is now favoured, but the earlier dates have not been abandoned: see Zorzi, "Cenni sulla vita e sulla figura di Bessarione", pp. 1 and 18.

134 ** See the text to n. 67 above.

Bessarion. Price was criticized in 1959 when he suggested that the angel might be a portrait of Regiomontanus himself.[135] Price knew it was not Bessarion, although, in fact, it is, but not the Cardinal. Price was perhaps right in supposing that the young man might represent Regiomontanus.

Bessarion was realistic enough to leave out some of the miracles of his namesake when he wrote his panegyric on the angelic saint,[136] but he surely was not naïve enough to believe that St Bessarion really was an angel.[137] Maybe Bessarion, the first time he saw Regiomontanus' angels on the Vienna calendars (see **Fig. I-11)**, said, jokingly, "They look a bit like you!". And maybe he was serious, and maybe this was an "in-joke" between the two men.

***Bessarion and Regiomontanus* conveniunt in unum**

The two names come together in the same epigram, once one realizes that IOANNES is to be associated with the letters RGO.

***Sergios and Regiomontanus* conveniunt in unum**

Regiomontanus has ingeniously used the verb SVRGO to combine the name of the dead SeRGiOs and his own name, (Ioannes de Monte) ReGiO. This was noted by Robert Dietz of the Frankfurt Seminar in April 2005, in the early days of our encounters with the two astrolabes together.[138]

Miscellaneous other hidden meanings

Favourite cities

The city of ROMA is specifically mentioned, but hints of VI(enna) and VE(netia), Bessarion's favourite city in Italy, are also identifiable at the tops of axes I ("V"), VI ("I"), and VIIIa ("E").[139] This appears to be one reason for Regiomontanus to have set the "S" of SVB outside the main rectangular area of the inscription.[140] Also VI for Vienna is hinted at in axis III, and to reach Rome one descends more or less by a north-south axis (*cardo*). I have already noted that Constantinople is implicit in the references to the 1062 astrolabe, and Trebizond from Bessarion's year of birth.

---

135 See nn. 107 above and 518 below.

136 See the study cited in n. 106 above.

137 There is always a wide gap between official religion and popular belief, but, in this case, we are dealing not with a popular belief but with an official doctrine that went over the edge. This is one reason why the veneration of certain saints as angels has been removed from the Eastern liturgy.

138 See also the text to n. 111 above.

139 Cappelli, *Lexicon abbreviaturarum*, pp. 389 (VI for Vienna) and 388 (Ven for Venetiis).

140 See also the text to n. 94.

Favourite authors (?)

It just happens that in the space roughly defined by the second and third quarters of the epigram one can find letters to form the names EVC(L)IDES, ARISTARC(H)OS, ARC(H)IMEDES, APO(LL)ONIOS, T(H)EODOSIOS, PTO(L)EMAIOS and his A(L)MAGEST, PROC(L)OS and T(H)EON. Obviously, this is pure coincidence, but it is nice.

And, of course, one does not have to look far for P(L)ATO and ARISTOTE(L)ES, as well as VIRGI(L) or VERGI(L), OVID or CICERO. The word SVRGO is nice because it contains the letters VRG of Vi/eRGil, who was possibly one of the sources of the Regiomontanus' inspiration for making an acrostic epigram. See further below.

Archimedes

Certainly Regiomontanus would have noticed the AR for ARchimedes in BessARion's name (this was first noted by Berthold Holzschuh – see between Parts I and II). Bessarion was deeply involved in the Italian rediscovery of Archimedes, and would have talked about it to Regiomontanus. Also Bessarion knew an Italian mathematician-artist who had recently made his own copy of Archimedes.[143] Furthermore, it was around 1462 that Regiomontanus copied the Latin texts of most of Archimedes' works (MS Nuremberg Cent. V 15), comparing them with Greek and Latin versions in the possession of Bessarion[144] So, yes, Archimedes was on his mind. He had copied extracts from Archimedes and Apollonius already in Vienna,[145] and his main duty to Bessarion was to do justice to Ptolemy and Theon. So I am confident that Regiomontanus would have seen in the letters B / AR / IO (axis IIb) a reference to his (IO) indebtedness to Bessarion for introducing him to the most recent manuscripts on Archimedes. Also, the vertical segment of the R in BESSARION is aligned with the vertical segment of the T of DICTI below. With some stretching of the imagination we might dare to suggest that the two letters R and T when rotated about this axis defined by their vertical segments generate a sphere, a cone, and – with more stretching – a cylinder (this was first suggested by Berthold Holzschuh). To demonstrate that this might be more obvious to the student of Archimedes than to a student of anything else, we remind the reader that

---

143 ** See the text to n. 177.

144 Rose, *The Italian Renaissance of Mathematics*, p. 94, and Zinner, *Regiomontanus*, p. 91.

145 Folkerts, "Die mathematischen Studien Regiomontans in seiner Wiener Zeit", pp. 182 and 184-186.

Archimedes requested that a cylinder circumscribing a sphere be used to decorate his tomb, together with a statement of the ratio of their volumes. This was recorded by Plutarch (*Marcellus*, ch. XVII) and the tomb with these features was actually seen by Cicero (*Tusculan Disputations*, V, xxiii, 64-66).[146]

Configured letters

There are several *intended* combinations of configured letters. The letters BE-NE-DICTI ROMAE, with the "B" at the left principal axis (IV) call to mind the Papacy, and BE-NE-DICTI perhaps Christians in general. Also, the configured letters DEO with DEI at the telestych (axis VIIIa), read in the two directions defined by the central "S", the combination defining a "Y", symbol of the Trinity and favourite letter of the Pythagoreans (as 10 was their favourite number).[147] The five letters D-E-I-O-E define a pentagon, and the geometry of the regular pentagon is intimately linked to the divine proportion.[148]

Potential letter configurations

The dedication is even richer in potential religious statements, achieved by many accidental letter combinations. The most obvious of these is:~ IN and RI in two vertical axes not far removed from the left principal axis, which would call to mind:~ I(esus) N(azarenus) R(ex) Iudaeorum). Also, we can find MA(ria) V(i)RGO and MA(ter) DE(i).[150]

In fact, one can read all sorts of words and hints of names in different parts of the text. To give just one example:~ from the words B**ESSARI**ON and the words DIC**T**I, **S**V**R**GO and **OPVS** below it, one can extract the letters to form the word TRAPESVNTIVS, "from Trapezunt or Trebizond". Did either Regiomontanus or Bessarion notice this? Given that Bessarion's year of birth is

---

146 See Marshall Clagett in the article "Archimedes" in *Dictionary of Scientific Biography*, I, p. 213.

147 On the significance of the number 10 see n. 33 above.

The importance of "Y" to the Pythagoreans is noted, for example, in Felice Feliciano's *Alphabetum Romanum*, Verona, 1460 (see the text to n. 495 below), *sub* Y. On "Y" as symbol of the Trinity see, for example, Metford, *Dictionary of Christian Lore and Legend*, p. 268.

148 See n. 99 above.

150 * On MA for Maria, VGO for Virgo, and MA DI for Mater Dei, see Cappelli, *Lexicon abbreviaturarum*, pp. 209, 389, and 97 (DI), also Metford, *Dictionary of Christian Lore and Legend*, p. 165 *sub* "M".

152 See the references in n. 14 and also the text to n. 44 above.

revealed by letters in the same words, I suspect that it is highly likely that they did.

Regiomontanus and Vergil

In order to accompany Bessarion to Rome, Regiomontanus was forced to cancel his announced lectures on the *Georgics* of Vergil at the University of Vienna.[152] Vergil (b. 70 BCE, d. 19 BCE) was regarded by the Romans as their greatest poet, and his influence in Renaissance Italy is reflected in his role in Dante's *Divine Comedy*, guiding the poet through Hell and Purgatory to the gates of Paradise. The poems focus on one of the most devastating effects of civil war, namely, the depopulation of rural areas, with farms succumbing to neglect and ruin, and Vergil pleads for the restoration of agrigultural life in Italy.

Now there is a section in the *Georgics* (2:322-339) in which each element of the poet's name occurs on the first and then on every third line:

P(ublius) / VER(gilius) / MA(ro) / PART(henopeius) / CRE(monae).[153]

One could hypothesize that Regiomontanus was aware of such acrostics, and that it was Vergil who inspired him to compose an acrostic epigram in which the name of his new teacher, Bessarion, and his own name, were hidden. I leave this conjecture to the specialists.

How did Regiomontanus conceive his acrostic epigram?

We shall never know how Regiomontanus achieved this masterpiece, but we are free to speculate.

Regiomontanus would have needed about 60 letters for a hexameter followed by a pentameter, and each step would require consideration of the meter. From the outset, he wanted to include the dates 1462 and **IO**62. Maybe he could use the **IO** of his name! Perhaps form a chronogram giving both dates. Also he wished to include the CD ANNI passed since 1062. Maybe he could use the ANNI in his name (IO)ANNI(S).

Somehow he had to include Sergios and at least part of de Regio Monte. Maybe Bessarion had pointed out to him that REGIO is embedded in the name SERGIOS. The whole operation was SVB PRAESIDIO of Bessarion: the SVB would go rather nicely with the phrase SVB CD ANNIS. Maybe the expression OPVS DICTI DE CARDINE DIVI was already floating about in his head, inspired by the Greek inscription on the Byzantine astrolabe. And maybe he had thought of the Cardinals as DIVI DE CARDINE DICTI?

---

153 Zarker, "Acrostic 'Carmina Latina Epigraphica'", p. 130.

He probably started with a few words and phrases like BESSAR**IO**N, DIVVS, OPVS **IO**ANNIS, the word IOANNIS conveniently containing IO and ANNIS, ROMAE, CARDO, SVB PRAESID**IO**, ... Then might have come SVB DIVI BESSAR**IO**NIS ... PRAESID**IO**. With the angel at the CARDO, then the Cardinal was DE CARDINE DICTVS in more than one sense. How about **SVB** DIVI BESSARIONIS DE / **C**ARDINE DICTI PRAESI / **D**IO, broken at the slashes to form SVB C D in a potential akrostych. As luck would have it, the letters just to the right of this can form a second axis with B A IO and ANNIS referring to the two names of Bessarion, BAsileIOs and IOANNIS (the Greek Ιωαννησ can be read IOANNIS or IOANNES. If we somehow put the R at the middle of BESSARIONIS above the CT of DICTI we can have another axis, for the words RCT suggest a straight line. Now we need the ANNIS of IOANNIS at the beginning of the fourth line and we have SVB CD ANNIS in the akrostych. Move the S of SVB to the left and we have a second axis with B A IO and ANNIS. Also the letters BE of BESSARION are above NE DICTI, so add ROMAE to the third line and we have BENEDICTI ROMAE in another axis defined by the B. We need a verb before IOANNIS OPVS. Also, we need to refer to at least part of the name de Regio Monte. SVRGO would refer to both himself, ReGiO, and to SeRGiOs. An axis through the second I in BESSARIONIS could lead to the R in PRAESI and the GO in SVRGO

The result was schematically something like this:~

S	V	B	DIVI	**B**	**ESSA**	R	<u>**IO**</u>N	**I**	S	DE
	C	A	RDI**N**	**E**	**DIC**	**T**	**I** P	**R**	A	ESI
	D	<u>**IO**</u>			**ROMAE**	SVR		**GO**		<u>**IO**</u>
	A	N	NIS		OPVS					14**62**

The meter is almost perfect, the only problem being with DIVI, but the compromise was worthwhile.

Now we notice another potential axis with the VI in DIVI for Vienna and CARDINE and ROMAE below. The letters could be aligned to achieve the axis from the V downwards leading to the I of CARDINE. We need something else for Bessarion, and - lo and behold – we have the I of BESSARION which could lead to the second I of DICTI and then to the V of SVRGO: "7" and Bessarion was born in 1407 (we assume). We need to connect it to the 14. Remember the colon and tittle of the Byzantine astrolabe! At the end of the first three lines we can find DEI and DEO, which can be made into a *carmen figuratum* around a central "S". It would all take some squeezing and stretching of letters and spaces.

The result can be represented schematically something like this:~

	I	II		III		IV		V	VI		VII		VIII
S	**V**	**B**	DI	**V**	I	**B**	ESSA	**R**	**I**	ON	**I**	S	**DE**
	C	**A**	RD	**I**	N	**E**	DI	CT	**I**	P	**R**	A	**ESI**
	D	**IO**		**RO**		**M**	AE	**S**	**V**	R	**GO**		**IO**
	A	**N**	NIS			OPVS			**:**	~		14	62

Clearly we have no proof that the text and the axis of the epigram were conceived in this way. However, Regiomontanus must have spent a long time preparing this brilliant epigram.

The epitaph

The epigram is as much an epitaph for a dead Sergios, his now-useless astrolabe, as well as for a lost Byzantine Empire, as it is a declaration that a European could finally make a proper, if standard, astrolabe.[154] I propose a rendition to accommodate most, but by no means all, of the meanings in the short epigram:~

> "Under the protection of the divine (*divus*) angel Bessarion depicted by the pole (*cardo*) of the markings directly above (*cardo*), who is the saint who lived some 400 years after Our Lord, and who looks like one of the angels by the hinges (*cardo*) of the astronomical calendar made in Vienna in 1461, and whose name was adopted by Basileios, born in Trebizond in 1407, who now sometimes calls himself "Ioannes" and who is serving as one of the hinges (*cardo*) of the Church (of Jesus of Nazareth, King of the Jews, crucified in Jerusalem), whose adopted name can be abbreviated IO like 10, the favourite number of his favourite Pythagoreans, with parts of his name dividing this epigram almost in the divine proportion, both on the right and on the left, at about 62 units out of 100 from either end, I arise in Rome under the protection of the godly (*divus*) Cardinal "Ioannes" BessarIOn as a

---

154 For Euro-centric readers (forewarned in n. 42) it should be pointed out that already in the 10th century the Muslim astronomer al-Khujandî (see the text to n. 420) made a far more sophisticated astrolabe than that of Regiomontanus. In the 14th century the astronomer Ibn al-Sarrâj produced a quintuply-universal astrolabe, the like of which was never produced thereafter, either in the Islamic world or in Europe (see the text to n. 423 below). Regiomontanus would have had a very hard time trying to figure out how it worked.

European astrolabes from the 10th to the early 15th century were often marred by incorrect star-positions, false identifications of stars, and incorrect terrestrial latitudes.

gift for him made by IOannes de RegIO Monte in the year:~ 1462, which is 400 years after the year 1062:~ the date of the astrolabe made for Sergios that Bessarion brought from Constantinople, the hinge (*cardo*) of the world, to Italy and then from south to north (*cardo*) from Venice, the hinge (*cardo*) of Europe,[155] to Vienna, thence from north to south (*cardo*) to Rome, and which bears a Greek inscription discussed in Vienna in 1460-61 and referring to the astrolabe as:~ an icon demonstrating the axis (*cardo*) of the sky (*divum*), *i.e.*, the rotation of the heavens."

We may presume that Bessarion was impressed by this monumental achievement reflecting the past and the present and also embodying hope for the future. In particular, he must have marvelled at the ways in which:~

- ❖ there were two principal vertical axes vaguely associated with the divine proportion, giving prominence to the letters "B" and the "I" of the "IO" in the name **B**ESSAR**IO**N;
- ❖ the vaguely DIVVS Bessarion is thus vaguely placed in this way by means of the *proportio DIVINA*;
- ❖ there were altogether eight vertical axes with hidden meanings relating to topics of personal interest to the two men;
- ❖ the two first names of Bessarion, his baptismal name BA(sileios) and his "adopted" name, Ioannis (Greek) or Ioannes (Latin), are combined in a single vertical axis;
- ❖ two Bessarions, one dead and one still living, the first even an angel, are represented in an image and the single name, BESSARION;
- ❖ the names of two persons, one dead and one still living, are combined in a single word, SVRGO;
- ❖ two IOannes are mentioned;
- ❖ the two dates 1462 and 1062 are represented by a single chronogram;
- ❖ two astrolabes are mentioned in a single inscription;
- ❖ five words of the Latin constitute a rendition of one verse of the Greek poem on the 1062 astrolabe;
- ❖ the two years of birth of Bessarion, 1407, and Regiomontanus, 1436, can be extracted with a little imagination;
- ❖ several concepts are suggested by the word *cardo*;

---

155 See McNeill, *Venice: The Hinge of Europe, 1081-1797*, although the origin of the appellation is not clear.

- ❖ Rome is specifically mentioned but memories of Constantinople (1062), Vienna (1460-61), and Jerusalem (INRI) are evoked; and, last but not least,
- ❖ the epigram is almost a perfect elegiac couplet.

Some of these features would certainly have reminded Bessarion of the biblical phrase *convenerunt in unum* that could equally well apply jointly to the Byzantine Emperor and the Ottoman Sultan (see also Part III) as it was applied in the *New Testament* to Herod and Pontius Pilate. In fact, we can be fairly certain that Regiomontanus had heard this from Bessarion already in Vienna, and that the phrase had given him the idea to make the two astrolabes, their two makers, their two dates, the two Bessarions, and a lot more, come together in one. *Quite by chance*, but surely not unnoticed by Regiomontanus and Bessarion, the letters

			V		**I**	**V**	
C				**NE**	**N**	**N**	
	O		**R**				**M**
		N					

are to be found in the left-hand three-eighths of the epigram, and, if one does not object to using a given letter twice or more when necessary, and one does not mind too much that the T is rather far away from the other letters, they can be thought of as spelling CONVENERVN(T) IN VNVM. We do not need all of the letters to make the text legible: the reader should realize that in the Middle Ages abbreviations such as *9ire*, *9it* and *9iût*, where "9" is a letter of the extended medieval alphabet (like "&" for *et*, "and", which has survived to this day), often used for the prefix *con-*, would have been understood as *convenire*, *convenit* and *conveniunt*.[156] However, the negative Biblical connotations of this phrase are now replaced by the playful and more positive representation of more than one concept – be it astrolabes, cardinals and angels, people living and dead, names, or dates – in a single one.

---

156 Cappelli, *Lexicon abbreviaturarum*, pp. xxv-xxvi and 68-85, especially p. 69.

On an astrolabe from medieval Spain (Toledo? 14th century), to which we shall return below (see the text to n. 424 below), the letter "9" was used as a hard "c" or "q": see King, *In Synchrony with the Heavens*, XV-2.2, 2.8, 3.2, *etc*. Contemporaneous literary or epigraphic sources with this feature have not been identified; these would help to localize the environment of the astrolabe (Toledo is just a first guess).

What happened to the two astrolabes before the mid-19th century?

No astrolabes are listed amongst the personal belongings of Bessarion when he died in 1472.[157] We can imagine that he had already given the 1062 and 1462 instruments to someone or some persons who could appreciate them. Here I might mention the three astrolabes depicted in the back corner of the painting "St Jerome in his Study" by Vittore Carpaccio (1502).[158] The person depicted as St Jerome (or St Augustine?) has been identified in some sources as Bessarion, not least because he has a beard and there is a cardinal's hat on the floor in front of his desk. The astrolabes are not the correct relative size for us to state categorically that two of these *must* be the astrolabes of 1062 and 1462.[159] Also, there are two astrolabes depicted in the splendid frontispiece to Bessarion's manuscript of the *Geography* of Ptolemy.[160] Here the most influential astronomer of Antiquity is depicted as a king – a charming confusion with the Ptolemaic dynasty[161] – and he is holding up an astrolabe, a large one, large

---

157 Mohler, *Bessarion*, I, p. 324.

We may compare the medieval French astrolabe (#202) that in 1522 was given to Hadrianus Amerotius, Professor of Greek in Louvain, by his grateful student, the Benedictine monk of Liège, Paschasius Berselius. (This is the second of the two astrolabes mentioned in n. 84 above.) The piece is of Picard provenance and datable *ca.* 1350, and it bears a later inscription with the dedication. All of the numbers on it are engraved in monastic ciphers, a system quite distinct from the Roman and Hindu-Arabic numerals, and alas Amerotius did not mention these ciphers in his *De Græcorum notis arithmeticis compendium*, a survey of Greek number notations published in Paris in 1530 and appended to his *Lexicon Græco-latinum* in over 25 editions published during the next 50 years. He omitted them probably not least because he did not realize that the prototype of such ciphers was of ancient Greek origin. Amerotius died in 1560 and there is no mention in his will of this particular object, whilst other more mundane objects are indeed specifically mentioned. We do not know how this astrolabe came to be in an English private collection *ca.* 1920. See further King, *Ciphers of the Monks*, pp. 141-151, especially pp. 149-151. It is a charming coincidence that Berselius learned Greek from Amerotius, as Regiomontanus improved his Greek in the company of Bessarion.

158 See, for example, Fortini Brown, "*Sant'Agostino nello studio* di Carpaccio", and the most recent discussions in Bedini, "Falconi, Renaissance Astrologer", pp. 58-59 and figs. 5-6, and van Cleempoel, "Representations of Astrolabes in Western Art", p. 106.

159 Ronchey, *L'Enigma di Piero*, p. 239 ad fig. 144, states categorically that the asrolabe on the right *is* the 1462 piece. Alas, all of the images available to me are too small to reveal whether this is true or false.

160 See the illustration from MS Venice Marciana cod. gr. 388 and a description by Antonio Rigo in *Venice 1994 Exhibition Catalogue*, p. 423.

161 This was one of the favourites of Otto Neugebauer, gathered in his *Lesefrüchte*, privately distributed (see also n. 286 below). He cited the following quote (source withheld):

enough to be the Byzantine piece, although the rete design is different. Also, Ptolemy is not observing with it; indeed, it looks as though he is reading an inscription on its lower back! Another astrolabe of similar size hangs in a pavilion to the side, but this also is neither the 1062 piece nor the 1462 one.[162]

In fact, we do not know of any response to the 1462 astrolabe other than a painting by Piero della Francesca (see Part III).

The 1462 astrolabe remained unknown until 1848, when it resurfaced in Rome, and the 1062 astrolabe had resurfaced in 1844 in Brescia. See further Appendixes C and E.

Regiomontanus' astrolabe and his other gifts to Bessarion

Regiomontanus' astrolabe was the first of a series of gifts to his patron. Perhaps he presented the completed object to Bessarion on June 6 of the year 1462, which was not only the day when he became 26 and the numerical information in the inscription would become valid, but also, by a charming coincidence, the feast-day of St Bessarion in the Byzantine Church.[163] Perhaps the instrument was completed in Vienna in 1461, and Regiomontanus was still working on the dedication as they travelled to Rome in late 1461. In any case, in 1463 Bessarion received Peuerbach's *Epitome* of the *Almagest*, which Regiomontanus had apparently completed in 1462 (both years are mentioned in the preface); in 1464, a spectacular treatise on plane and spherical trigonometry, the first of its kind ever written; and in 1465, a new treatise on the armillary sphere.[164]

> "The idea (of Aristarchus that the sun was in the center of the universe) was soon outlawed by Ptolemy, who favored the geocentric view, and was in a position to back it up with political force."

The idea of the astronomer-king actually circulated already in medieval Europe. See Burnett, "King Ptolemy", pp. 340-343, and also, on this miniature, van Cleempoel, "Representations of Astrolabes in Western Art", pp. 99-100 and 101.

162 Only the back is shown, and this bears a double universal horary quadrant and a double shadow square. On such features see the references cited in nn. 390 and 391 below.

The minature also features a quadrant and two different varieties of *torquetum*. Since the history of the latter is by no means clear, especially its transmission from al-Andalus to Northern Europe, these illustrations are of considerable interest and should be studied in detail. See Hartmann, "Die astronomischen Instrumente des Kardinals Nikolaus Cusanus", pp. 17-28 and pls. 1-3; Zinner, *Astronomische Instrumente*, pp. 177-183; Lorch, "The Astronomical Instruments of Jâbir ibn Aflah and the Torquetum"; and North, *Richard of Wallingford*, II, pp. 297-300, and *idem*, *The Ambassadors' Secret*, pp. 143-149; and the literature there cited.

163 See the text to nn. 104 and 116.

164 Zinner, *Regiomontanus*, p. 88. The treatise on trigonometry is published as Hughes, *Regiomontanus on Triangles*.

(Regiomontanus also made a pocket sundial for Pope Paul II[165] and, later, *ca.* 1470, a large astrolabe for Matthias Corvinus, King of Hungary.[166]) Bessarion gave Regiomontanus several important manuscripts.[167] When Regiomontanus died, he left some 145 books and various instruments: an inventary of 1522 survives, but not more than a dozen of the books and a few of the instruments.[168]

In what follows, the reader should not lose sight of the fact that Regiomontanus' agenda – a new astrolabe for Bessarion with its brilliant epigram and its image of St Bessarion – is modest when compared with Bessarion's agenda to revive the heritage of ancient Greece, or at least to make it available in Latin, and to save the few surviving fragments of the Greek world, let alone Europe, from the Turks. For over twenty years, nobody had seriously heeded or acted upon his pleas for a crusade.

---

165 This, which bears a profile of the Pope and an inscription "Paulo Veneto Pape II Italice Pacis Fundatore. Roma", survives in Nuremberg. Since it was part of Regiomontanus' legacy, either it was never presented to the Pope, or what we have was a model for another that was given to the Pope, but which is not known to survive. See further item VII-8 in Appendix D.

166 On this, alas known only from a written source, see Zinner, *Astronomische Instrumente*, p. 480, and King & Turner, "Regiomontanus' Astrolabe for Bessarion", p. 186.

167 Zinner, "Einige Handschriften des Regiomontanus", also *idem*, *Regiomontanus*, pp. 297-371.

Regiomontanus' copy of Theon's commentary on the *Almagest* (MS Nuremberg Gr. Cent. V. APP. 8) was given to him by Bessarion. See also Monfasani, *Collectanea Trapezuntiana*, p. 673.

A manuscript of the *Aphorismi astrologici* dedicated to Bessarion by Ludovigo Ricci Corner was passed on by the Cardinal to Regiomontanus. See Rigo, "Bessarione e Regiomontano", p. 70, and *idem*, "Gli interessi astronomici del cardinal Bessarione", p. 109.

168 See Petz, "Urkundliche Nachrichten über den literarischen Nachlaß Regiomontans und B. Walthers 1478-1522", especially pp. 247-262 on the inventory of manuscripts, and King, "Die Astrolabiensammlung im Germanischen Nationalmuseum", I, pp. 109-110, on the instruments.

INTERLUDE II:~ On esoteric symbolism

In 1975, the distinguished Renaissance scholar D. P. Walker published an article entitled "Esoteric Symbolism" which dealt with the connection between games and verse in 16th-century French poetry. In the introduction he made the following observations:~[169]

> This essay deals with symbols having a hidden meaning or meanings which can be grasped only by interpreters who have some special skill or fitness. I shall confine myself largely to verbal symbols, to ordinary language. My examples will mostly be taken from the sixteenth century, a period when esoteric language was much in favour, but when the principles behind its use were not often discussed. Among the many fields in which esoteric language was and is used, I shall concentrate on one where it has always been prevalent: religion. But I want first to mention some of these other fields, which may turn out to be relevant to religious esoteric language.
>
> In military and diplomatic life, one often conveys a hidden meaning by means of a code or cipher. This brings us to one basic distinction by which all esoteric language can be divided into two classes. A cipher may be so devised that either, as in most present-day ones, the enciphered text presents an apparently unintelligible series of letters, Class I, or, as in most of the methods proposed by (the German monk) Trithemius (in his *Polygraphia* and *Steganographia*) in the early sixteenth century, the enciphered text reads like an ordinary message, but the clear[171] is a quite different message, Class II. In more general terms: a meaning may be hidden, either by using obscure, difficult, apparently unintelligible language, but susceptible of only one correct interpretation, or by using language that conveys two or more meanings, the less evident of which is usually the more important. … … (To) take (other) examples from the

---

[169] Walker, "Esoteric Language", pp. 218-220. See already the text to n. 10 above.

[171] * The term "clear" here signifies "clear text", in the sense of what the text was intended to convey. This is the sense of the German *Klartext*.

> Bible: any parable is in Class II, and the beginning of the Gospel of John is in Class I.
>
> A vast field of esoteric language of Class I is that of technical jargons, ranging from medicine to music, from logic to law. Here I want to make another distinction, applying also to Class II, which seems to me fundamental. Either the jargon may be used, or even devised, in order to be incomprehensible to all but the initiate, Class A; or it may just happen to be so, Class B. This distinction rests, of course, primarily on the intention of the user, so that the same jargon might come into both classes. … …
>
> Finally, before I come to religion, there is an important field of esoteric language, Class IA or IIA: various kinds of games – guessing games, in which players invent deliberately obscure symbolic expressions of which their opponents must guess the meaning (charades, riddles, anagrams, picture-messages), or games which involve expressing two meanings with one set of symbols (acrostics, enigmas, puns). I have called these important because I want to suggest that these games are closely allied to certain kinds of verse, music, and picture, and, more generally, that the delight in sheer ingenuity of expression, in difficulties overcome, which appears in a pure form in such games, is an important element in many forms of art, perhaps in all art (both from the artist's and the recipient's point of view), though its importance varies, I think, considerably at different periods.
>
> There is a striking example of the connexion between games and verse in the dominant school of French poetry of the early Renaissance, the *Grands Rhétoriqueurs*, as they have been called. First, there is the obvious connexion: they are very fond of acrostics, puzzle-poems, complicated puns, etc. Secondly, their attention is often concentrated on feats of technical ingenuity, especially on difficult rhyme schemes. … …

In Part II, I investigated an enciphered poem composed by Regiomontanus that reads like a normal text. Its full meaning can only be understood by the initiated. **Regiomontanus' epigram belongs to Walker's Class IIA.** The game I played to decipher it was surely played by Bessarion when he first saw it. But as he played the game that Regiomontanus had planned for him, the two of them, not necessarily alone, saw that the epigram, even deciphered, just happened to have even more hidden meanings than Regiomontanus had realized. For, as luck would have it, **Regiomontanus' epigram also belongs to Walker's Class IIB.**

In Parts III and IV, we shall be playing another, quite different game with the now deciphered epigram, which, *mainly quite by chance*, and not least thanks to the nature of Latin, contains a completely different set of potential "messages". The evidence that somebody – surely Bessarion and Regiomontanus together, if only at the outset of the venture – already played that game is a painting by Piero della Francesca.

The operation that Bessarion and Regiomontanus conducted probably had no counterpart in the Renaissance, just as my presentation of it more or less defies all of the rules of modern scholarship on classical and medieval texts. I am happy to acknowledge that I feel more comfortable working with horizontal texts in medieval scientific Arabic or with vertical astronomical tables than jumping around inside a neo-Latin epigram. However, what we shall find are some 60 "monograms" for names and words more or less in the right places that are relevant to Piero's painting. A statistician might be able to calculate the probability of 62 letters arranged in four lines yielding some 60 meaningful combinations in Latin. If I had found just one or two possibly relevant names, this study would end right here.

PART III:~ PIERO'S PAINTING AND REGIOMONTANUS' ACROSTIC

"Time present and time past
Are perhaps both present in time future,
And time future contained in time past."
T. S. Eliot (1888-1965).

"No picture could exude a more pronounced air of geometric control and no painting was ever more sumptuously planned." Martin Kemp, *The Science of Art* (1997), p. 30, on Piero's "Flagellation".

"The "obsessive" correctness of its geometry makes (Piero's "Flagellation") a most rewarding painting for detailed mathematical analysis." Antonio Criminisi, "Getting into the Picture" (website).

"The identity of the three principal persons and the purpose of their presence remains a mystery in spite of all the studies and investigations of specialists." Translated from Stefano Zuffi, *L'art au XVe siècle* (2005), pp. 168-169.

"The raw materials with which the historian of Renaissance art works are of two kinds: the artefacts which are the central subjects of study; and written evidence which may provide indications of why the artefacts look like they do. In the standard histories of Renaissance art, the 'simple' business of juxtaposing the works with the documentation has become overlain with heavy veils of modern interpretation – most recently by incestuous discussions of theory and methodologies. I am not suggesting that artefacts and documents can simply be left to speak for themselves. Rather I will be asking what we may consider to be the relevant sources and how their nature as cultural products in their own right can be respected." Martin Kemp, *Behind the Picture* (1997), p. vii.

Piero della Francesca: the man and his works

Piero was born around 1412 in the small town (Italian *borgo*) of Sansepolcro, situated at the centre of an agricultural basin in the upper valley of the Tiber, some two-thirds of the way between Urbino and Arezzo. The town apparently takes its name from a story that a pilgrim brought back some stones from the Holy Sepulchre in Jerusalem that were then built into the wall around the town. Piero's father was Benedetto di Pietro, a shoe-maker, and his mother was Romana di Renzo di Carlo di Monterchi; their eldest son was called Piero after his grandfather and, again after grandfather and father, also dei or de' Franceschi, di La Francesca or della Francesca.

Piero was a mathematician of consequence, not only in the mathematics of Euclid and Archimedes, but also in the new mathematics of perspective. As an artist, as well as working in his hometown of Sansepolcro, he accepted assignments in Florence, Rome, Ferrara, Rimini, Arezzo, Monterchi, Loreto, Perugia and Urbino. Piero died and was buried in his native town in 1492.

A large full-size painting of Piero attributed to Santi di Tito (b. 1536 in Sansepolcro, d. 1603 in Florence) hangs in the Museo civico in Sansepolcro. Nobody seems to like this painting, one reason surely being that its attribution precludes it being a "real" portrait.[172] But art historians surely do not like paintings of artists like this one. If Piero had been painted "in action", working on a fresco, modern interest might be greater. In the numerous works on Piero's art, this portrait features very seldom, perhaps only twice.[173] However, the painter certainly captured the mind of his subject, for he is shown in humanist dress with his hand on two books, the names of their authors clearly visible on their spines: EVCLIDES and ARCHIMEDES. The inscription appropriately reads:

PETRVS DE FRANCESCA
EX NOBILI FRANCISCOR[VM] FAMILIA
PICTVRÆ ARITMETICÆ GEOMETRIÆ
AMPLIFICATOR

"Piero della Francesca, of the family of the noble Franciscans,
who greatly contributed to the arts of painting, arithmetic and geometry."

---

172 It is not mentioned in Lecchini & Collareta, eds., *Disegni di Santi di Tito*, or Spalding, *Santi di Tito*, or even in Maetzke & Gallopi Nappini, *Il Museo civico di Sansepolcro*!

173 The only exceptions seem to be "Evelyn", *Piero della Francesca* (1912), where the portrait features on the cover, and Brilli, *Alla ricerca di Piero della Francesca* (1994), p. 15.

A related wood-cut was used by Santi's colleague, the historian Giorgio Vasari (b. Arezzo, 1511, d. Florence, 1574), to preface the several pages devoted to Piero's life in the 1568 edition of his *Le vite de' più eccellenti pittori scultori e architettori*. The face could be that of the same person, somewhat younger, and a Grace is painting above him, and two *putti* on either side are brandishing geometrical instruments. The inscription reads:

PIERO DELL FRANC. PIT.
DAL BORGO A S. SEPOL.

"Piero della Francesca, painter, from the *borgo* at San Sepolcro."

(See **Figs. III-1 and III-2** for both images.) Vasari, who worked with Santi in Florence, writes enthusiastically of Piero's life and works, sometimes quite specifically, and from his detailed and often well-informed accounts, it is clear that a substantial number of Piero's works are lost, some already in Vasari's time. In fact, it seems that only six out of the 18 works attributed to him by Vasari are still extant,[174] although the survival of the Arezzo frescoes means that the proportion of his works surviving is substantially higher than this.

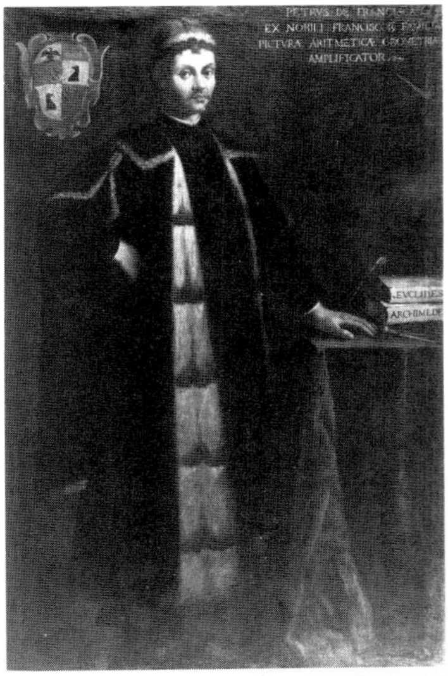

[174] Field, *Piero della Francesca*, p. 184.

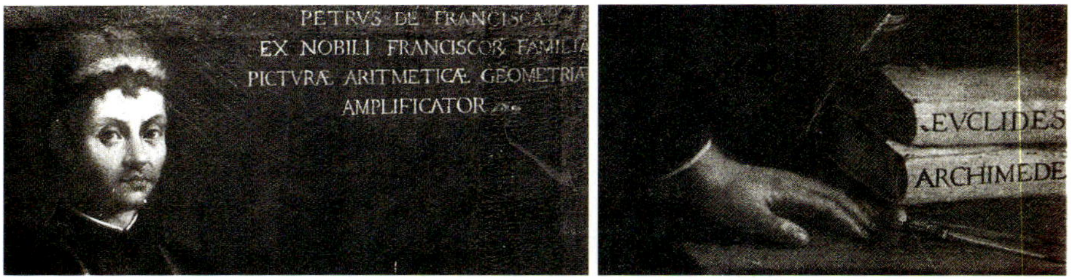

Figs. III-1a-c:~ The “portrait” of Piero (180 × 118 cm) attributed to Santi di Tito (1536-1603) and details of his face and the two books near his left hand. [Courtesy of the Museo Civico, Sansepolcro.]

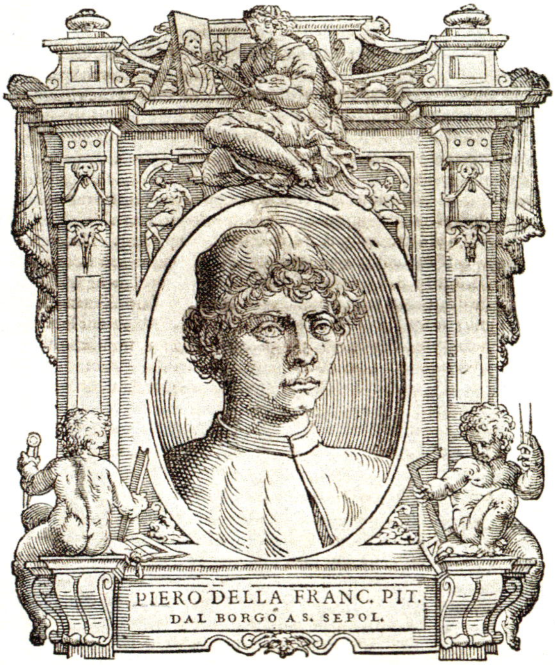

Fig. III-2:~ The woodcut “portrait” of Piero in the 1568 edition of Giorgio Vasari’s *Le vite de’ più eccellenti pittori ...* (first printed 1550). [Courtesy of the Accademia Raffaello, Urbino.]

Piero – the mathematician

Piero belongs to two modern worlds that seldom meet: besides being one of the best-known painters of the Quattrocento, he features in the authoritative *Dictionary of Scientific Biography*. Vasari's remarks about Piero's mathematical prowess, which are borne out by his surviving writings, bear repeating:~[175]

> "(Piero) was regarded as a rare master of the difficult problems of regular bodies, of arithmetic and geometry, but he was prevented by the blindness that overtook him in his old age, and then by death, from making known his brilliant researches and the many books that he had written, which are still preserved in his hometown of Borgo. ... ... / In his youth Piero applied himself to mathematics, and although when he became fifteen it was decided that he should become a painter, he nevertheless kept up with his earlier studies. Indeed, it was because of his marvellous achievements both in these (mathematical) studies and in painting that he was employed by [Federigo][a] da Montefeltro, the earlier duke of Urbino, ... (S)ome of Piero's writings on geometry and perspective have been preserved (in Urbino), and in these he was in no way inferior (in those sciences) to anyone of his own or indeed of any other time. And this is illustrated by all his works, which are full of (figures showing examples of) perspective,[b] ... . / As I said earlier, Piero made an intense study of painting and practised copiously in perspective. He acquired an excellent knowledge of Euclid, understanding better than any other geometrician the nature of all of the optimal curves drawn on the regular bodies (when viewed in perspective?);[c] and the clearest elucidations of these matters come from his pen. ... ... / (Piero's) books, for the most part, are in the library of the second Federigo,[d] duke of Urbino; their qualities are such that because of them he has justifiably acquired the reputation of being the leading geometrician of his day."
>
> a Text has Guidobaldo, who was born only in 1472 and lived only to 1508!
>
> b Text has *piene di prospettive*.
>
> c Text has *tutti i miglior giri tirati ne' corpi regolari*.
>
> d Text has *del secondo Federigo*. It seems that Federigo della Rovere is meant here, the last duke of Urbino who died age 18.

Piero's own compilations are the following:

---

175 See Vasari, *Le vite de' più eccellenti pittori ... / Lives of the Artists*, in the bibliography. Translation based on the London 1987 English version (I, pp. 191 / 192 / 196 / 198, with minor changes incorporated from the Italian editions of Siena 1791 (III, pp. 247 / 250-251 / 256 / 259) and Novara 1967 (II, pp. 375 / 376 / 382 / 384).

De prospectiva pingendi, *On Perspective in Painting*, extant in four Latin and three Italian manuscripts, each containing about 80 illustrations.

Del abaco, *On Computation*, extant in a single Italian manuscript in Florence, with 134 autograph illustrations. (Works with this kind of title usually have nothing to do with the abacus as such.)

De quinque corporibus regularibus, *On the Five Regular Solids* (that is, the tetrahedron or pyramid, hexahedron or cube, octahedron, dodecahedron and icosahedron, each with geometrically similar sides), extant in a single Latin manuscript (from Urbino) in the Vatican, with 106 autograph illustrations. This was published in 1995 with a facsimile together with a critical edition of the Latin text accompanied by the Italian version by Luca Pacioli (see below), and a critical edition of the drawings.[176]

I have no intention of discussing these works further, not least because they have already been studied in detail, but also because they do not seems to cast any light on the basic geometry of that one painting of his that concerns us here. Rather I refer the reader to the recent secondary literature, especially the article by Menso Folkerts on Piero and Euclid, now augmented by a book by J. V. Field on Piero's mathematics, as witnessed by his treatises and by his art, and by a new book on the history of perspective by Kristi Andersen.

Piero's own copy of Euclid may be preserved in some library somewhere (though Menso Folkerts did not find it). On the other hand, in 2005, one of the dozen or so Latin manuscripts of various works by Archimedes – MS Florence Riccardiana 106 – was recognized by James Banker as being copied in Piero's hand – a hand known, for example, from MS Florence Ashburnham 359 of the *Del abaco* (see below).[177] (Another of the surviving Latin manuscripts of Archimedes is in the hand of Regiomontanus, for which he had consulted Greek

---

176 Namely, as part of a project initiated in 1992 to publish all of Piero's works, organized by Cecil Grayson, Marisa Dalai Emiliani and Carlo Maccagni, who edited this treatise. See Piero in the bibliography.

177 See Banker, "A Manuscript of the Works of Archimedes in the Hand of Piero della Francesca". On this manuscript, not used by Marshall Clagett in his *Archimedes in the Middle Ages* because he apparently did not realize what was in it, see his note 23 on p. 12 of his vol. I.

The library website (www.riccardiana.firenze.sbn.it/eventi_archimede.html) states that the works of Archimedes in the manuscript are *De sphaera et cilindro* (Book I, fols. 1r -29v, Book II, fols. 10r-17r); *Circuli dimensio* (fols. 17v-18v); fol. 19 blank; *De conoidalibus et sferoidibus figuris* (fols. 30r - 51v); *Archimedis inventa circa elicas hoc est spirales lineas et spatia dictis li*

neis contenta (fols. 52r-65r); *Archimedis Planorum aeque ponderantium inventa vel centra gravitatis planorum* (fols. 65r-68v); *Archimedis de his que aeque ponderarti* (fols. 69r-73r); *Archimedis quadratura parabule* (fols. 73r-78r); and *Archimedis tractatus de arene numero* (fols. 78r-82r).

and Latin manuscripts of Bessarion himself.[178]) Of course, Piero's mathematical writings are replete with the standard abbreviations of Latin words. He also used them in the captions to his paintings, for example, SCI for **S**AN**C**T**I** on "The Flagellation", and MIA for **M**iser**I**cordi**A** on both lower corners of the polyptych displaying the Madonna della Misericordia.[179]

Vasari's ungenerous statements that the Franciscan mathematician Luca Pacioli (1445-1517), also a native of Sansepolcro, who was born and died there, plagiarized Piero's works were unjustified and have long been disproven.[180] Luca began his *Proportione divina* in Milan in 1496-97, and the work was published in Venice in 1509, together with his Latin edition of Euclid's *Elements*. Luca praises Piero in several places in his various writings. And if anyone deserves praise and repeated mention, it is Euclid and Archimedes, whom both Piero and Luca rightly revered. To the history of art, Luca is perhaps best known from the portrait of him holding a pointer over a geometrical diagram, with a copy of Euclid by his side, and a young man behind him who is generally thought to be Guidobaldo da Montefeltro (b. 1472). This painting by Jacopo dei Barbari (Venice? *ca.* 1445 - *ca.* 1515) housed in the Museo di Capodimonte, Naples.[181]

Piero – the artist

For readers who do not already know this, Piero's paintings and frescoes fall into the following main categories:~

- ❖ Major events and images of supreme religious significance, such as the Annunciation; the pregnant Madonna; the birth of Christ; the Baptism of Christ; the Madonna and Child; the Madonna of Mercy; the Passion of Christ; the Crucifixion; and the Resurrection of Christ.
- ❖ A series of such events relating to the so-called "Legend of the True Cross", exemplified by the magnificent Arezzo frescoes. The story, adapted from the colourful version of the *Legenda aurea*, runs from the time of Adam to "the present", when the topic was very much an issue, not least

---

178 See the text to n. 144.

179 See, for example, Calvesi, *Piero della Francesca*, pp. 207 (SCI SEPVLCRI) and 199 (MIA), and on these abbreviations, Cappelli, *Lexicon abbreviaturarum*, pp. 243 and 218.

180 *Ibid*., pp. 191 and 196. For details see the article "Pacioli, Luca" by S. A. Jayawardene in the *Dictionary of Scientific Biography*.

181 See, for example, *Sansepolcro 1994 Conference Proceedings* and *Sansepolcro 1994 Exhibition Catalogue*, also Brilli, *Alla ricerca di Piero della Francesca*, p. 234. The painting is appropriately used on the cover of Crosby, *The Measure of Reality – Quantification and Western Society, 1250-1600* (1997).

because Bessarion had brought a Byzantine reliquary of the "True Cross" to Italy.

- ❖ Frescoes or paintings including portraits of individuals, such as Sigismondo Malatesta, Federigo da Montefeltro and his wife, Battista Sforza.
- ❖ Paintings of individual religious personalities, such as St Jerome, various other saints, and various angels.
- ❖ The anomalous and much restored "Hercules" fresco of a barely-clad young man with a club.[182]

In many of Piero's paintings, contemporary personalities are included in brilliant time-warped representations of historical scenes, and contemporaneous scenes with historical personalities. Thus, for example, Bessarion (nine times!)[183] and Ioannis VIII,[184] King Sigismund of Hungary (elected "Holy Roman Emperor" in Rome in 1433), and various members of the Bacci family appear in "The Legend of the True Cross", and another Bacci perhaps appears in the "Madonna del Parto". Or early contemporaries are used for the images of historical personalities, such as King Sigismund of Hungary for St Sigismund, shown with Sigismondo Malatesta venerating him.[185] Also, Federigo's wife, Battista Sforza, and their son Guidobaldo (b. 1472) are embodied in the scene of the Brera Madonna and Child, for it is certainly Federigo kneeling by their side and Bessarion, who baptized the child, behind him.[186] Extra-Biblical legend – from the *Legenda aurea* – played a leading role in the Arezzo frescoes, but also in

---

182 On the fresco see, for example, Calvesi, *Piero della Francesca*, pp. 48 and 207. Connections with Hercules were suggested by Piero's mother's family home of Monterchi, *Mons Herculis*, said to have been delivered from a dragon by Hercules, or by Piero's contacts with Rome.

183 Page-references to images in Centauro & Settesoldi, *Piero della Francesca – Committenza e pittura*, are given in square brackets. Bessarion appears in Piero's "Legend of the True Cross":

four times in *The Battle of Heraclius against Chosroes* [150];

twice in *The Exaltation of the Cross* [130-133];

once as the sleeping Constantine in *The Dream of Constantine* [154];

once as the fleeing Maxentius in *The Victory of Constantine* (where Constantine is represented by Ioannis VIII) [155-157]; and

once as Solomon in *The Meeting of King Solomon with the Queen of Sheba* [140].

184 Ioannis VIII is shown as Constantine in Piero's *Victory of Constantine* in Piero's "Legend of the True Cross". See Centauro & Settesoldi, *Piero della Francesca*, p. 155.

185 Calvesi, *Piero della Francesca*, pp. 26-28.

186 Calvesi, *Piero della Francesca*, p. 217, citing a report from 1822.

some of Piero's other works. And St Jerome is shown with his legendary cardinal's hat at his feet.[187]

Only two of Piero's works are dated, namely, "St Jerome Penitent", dated 1450, and "Sigismondo Malatesta kneeling before St Sigismund", dated 1451 (although the roundel featuring the Rocca of Rimini bears the date 1446). Others, such as the Madonna della Misericordia polyptych and the San Agostino polyptych and the frescoes in the pontifical chambers, are mentioned in dated archival materials relating to commissions or payments. However, it seems that there is neither general agreement amongst specialists about the dating of any of his works – the majority – that are not dated, nor any general agreement about the commissioning of those works – again the majority – for which there are no written commissions or notices of payments.[188] In the case of "The Flagellation", variously dated from *ca.* 1445 to *ca.* 1475, most of the estimates are highly speculative since there is no external evidence of a commission and it is not known either who gave the commission or where it was given. Vasari makes no mention of any painting that could be equated with "The Flagellation", confirming the hypothesis that it was a very private affair.

Table 2

Chronology of Piero and his works

(Based on Calvesi, *PdF*; Centauro & Settisoldi, *PdF*; and Banker, "Contributi alla cronologia della vita e delle opere di PdF"; page references given as [C x], [CS x] and [B x]; my own hypothetical suggestions are labelled [K].)

1412

P born in Sansepolcro. [B 248-252, C 12] (Previously, his date of birth was usually given as 1410-20.)

1431

In June P receives payment for painting candles or candle-holders in Sansepolcro. [C 12, 179]

1430s

P works on heraldic devices and standards in Sansepolcro with the painter Antonio d'Anghiari (none of whose works have survived). [C 179-180]

1439

187 Maetzke, *Piero della Francesca*, p. 76.

188 A useful overview of comparative datings of Piero's works to *ca.* 1995 is presented in Lavin, "*Monarca della Pittura*: Piero and his Legacy", pp. 14-16. It is healthy when historians survey and call into question the accepted chronology of the works of a given painter, as did Cecil H. Clough of Piero in 1970 – see his "Piero della Francesca – Some problems of his art and chronology" (see also n. 335 below).

In September P is cited in the record of a payment to his new teacher, Domenico Veneziano, and himself for work on frescoes in the chapel of San Egidio, Florence. [C 180, B 253-254] (Several of Veneziano's works have survived.) P would have seen Ioannis VIII and Bessarion in Florence, where the Council had moved from Ferrara.

1442

P is mentioned in a list of those eligible for election to the Consiglio del Popolo in Sansepolcro. (This does not mean he was present in the town.) [C 180]

1445

In June P receives from the confraternity of the Misericordia in Sansepolcro a commission for the Misericordia polyptych (now in the Museo civico, Sansepolcro) to be painted within three years and receives part-payment the following year. P is summoned to finish the work in 1454, which he apparently did finish around 1462. [C 16, 18, 28-33, 180, 181, 198]

1447

P receives a commission with Domenico Veneziano to decorate the sacristy at Loreto. [C 15, not in C 180]

1448 - early 1450s

P in Ferrara. [C 16]

1450

P paints "St Jerome Penitent", signed and dated, probably completed in Ferrara (now in Berlin). [C 22, 193] (An anachronistic Cardinal's hat is shown at the saint's feet.)

1451

Fresco of "Sigismondo Pandolfo Malatesta kneeling before San Sigismondo" completed in the Tempio Albertina in Rimini, signed and dated, still *in situ*. [C 24-28, 193-195]

1454

Commission of the Sant'Agostino polyptych by the Augustinians of Sansepolcro, paid for by 1469-70. [C 180-181]

mid 1450s (?), *ca.* 1465 (?)

"The Baptism of Christ", painted for the Church of San Giovanni Battista in Sansepolcro [C34-37, 197] (now in the National Gallery, London). Calvesi dates it to the mid 1450s, others to *ca.* 1465. This painting is important to the present study as it is "almost as geometrical" as "The Flagellation".

~1457 - ~1466

P involved in the frescoes of "The Legend of the True Cross", based on the *Legenda aurea*, in the Church of San Francesco, Arezzo, with occasional breaks of several months on account of harshly cold winters and excessively hot summers as well as other commissions.

1459

On April 12, P receives payment for frescoes in the pontifical chamber in Rome. [C 16, 37, 182] He would have met Bessarion in Rome.

On Nov. 6, P's mother Romana dies in Sansepolcro and is buried in the family crypt at Badia, Sansepolcro. [C 182]

late 1461 (??)

First possibility for a hypothetical encounter of P with Bessarion and Regiomontanus and/or Giovanni Bacci in Sansepolcro or Arezzo for the commission of the painting "*Convenerunt in unum*" [K].

1462

P's presence in Sansepolcro documented on three occasions (April 20, May 4 and August 12). [C 182-183]

1462-64 (??)

Possible dates for the execution of "*Convenerunt in unum*" [K]. The period 1463-1464 is suggested by Calvesi, who proposes a quite different motive behind the painting [C 20, 48, 56]. Others propose dates between *ca.* 1445 and *ca.* 1475.

mid 1463 (??)

Second possibility for a hypothetical encounter of P with Bessarion and Regiomontanus and/or Giovanni Bacci in Sansepolcro or Arezzo for the commission of the painting "*Convenerunt in unum*" [K].

1464

On Feb. 20, P's father, Benedetto, dies and is buried in the family crypt at Badia, Sansepolcro. [C 183]

P receives a commission from the Compagnia della Trinità in Arezzo for a gonfalon banner with the image of the Holy Trinity, to be completed by March, 1465. [C 183]

1466

P receives a commission from the Compagnia dell'Annunziata in Arezzo for a gonfalon banner with an image of the Annunciation. [C 183-184]

1468

From June, record of a payment for the Sant'Antonio polyptych to the convent of Sant'Antonio in Perugia from the municipal council. [C 184]

1468-1488

Numerous documents mentioning property and payments. [C 185-189]

1469

Urbino, April 8: Giovanni Santi (father of Raphael) receives a payment from the confraternity of Corpus Domini in Urbino for Piero to work on a painting. [C 185]

1487

According to his will dated July 9, 1487, Piero is sound of mind and body. [C 188]

***ca.* 1490**

Earliest date for Piero to have gone blind. [B 256-257] (According to Vasari, P went blind at age 60, which is impossible.)

1492

On Oct. 12, Piero dies in Sansepolcro and is buried in the family tomb at Badia, Sansepolcro. [C 64, 66, 189] (According to Vasari, he was 86 years old when he died, which cannot be correct.)

1992

Quincentennial celebrations in Sansepolcro, including an exhibition and a conference, also in Arezzo, Urbino and Florence.

****

Table 3

The extant paintings and frescoes of Piero and their locations

(The order is roughly chronological, and the numbering follows Calvesi, *PdF*, pp. 193-222.)

Abbreviations for collections: Arezzo D = Duomo; Arezzo SF = San Francesco; Berlin SM = Gemäldegalerie, Staatliche Museen; Boston GM = Isabella Stewart Gardner Museum; Florence GU = Galleria degli Uffizi; Lisbon MAA = Museu de Arte Antiga; London NG = National Gallery; Milan MPP = Museo Poldi Pezzoli; Milan PB = Pinacoteca di Brera; Monterchi MMP = Museo "Madonna del Parto", fresco formerly in the chapel of the cemetery; New York FC = Frick Collection; Paris ML = Musée du Louvre; Perugia GU = Galleria Nazionale dell'Umbria; Rimini TM = Tempio Malatestiano; Rome SMM = Santa Maria Maggiore; Sansepolcro MC = Museo Civico; Urbino GM = Galleria Nazionale delle Marche; Venice GA = Galleria dell'Accademia; Washington NGA = National Gallery of Art.

Notes: Only 1,2, 4 and 10 are signed. Only 1 and 2 are dated. The attribution of 3 is contested. See Calvesi, *PdF*, pp. 224-226, for other works that have been attributed to Piero, including five Madonnas and Child and San Ludovico. Names of locations, towns or cities are double underlined when they correspond to the places where the works were originally executed. An asterisk indicates that the author saw the work in 2006.

1) St Jerome Penitent, 1450 — ?? > Berlin SM *

2) Sigismondo Malatesta kneeling before St Sigismund, 1451 — Rimini TM *

3) Portrait of Sigismondo Malatesta — Rimini (?) > Paris ML

4) St Jerome and a devotee with a *becchetto* before Sansepolcro — ?? > Venice GA

5) The Baptism of Christ, with John the Baptist and three angels (Graces)
Sansepolcro > London NG

6) Madonna della Misericordia polyptych
(Madonna sheltering various moderns, one masked, with various saints and angels, various scenes, including a Flagellation, a Crucifixion, two Marys at the Tomb)
Sansepolcro MC *

7) "The Arezzo frescoes" = "The Legend of the True Cross"
including the Death of Adam; the Discovery of the Wood for the Cross; the Meeting of Solomon and the Queen of Sheba; the Battle between Constantine and Maxentius, the Transport of the Sacred Wood; the Dream of Constantine, the Torture of the Jew; the Annunciation; the Return of the True Cross to Jerusalem; the Discovery and the Testing of the True Cross; and the Battle between Heraclius and Chosroes
Arezzo SF *

8) St Luke — Rome SMM

9) "Hercules" (?) with a club — Sansepolcro > Boston GM

10) ***"Convenerunt in unum"*, called "The Flagellation of Christ"**
??????? > **Urbino GNM** *
11) The Resurrection of Christ (with four sleeping disciples) Sansepolcro MC *
12) St Julian (??) Sansepolcro MC *
13) St Mary Magdalene Arezzo D *
14) Madonna del Parto, the pregnant Virgin Monterchi MMP *
15) San Agostino polyptych (6 saints, angel, Crucifixion)
Sansepolcro > Lisbon MAA; London NG; New York FC (Crucifixion and 2 saints); Milan MPP; Washington NGA
16) San Antonio polyptych
(Madonna and Child, 3 miracles, Annunciation added) Perugia GNU *
17) "Brera altarpiece" (Madonna and Child, with angels, saints) Urbino > Milan PB
18) Federigo da Montefeltro and Battista Sforza Urbino > Florence GU
19) Madonna di Senegallia (with Child and two angels) Urbino GNM *
20) The Nativity (Madonna and Child, with five angels, two shepherds and St Joseph)
Piero's estate in Sansepolcro > London NG

The most enigmatic and controversial painting of the Quattrocento

Piero's painting known as "The Flagellation of Christ" – see **Fig. III-3** – is the most enigmatic and controversial work of art of the Quattrocento. The flagellation scene on the left is a background to three "moderns" on the right, of whom the middle one, in cardinal red, looks rather like a typical Piero angel. It is unclear who commissioned this remarkable painting, possibly the man on the far right, possibly the bearded man, who holds pride of place. The "flagellation" scene is somehow related to the Byzantine Emperor Ioannis VIII, depicted on the left sitting on a throne in front of the *scala Pilati*, the stairs that Christ ascended when his death sentence was passed. It is generally accepted that two of the figures have double "identities". The Christ figure also represents the Church and perhaps Byzantium, and Ioannis VIII is "sitting in" for Pontius Pilate. Watching the Flagellation is a figure in oriental garb, surely the Ottoman Sultan Mehmet II, who was responsible for the Fall of Byzantium.

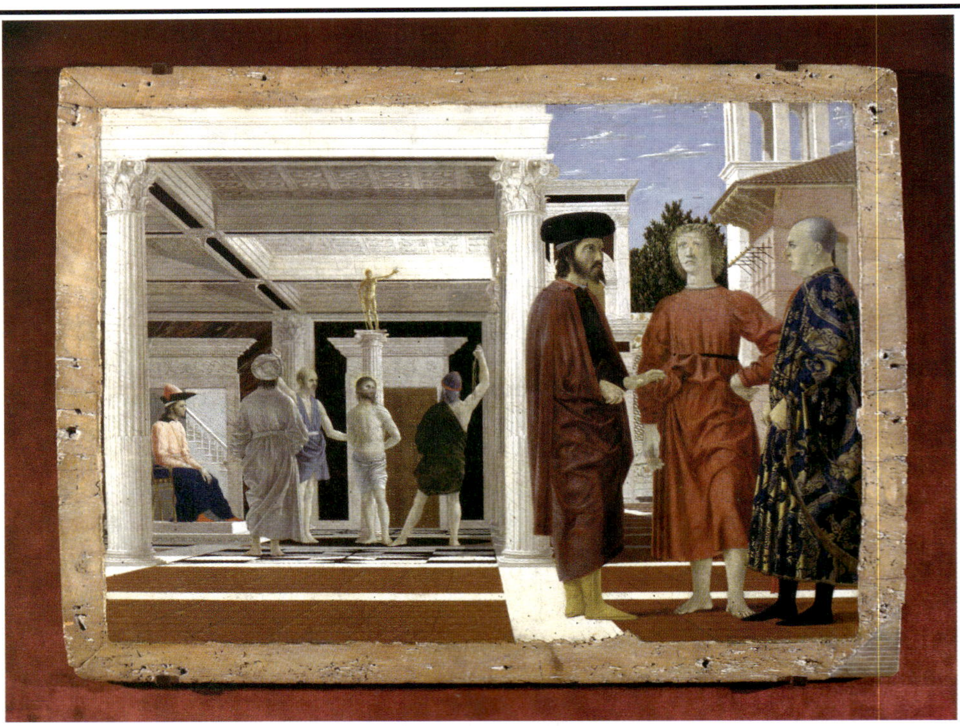

Fig. III-3:~ The spectacular painting by Piero della Francesca, which has hitherto defied interpretation. The key is in its original title: *Convenerunt in unum*, with the multiple identities of the persons depicted provided by Regiomontanus' epigram. The painting is surprisingly small, which indicates that it was intended for a private individual, not for public display. [Courtesy of the Soprintendenza per il Patrimonio Storico, Artistico ed Etnoantropologico delle Marche, Urbino.]

Who are the *eight persons* – not just three – represented across the painting? Why are they depicted in the way they are? Why are the three "moderns" depicted to the side and forefront of a symbolic flagellation scene? Why is there an illuminated statue on top of the column behind Christ? Why does the man sitting on Pontius Pilate's throne look like a Byzantine emperor? Is it the Ottoman Sultan standing next to him? (Some have proposed King Herod.) Why does one of the two men "flagellating" Christ have a distinctive face, and why is he touching Christ? Who are the three "moderns" and why does the one in the middle have an angel-like face? Why is the painting so static? Nothing is

---

190 * Passavant, *Rafael von Urbino*, I, pp. 432-433, wrote (my emphasis in bold):

happening; nobody is moving; nobody is talking; nobody has his eyes fixed on anybody else. What is it all about? Who conceived this remarkable double scenario? And why?

A comment at the outset: Piero has not put the Flagellation scene in the background because it was less important than the three "moderns". He has given the Flagellation scene exactly one half of the painting, and the three men the other half. As I shall show, this was a symbolic Flagellation, for the Christ-image represents other persons and "things" – such as St Peter and the Church – that had been "tortured" in the minds of at least two of the men on the right. Also, the image of the classical god represents as well as the god of Rome various concepts – such as Byzantium and Constantinople – of which the same could be said.

The title "The Flagellation of Christ" is a modern title that is perhaps not fully appropriate, for there is far more to the picture. A report dated in 1839 by the Frankfurt art-historian Johann David Passavant (1787-1861) implies that the words *convenerunt in unum* were written *on* the painting *near to* the three men on the right.[190] Alas the German is rather imprecise, and Passavant was not to know that the inscription would soon therefter disappear. Although Passavant did not mention any frame, it seems that there was at some time a frame around the painting, again long since removed. That frame could have been original – a surviving painting by Piero with an original frame is Piero's "St Jerome Penitent"[191] – but if it was, then it surely covered not a square millimetre of the painting itself. There are holes around the wooden board behind "The Flagellation" that serves as its base, and these surely indicate where the frame was originally attached. Most modern investigators have assumed that the inscription was on the frame. Recent radiographical analyses reveal no trace of any such inscription on the painting itself.[192]

Some recent interpretations

> "Every now and then it is useful to review the history of a discipline in order to gain insight into the origin of opinions prevailing in it. In doing so, one sometimes finds that extraneous factors played a significant role in the development of theories which have come to be accepted as dogma and have been passed down unquestioned from generation to generation in the firm belief that these

---

191 Calvesi, *Piero della Francesca*, p. 193. The frame is shown in Field, *Piero della Francesca*, fig. 3.2 on p. 80. I saw this painting in September, 2006, and noted that it is painted gold and that the surface of the painting is slightly convex outwards, but vertically.

192 Seracini, "Piero: *Flagellazione*". See also the text to n. 224 below.

theories are grounded solely on 'scientific' considerations. A case in point is provided by the history of views concerning the phonological value of the 15th rune in the Germanic fupark, the so-called algiz-rune, ..." Elmer H. Antonsen, "15th Rune" (1980), p. 1, cited in King, *The Ciphers of the Monks* (2001), p. 16, n. 2.

The reader cannot expect to find here a survey of all of the identifications that have been made over the past 150 years of the main figures in the painting. Every single one of them was based on conjecture, since documentary evidence was (thought to be) lacking. Most writers had no problem seeing the Byzantine Emperor on the throne on the left, or supposing that the Christ-figure represented more than the historical Christ. The man with the beard has been associated with at least a dozen different historical figures. See Table 4 in which those suggestions that this author thinks are the most reasonable (in view of the new evidence) are in bold font.[193] It should also be kept in mind that some of the faces have been restored, and, anyway, we are probably not dealing with portraits.[194]

Table 4

Previous identifications of the three "moderns" in Piero's "Flagellation"

Source	**Date**	**Bearded man**	**Youth**	**Man on right**
Inventory	1744	Guidantonio	**Oddantonio**	Federigo
Passavant	1839		three enemies of Federigo	
Dennistoun	1851	Manfredo de' Pii	Oddantonio	Tommaso dell'Agnello
Ugolini	1859	<	*ditto*	>
Pichi	1892	Serafini	Oddantonio	Ricciarelli
Witting	1898	Caterino Zeno	missionary / Christ	Federigo
Graber	1922	evil councellor	Oddantonio	evil councellor
Longhi	1927	<	*ditto*	>
Clark	1951	Thomas Palaiologos	Allegory	Guidantonio
Gombrich	1952		anonymous bystanders	
Gilbert	1952		*ditto*	
Running	1953	Gentile	soldier	Joseph of Arimathea
Siebenhüner	1954	Ioannis VIII Pal.	Oddantonio	Guidantonio

193 This is considerably expanded and updated from the information in van Waadenoijen, "La 'Flagellazione' di Piero della Francesca" (1993), p. 186, and Lavin, "*Monarca della Pittura*: Piero and his Legacy" (1995), p. 17.

194 See the old images published by Longhi and Clark.

Gombrich	1959	Judas	Christ	Sanhedrin
De Tolnay	1963	Jew	pagan Greek	Aryan
Murray	1968	Jew (?)	angel	Western layman
Lavin	1968	Ottaviano Ubaldini, father of Bernardino	**allegory of beloved son**; **theme is consolation**	**Ludovigo Gonzaga, father of Vangelista**
Hartt	1970	Palaiologos	David	judge
Gilbert	1971	Gentile	soldier	Joseph of Arimathea
Battisti	1971	Byzantine	Oddantonio	F M Visconti / Fr Sforza
Gouma-Peterson	1976	Greek ambassador	"athlete of virtue"	western prince
Clough	1978		Oddantonio	
Borgo	1979	Sanhedrin	gardener	Sanhedrin
Salmi	1979	Agnolo Bacci	Andrea Bacci	Francesco Bacci
Ginzburg	1981	**Bessarion**	**Buonconte**	**Giovanni Bacci**
Hoffman	1981	Jew	youth	Christianized Jew
Turchini	1982	**Bessarion**	**Buonconte**	**Ottaviano Ubaldini**
Pope-Hennessy	1986	"*The Dream of St Jerome*"		three literary critics
Lollini	1991	rabbi	young Jew	Jew
Tessari	1991	establishment of the Rule of Christ on earth and in heaven (angel)		
Lightbown	1992	-	-	Francesco Sforza
Ciardi Duprè	1992	Ioannis VIII	Oddantonio	Guidantonio, father of Oddantonio and Federigo
van Waadenoijen	1993	prophet Isaiah	St John Evangelist	Ottaviano Ubaldini (?)
Calvesi	1995	**Thomas Palaiologos**	**Matth. Corvinus**	Italian prince
Lightbown	1997	high-ranking Byzantine	angel	Francesco Sforza
Ronchey	2000	Bessarion	Thomas Palaiologos	-
Shank	2000		connection with the controversy between Bessarion and George of Trebizond	
Lavin	2002		as in Lavin 1968	
Field	2005	scholar	generic saint / angel	merchant
Roeck & Tönnesmann	2005	Judas (Federico)	Oddantonio	Reuben, father of Judas (Federico's father)
Holzschuh	2005	**Bessarion**	**Regiomontanus**	**Ottaviano Ubaldini**
de Rycke	2005	Io. Argiroupolos	David / Oddantonio	Francesco Accolti
Ronchey	2006	Bessarion	Thomas Palaiologos	Niccolò III d'Este
Roeck	2006	Judas (Federico)	Oddantonio	Reuben, father of Judas (Federico's father)

In fact, as we shall see, all attempts to identify single individuals amongst the principal three images on the right of the painting were inevitably futile or could only be partly successful. Nobody ventured to think that every single one of the eight persons depicted could represent more than one single person. However, some previous identifications are relevant if the text that I propose as the inspiration for the painting is indeed that. These we now consider.

Already in the earliest identification of the three principal moderns – a notice in the inventory of the sacristy of the Cathedral at Urbino dated 1744 – we encounter the suggestion that the painting focuses on the death of Oddantonio (d. 1444), Count of Urbino, the murdered brother of Federigo da Montefeltro, featured in the middle between the latter's son Guidobaldo on the left and Federigo himself on the right.[195] The young "tyrant" Oddantonio was assassinated under mysterious circumstances, opening up the way for his "brother" to take control of Urbino and its territories. Guidobaldo need not concern us further for he was not born until 1472. Also, this is not Federigo on the right, or is it?[196] And perhaps Oddantonio should not be forgotten.[197]

Thalia Gouma-Peterson in 1976[198] recognized two Greeks, Ioannis VIII on Pilate's throne, and the man with the beard as "a generic or representative type of Greek ambassador, an ideal mediator". She dismissed Bessarion as a candidate for physiognomic reasons and because he was wearing neither a monk's habit nor a cardinal's attire. She suggests that the young man in red was an "allegorical representation of the Christian athlete of virtue".

Marilyn Aronberg Lavin in 1968 and 1972 and again in 2002[199] proposed that the bearded man on the left was Ottaviano Ubaldini dalla Carda, nephew of Federigo da Montefeltro, his chief counsellor and treasurer of state; and that the man on the right was Ludovigo Gonzaga, Marquis of Mantua, who hosted the *Conventus* of Mantua in 1459. His profile is identifiable from a bronze statue attributed to Donatello made *ca.* 1450. Both men were life-long friends. The handsome figure in the middle personifies the 'beloved son', who like Christ, will rise to glory. The scene was related to the recent deaths of Ottaviano's son, Bernardino, in 1458, and Ludovigo's nephew and adopted son, Vangelista, some

---

[195] See, for example, Ginzburg, *The Enigma of Piero*, pp. 48 and 52-53; Turchini, "Un'ipotesi per la "Flagellazione" di Piero", pp. 66-67; and Calvesi, *Piero*, p. 207.

[196] See the text to n. 307.

[197] See the text to n. 294.

[198] Gouma-Peterson, "Piero's Flagellation".

[199] Lavin, "Piero della Francesca's "Flagellation": The Triumph of Christian Glory" (1968), *Piero's 'Flagellation'* (1972), and the summary in *eadem*, *Piero* (2002), pp. 93-112.

time between 1456 and 1460.[200] The theme of the painting is consolation, and its purpose private devotion. Lavin still insists on this interpretation.

Joseph Hoffman in 1981[201] argued that the man with the beard, although surely not Italian (Italians did not wear beards at that time), need not necessarily be Greek; indeed, he could be a Jew. We could then have a scenario in which a Jew on the left and a Christian on the right debate before the young man in red who "appears to adjudicate".

Carlo Ginzburg in 1981[202] proposed that the bearded man was Bessarion, and the man on the right was the jurist Giovanni Bacci. Furthermore, the barefoot young man in the middle must be Buonconte da Montefeltro, Federigo's natural son, who had been loved and admired for his precocious intellectual gifts by Bessarion, and who – along with Bernardino – had died of the Plague in 1458.[203] But is it really Bessarion who is depicted in the painting?[204] And is it really Bacci on the right?

Angelo Turchini in 1982[205] proposed that the three men were Bessarion, Buonconte and Ottaviano Ubaldini dalla Carda. He also expressed the doubt that the painting might have nothing to do either with Urbino or with Federigo da Montefeltro.

Maurizio Calvesi in the 1990s convincingly argued that the three men were Thomas Palaiologos, brother of Ioannis VIII and pretender to the Byzantine throne (after Kenneth Clark), an Italian prince or his representative, possibly Ludovigo Gonzaga, with military interests in a crusade, and between them the young king of Hungary, Matthias Corvinus, who ascended the throne aged 15 in 1458.[206] He is portrayed in red elsewhere, and wearing laurels, and his favourite flower was the rose. The background to the painting was the Council (*conventus*!) of Mantua in 1459, when Pope Pius II convened a meeting of sovereigns and princes. Federigo was appointed head of the papal forces, and the young Hungarian king was assigned to lead the crusade himself.

---

200 On these two young men and their deaths see Lavin, *Piero's Flagellation*, pp. 70-71.

201 Hoffman, "Piero's 'Flagellation'". See the text to n. 317 below.

202 Ginzburg, *The Enigma of Piero*, pp. 60-101, with later reflections on pp. 116-128.

203 See the text to n. 38.

204 On the basis of the physiognomy of the face in the image, Fabrizio Lollino ("L'iconografia di Bessarione", p. 282), who collected all available representations of the Cardinal, is adamant that this cannot be a picture of Bessarion.

205 Turchini, "Un'ipotesi per la "Flagellazione" di Piero".

206 Calvesi, *Piero della Francesca*, pp. 48-50, 109-116 and 155, also citing earlier studies in 1992 and 1996. See also his 1995 study "*La Flagellazione*". We shall return to his theory in the text to n. 310 below.

In 2000, Ginzburg commented that virtually no one had accepted his interpretation.[207]

That same year, the American historian of science Michael Shank gave a lecture[208] in which he argued that:~

> "the painting reflects in part the aftermath of an important astronomical and patronage controversy that pitted George of Trebizond against Cardinal Bessarion."

As we shall see, if my interpretation of the painting is correct, all of these scholars were very much on the right track. Yet even scholars intimately familiar with the works of Piero cannot begin to accept one or other of these identifications,[209] and new ones, wilder than before, are still being formulated.[210]

The collection of essays in *The Cambridge Companion to Piero della Francesca* (2002) gives the impression that art historians have nothing more of consequence to say on "The Flagellation". The editor, Jeryldene M. Wood, comments in her introduction that the painting "exemplifies the perplexing iconographical problems faced by scholars", and that the various interpretations

---

207 Ginzburg, *The Enigma of Piero*, pp. 116-127.

208 From the description of a lecture delivered by Michael Shank at the Institute of Medieval Studies in Kalamazoo, Michigan, in April, 2000: see Shank in the bibliography. It is a pleasure to acknowledge some fruitful phone-calls with Michael Shank.

209 In Field, *Piero della Francesca* (2005), p. 178, we encounter "a scholar of some kind, possibly a theologian or a jurist; associated with the judgement hall", "a merchant associated with the town house", and the youth with "the kind of dignity Piero gives to figures such as saints and angels", "the deep red colour of his clothes also suggest(ing) some title to spiritual nobility".

210 In Roeck & Tönnesmann, *Die Nase Italiens* (2005), pp. 67-72, we are confronted with a completely new theory, formulated virtually without any concession to previous ones.

The painting supposedly has a connection to the fictitious biographies of Pontius Pilate and Judas in the *Legenda aurea*, "The Golden Legend", a collection of stories about saints that was popular in the Late Middle Ages. Pilate is represented by Ioannis VIII. The young man is Oddantonio, murdered in 1444. The man with the beard is Judas, to be identified with Federico da Montefeltro, and the man on the right, Reuben, Judas' father, here identified with Federico's father, but also – because he has thistles on his gown – with Bernardino Ubaldini dalla Carda. The story is about apples from Reuben's garden that Judas takes for himself. Then Judas, when confronting his father, kills him. Pilate gives Reuben's possessions to Judas, and his wife, so that Judas ends up marrying his own mother. The statue is of Paris, with an apple in his hand.

What the two authors do not discuss is why the "apple" is painted silver and how Piero could ever have been convinced to paint such a picture and then sign it.

"circle around a pair of assumptions: that the *Flagellation* was always at Urbino and that the three figures portray actual individuals".[211]

In 2005, the Belgian art historian Jean-Pierre de Rycke made a completely new proposal for the three figures on the right:[212] the young man embodies the Psalmist David, who first coined the equivalent of the phrase *Convenerunt in unum*, and the dead Oddantonio. The bearded man is Ioannis Argiropoulos (b. Constantinople? between 1393 and 1405, d. Rome, 1487),[213] supposedly a member of the Greek delegation to Ferrara-Florence, who was an avid supporter of the Union and who converted to Catholicism. In 1456, he emigrated to Florence, where for 15 years he taught Greek philosophy. The man on the right is the Italian Humanist Francesco Accolti.[214] The painting dates from *ca.* 1450 and relates to the Union of the Eastern and Western Churches.

In an article published in 2000, but more especially in April 2006, the Italian Byzantinist Silvia Ronchey presented a brand new interpretation of the painting revealing it as a political statement of Bessarion.[215] With this conclusion I can only agree. The three figures on the right are now, according to Ronchey, Bessarion, Thomas Palaiologos and Niccolò III d'Este, the host of the Council of Ferrara. How they all come to be represented by Piero *ca.* 1450 is explained in over 500 pages of scholarly text, hailed by the Italian press as the final solution to Piero's enigma. The reaction of the scholarly world will come later. In any case, the work is exciting and timely. Various relevant events in the life of Bessarion and his connection to the last Byzantine rulers are clearly outlined, but inevitably the hypothesis of their relationship to Piero's painting is highly speculative. Nevertheless, this study of the painting is the first to mention all of the principal players who I think are relevant to it, as well as Regiomontanus' 1462 astrolabe.

All of these interpretations were made without one *iota* of documentary evidence as far as the painting itself was concerned. So it is not unreasonable to

---

211 Wood, ed., *The Cambridge Companion to Piero della Francesca*, pp. 9-10.

212 De Rycke, "La 'Flagellation' d'Urbino". See further the text to n. 311 below. By coincidence, copy of this arrived on the desk of Dr. Agnese Vastano in Urbino on the day of my lecture there. Dr. Vastano kindly provided me with a photocopy.

213 See the article "Argyropoulos, John" in the *Oxford Dictionary of Byzantium*, also Monfasani, "The Averroism of John Argyropoulos", pp. 157-176.

214 See the article "Accolti, Francesco", in *Dizionario biografico degli Italiani*, I, pp. 104-105.

215 Ronchey, "Malatesta/Paleologhi", pp. 543-557, and *eadem*, *L'Enigma di Piero*. The full set of bibiographical notes has been reduced by the publishers but, as indicated *ibid.*, p. 513, it is available at:
www.rizzoli.rcslibri.it/enigmadipiero .

assert that, until the documentary evidence was discovered, not a single one of the three "moderns" had been definitively identified in a way that might satisfy all interested scholarly parties. John Pope-Hennessy was right when he wrote in 1986 that:~[216]

> "The case requires to be reconsidered in the light of the true subject of the painting, as a humanist embodiment of one of the most controversial intellectual problems of its time. Ironically the nature of the subject should have been evident from the start, for at Urbino the painting bore on its frame the phrase: '*Convenerunt in unum*,' … … "

Nevertheless, the subject escaped him, as it has most others, and so did the real significance of the Latin phrase.[217]

The painting has been traditionally thought to have been dedicated to the Count Federigo of Urbino, whom Bessarion urged, not just on one occasion, to undertake a crusade against the Turks, who now had their eye on treasures like Venice. However, as we shall see, it is Bessarion that is the focus of the painting, and it seems highly probable that it was intended as a very personal gift for him or even a personal commission. After all, it would not have been a good idea to threaten Federigo with being likened to those who had, in Bessarion's opinion, "betrayed" the Church as Herod and Pilate *convenerunt in unum*, "come together in one" against God and His Christ.

A severe problem confronting previous investigators in coming to terms with this painting is the apparent total lack of any mention of it in the archival materials relating to Piero's paintings.[218] The only explanation of this seemed to be that the work was intended for a very select audience; in other words, it was perhaps achieved in secret with no paper trail. Now that a relevant contemporaneous document has been identified, this author is confronted with the unenviable task of interpreting it to a background of relevant snippets from earlier modern authors on what is, for him, unfamiliar territory. We are indeed

---

216 Pope-Hennessy, "Whose Flagellation?", especially p. 165.

217 In fact, he later completely misrepresented and even renamed the painting as *The Dream of St Jerome*. See Pope-Hennessy, *op. cit.*, and *idem*, *The Piero della Francesca Trail*, especially the caption to fig. 2 on p. 9, and p. 10 on the unduly harsh criticism of Ginzburg. Cole, *Piero della Francesca*, p. 73, n. 38, calls this "the one plausible alternative explanation of the subject (of *The Flagellation*)". (A less drastic renaming of the painting, after the interpretation of Kenneth Clark, was by Philip Hendy in 1968, who proposed: *An Allegory, with the Flagellation of Christ* – see his *Piero della Francesca*, p. 115.)

218 On these see Calvesi, *Piero*, pp. 177-189, and also Centauro & Settesoldi, *Piero della Francesca – committenza e pittura*, pp. 166-170, on the Arezzo frescoes.

fortunate that much that was written on "The Flagellation" before Ginzburg is already discredited and need not even be mentioned here.

A contrived polysemous scenario

Most of the authors mentioned above give the impression that they alone are right. In 1987 David Carrier observed:~[219]

> "Presumably at most one interpretation (of Piero's "Flagellation") can be correct,"

and he qualified this with a footnote:~

> "This is to assume that the image is not polysemous [*i.e.*, having many meanings]."

As we shall see, this sums up the problem with all previous research, for the painting now known as "The Flagellation of Christ" is indeed polysemous. Indeed, I would maintain that each of the nine images – eight persons and one god – is polysemous. And the epigram of Regiomontanus explains how this might have come about.

The polysemous nature of the painting has already been captured by the Italian journalist, essayist and author Edgarda Ferri in 2001.[220] In a novel about Piero, Ferri ingeniously invents a conversation concerning the painting between Federico and the artist that took place late in 1461. Both men are on the point of leaving Urbino. But first the Count takes the artist down to the Sacristy of the Duomo, not yet completed, where Piero had prepared the outline for "The Flagellation" but not the faces of the men to be depicted. Federigo says to Piero:~

> "Your works exude mystery; this one, however, seems to me to be the most enigmatic and the most intriguing of them all. Whoever looks at it, will have to rack their brains in order to decipher the significance of the two scenes. Who are the three persons in the front? I would be truly happy if the blonde youth, most handsome and barefoot, might be my beloved Buonconte; that the gentleman wearing the dark blue gown, woven with golden thorns, whose strong resemblance makes me suspect that he might be my dear friend Giovannni Bacci of Arezzo; and that the bearded man might be Cardinal Bessarion."

Over the next two pages, Federigo tells Piero the other scenes that he could imagine. (Ferri more or less runs through a selection of the published hypotheses, having started with Ginzburg's.) The bearded man could be Judas

---

219 Carrier, "Piero and his Interpreters", p. 157.

220 Ferri, *Piero della Francesca – Storia e misteri del maestro della luce*, pp. 209-211.

dealing with the Sanhedrin. The three men could be a soldier, a gentile and Joseph of Arimathea. The scene could represent Ottaviano Ubaldini, who had lost his son Bernardino, consoling Ludovigo Gonzaga, who had lost his son, Vangelista. Federico points out that the most simple-minded inhabitants of Urbino would see his own dead half-brother Oddanonio in Piero's angel. Piero is speechless. Finally, the two men part company, with Piero leaving for his house in Sansepolcro, promising to return when he has finished the faces.

I dare say that no academic could ever have devised this scenario: it would have required too much imagination. In the sequel, I shall try to show that Edgarda Ferri's contrived scenario comes closer to what Piero actually painted than any of the scholarly proposals of the past 150 years, for not one has seriously explored the possibility that the painting "*Convenerunt in unum*" might be polysemous, all those meanings and all those scenes coming together on one canvas.

On the geometry of the painting and its two principal axes

The painting is attached to a wooden board of dimensions 67.5 × 91 cm.[221] Presumably the frame that was once attached to it at one time was removed because it covered part of the painting.[222]

The painting is not flat, but rather, slightly cylindrical, following the contours of the wooden board about 1 cm thick to which it is attached. The central horizontal axis of the painting itself protrudes about 2 cm from the plane defined by the back corners of the wooden board. This could be the result of the wood having warped, but it should be kept in mind that a few contemporaneous paintings were deliberately painted in this way, as witnessed by their surviving frames. (For example, Piero's framed "St Jerome Penitent" is slightly curved, this time convex with respect to the central vertical axis.[223]) On the other hand, modern radiographic analyses do reveal several horizontal breaks, as if the painting had succumbed to the warping of the wooden board.[224]

Unlike the case of the epigram, there is no problem identifying the borders of the painting: it is painted on a rectangular white background on the wooden panel. The complete foreground of the painting should reveal that the floor in front of the feet of the man with the beard is damaged. The complete right side

---

221 *Urbino 2005 Exhibition Catalogue*, p. 183.

222 See the text to n. 190 above.

223 For another example, this time convex with respect to the central horizontal axis, see *Christ in the Tomb* by Bartolomeo di Tommaso of Foligno *ca.* 1430 – see Garibaldi, *The National Gallery of Umbria (at Perugia)*, pp. 48-49.

224 Seracini, "Piero : *Flagellazione*". See also the text to n. 192 above.

should reveal the white background at the back of the right hand figure where a small area of the paint has fallen off. Most published versions of the painting tend to have the left side and the right side and the bottom chopped off. In fact, the painting continues to the left of the column on the left, with parts of the ceiling of the hypostyle hall visible there. The purpose behind this is, as we shall see, to identify the central axis of the left-hand column.

Piero appears to have begun his painting by constructing a square with sides of one Florentine *braccio* or cubit. This unit was equivalent to 58.4 cm,[225] and the height of the painting is apparently 58.4 cm.[226] Arezzo became Florentine in 1384, so that Piero was born a Florentine citizen.[227] If he had made the painting for somebody in Urbino, he would not have used a Florentine cubit. On the other hand, 58.4 cm is one of several measures (second from the right) represented on a Renaissance gauge for lengths in the doorway of the Municipio at Urbino (directly opposite the Cathedral), but this is specifically applied to silk (labelled SETA).

It has been proposed that Piero would have produced the base of the square beyond one corner. Using the diagonal of the square – an incommensurable √2 *braccio* – as radius, he would then have constructed the arc of a circle to cut the base produced, thus providing the base of the rectangular frame for the painting.[228] However, if we multiply 58.4 by √2 we obtain *ca.* 82.6 cm, and the width of the painting is apparently 81.5 cm. So this proposition needs to be checked. If we assume that 81.5 is correct, and that it was derived by multiplying the height by √2, then the height would have to be 57.6 cm. Rough estimates for the "height" of the convex surface (using Pythagoras' Theorem) using distortions of 2 cm then 3 cm then 4 cm would be 58.5, 58.7 and 58.9 cm. So we still have a width a few millimetres too large. Complicated measurements would be necessary before one could say more. See **Fig. III-4**.

---

225 See Zupko, *Italian Weights and Measures from the Middle Ages to the 19th Century*, pp. 40-49 on the *braccio* and p. 46 on this value.

226 This is commonly given in the literature. A document given to me by the Museum in 2006 suggests 58.3 and 81.5 cm for "la superfacie dipinta", as if the curvature had been taken into consideration.

227 See, for example, Banker, "Contributi alla cronologia di Piero della Francesca", p. 248.

Duccio Alessandri suggested to me that Piero may have used the *Florentine braccio* on account of his having been an apprentice of Domenico Veneziano in Florence in 1438. On Piero's relations with this teacher of his, see, for example, Focillon, *Piero della Francesca*, pp. 58-60 of the Italian translation, and Field, *Piero della Francesca*, pp. 90-94.

228 For example, this is proposed in:
www.dartmouth.edu/~matc/math5.geometry/unit13/unit13.html.

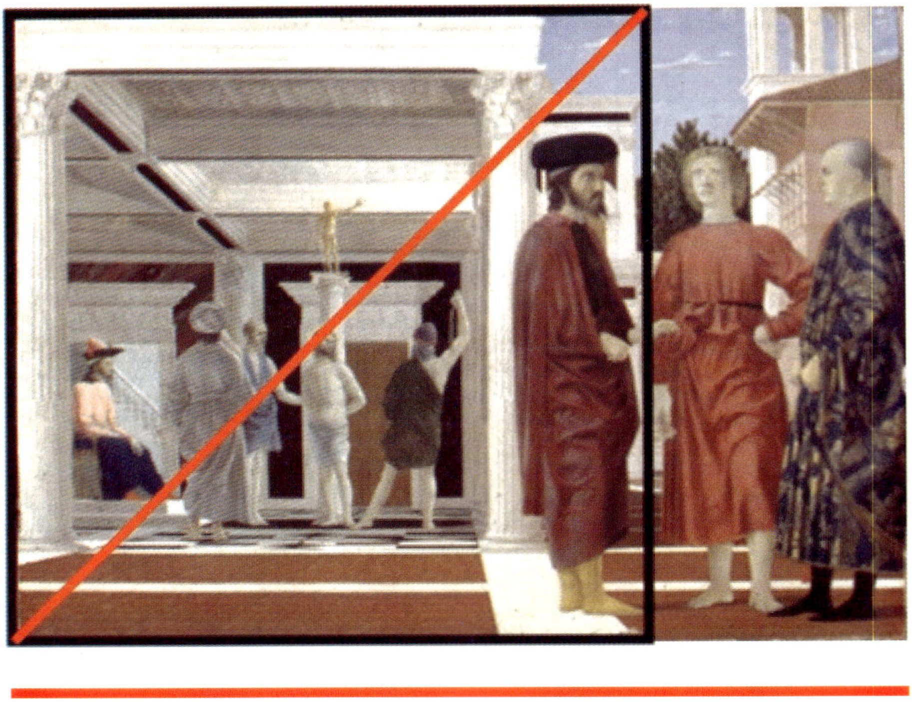

Fig. III-4:~ It seems that Piero prepared the space for his painting by first constructing a unit *braccio* square and then using the diagonal to construct the width of the space. However, with the measurements of the height and width currently available, or with slight adjustments to take into consideration the curvature of the painting, there seems to be a discrepancy of a few millimetres. The painting needs to be measured again, more carefully than before, taking into consideration the curvature.

In any case, Piero's rectangle can be thought of as composed of a left-hand square and a right-hand square. The centres of these two left and right squares are at the intersections of their respective diagonals. The vertical lines through these two centres are two principal vertical axes in Piero's painting. The first is the vertical axis between the eyes of Christ, which passes through the centre of the left-hand square. The second is the vertical axis between the eyes of the man with the beard, which passes through the centre of the right-hand square.[229] If the width of the painting *were* √2 *braccio*s, it is not difficult to show that these axes would theoretically divide the painting at ¼ √2 = *ca.* 0.354 and (1 – ¼ √2) = *ca.* 0.646 of its width. These values are, of course, incommensurable, but that would not have stopped Piero constructing lines to represent them. See **Fig. III-5**.

In fact, the axes are at *ca.* 0.354 and 0.649 of the width of the painting, which seems to confirm the supposition that he did use the diagonal of the left base square. These positions are not far removed from the axes that would divide the width of the painting at 3/8 (0.375) and 5/8 (0.625). Likewise, they are not far from the two axes that divide the width in the divine proportion (at *ca.* 0.382 and *ca.* 0.618).

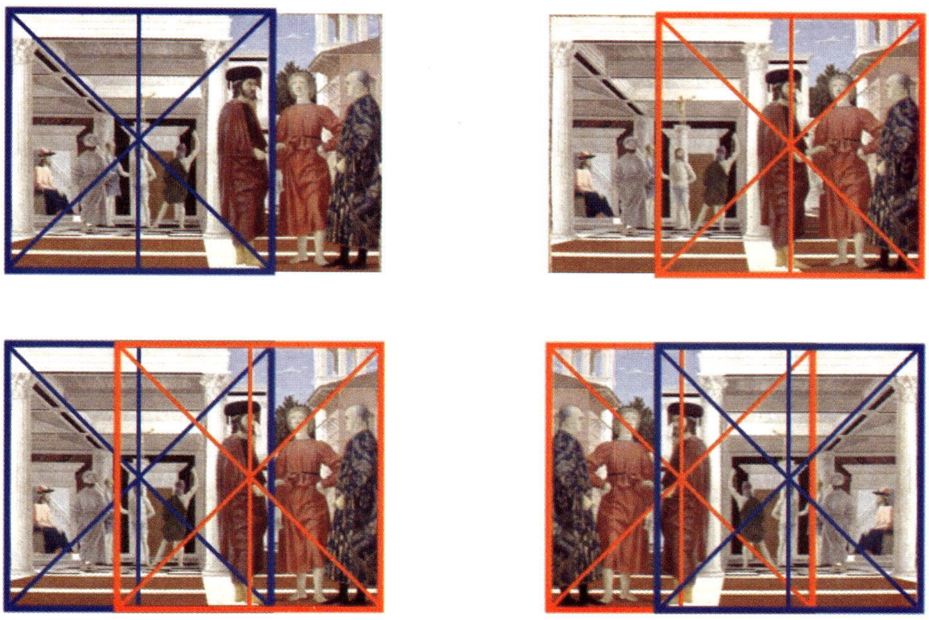

Fig. III-5:~ Preliminary investigations show that Piero constructed the left-hand and right-hand squares of his rectangle. The vertical axes bisecting each square

---

[229] The inspiration for these assertions derives from the image at:
www.artemis.jussieu.fr/hermes/hermes/actes/ac0102/flagellation.gif ,
alas based on a cropped image and without commentary. This site was brought to my attention in June 2005 by Nadine Faerber of the Frankfurt Seminar.
Not even these axes are mentioned in Field, *Piero*, pp. 174-181.

would serve for the Christ-figure on the left and for the Bessarion-figure on the right. These axes appear to pass between the eyes of each figure. The same holds, of course, for the picture in reverse.

The use of perspective in this painting by Piero has attracted a great deal of attention, and "Sam" Carter's reconstruction the floor-plan and Antonio Criminisi's animated representation of the three-dimensional scene portrayed by Piero are particularly impressive.[230] See **Figs. III-6a-b** and **III-7**. However, it is not the perspective that concerns us here, but rather the two principal vertical axes that we have identified, and Piero's brilliant use of the divine proportion.

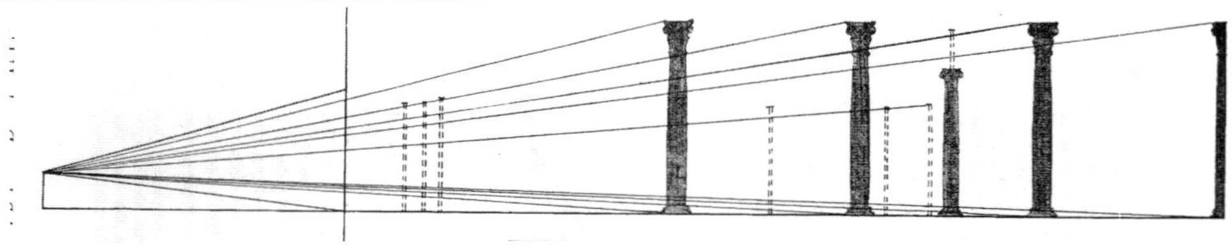

Fig. III-6a

Fig. III-6a-b:~ "Sam" Carter's 1953 reconstructions of the side view and the floor design of "The Flagellation". Curiously, this brilliant work did not consider the basic geometry of the painting (as revealed by Figs. **III-4** and **5**), which has also not been pursued in any other more recent publications. [From Wittkower & Carter, "The Perspective of Piero's 'Flagellation'", opposite p. 294.]

230 See the important study Wittkover & Carter, "The Perspective of Piero's 'Flagellation'", and, most recently, Andersen, *The Geometry of an Art*, pp. 34-80. For the most recent three-dimensional reconstructions, see Criminisi, *Accurate Visual Metrology*, pp. 140-142 and colour pls. 4-5, and his various websites, including "Getting into the Picture", also Field, *Piero*, pp. 176-177. A reconstitution of the floor using computer graphics is in Fondelli, "La restituzione della pavimentazione". See also Kemp, *The Science of Art*, pp. 30-32.

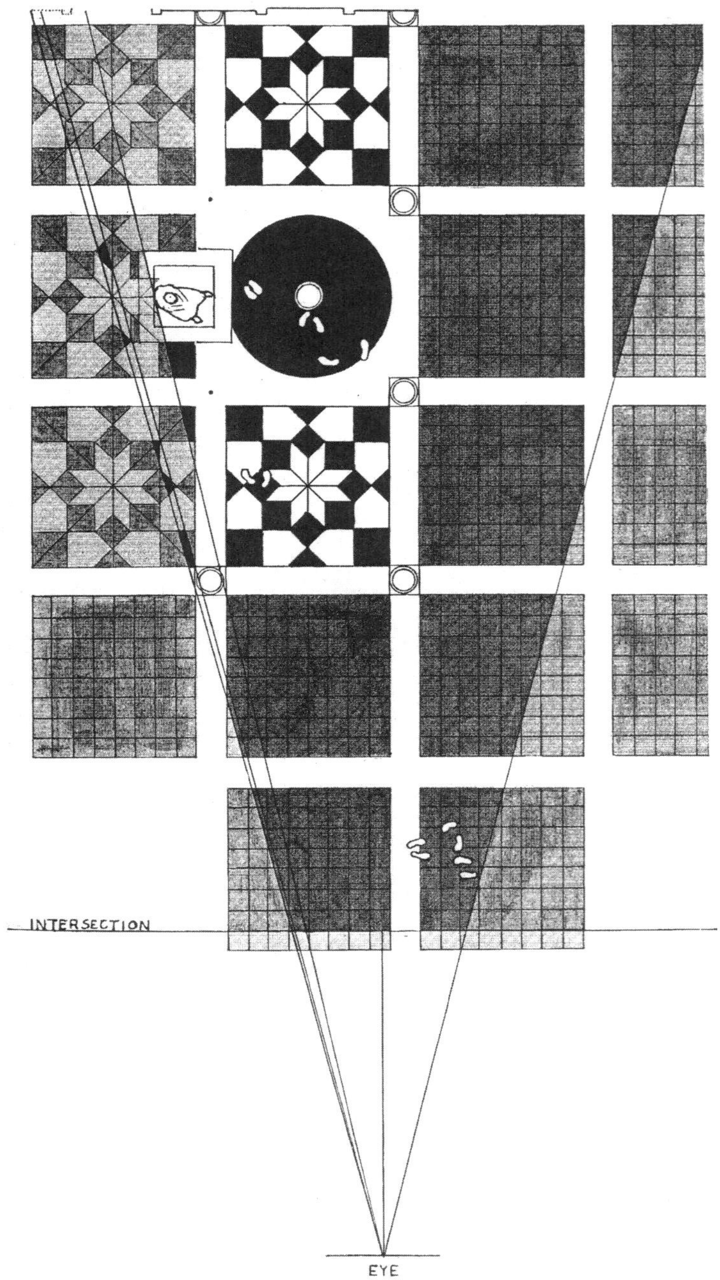

Fig. 6b

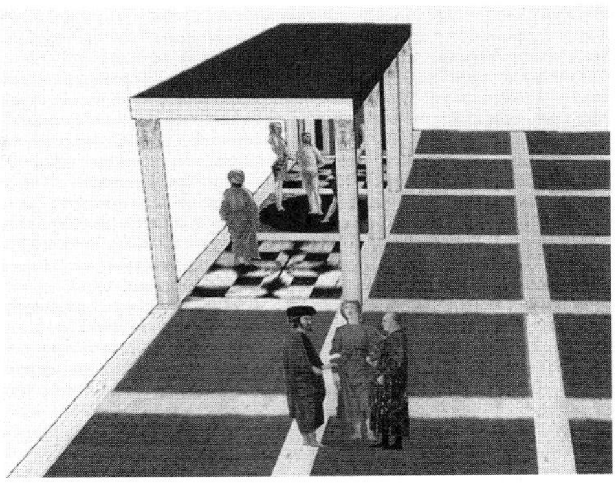

Fig. III-7:~ A three-dimensional reconstruction by Antonio Criminisi.

Piero's use of the divine proportion

In the painting there are two other axes defining a space that appears to be deliberately divided in the divine proportion.

Whilst we are not sure how Piero constructed the rectangle used for his painting, Holzschuh and I maintain that the vertical axis between the eyes of Christ

---

238 ** This does not seem to have been noticed previously, probably because most images of the painting are cropped at the sides, but also because it seems that there has never been a proper mathematical investigation of the actual painting, as distinct from studies of its perspective (see n. 230). Numerous websites accessible with the words "Flagellazione" and "sezione aurea" or "Flagellation" and "golden section" give the impression that Piero's painting is full of "golden rectangles" and the like, but I have not found a single serious discussion with an appropriate diagram.

divides the width of the flagellation scene, defined by the central axes of the two front columns, in the divine proportion: see **Fig. III-8**. This would explain why Piero deliberately included space to the left of the front left column, extending horizontal axes of the ceiling to the left of it. (This space is not shown in most modern reproductions.) On the other hand, the central axis of the front column on the right is partly hidden by Bessarion's cloak, although the full column is visible below and above it. Of course, the columns are not circular, so the axes we find may be only an approximation to the actual centres. In any case, I invite readers to measure for themselves the distance of the vertical axis between Christ's eyes from the left column and the distance between the columns, divide the former by the latter, and see if they get 0.62 or thereabouts. I admit that, even if they do, it would not prove that this was deliberate.

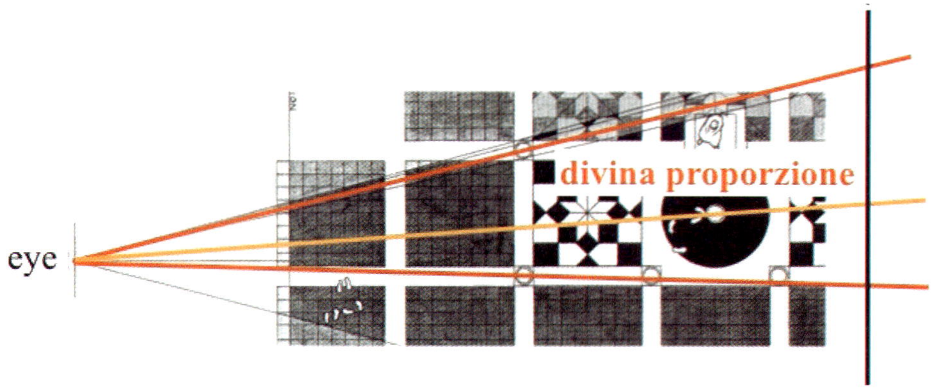

Fig. III-8:~ Sam Carter's diagram and the divine proportion in which the Christ-figure divides the space between the axes of the front two columns.

In addition, this feature explains, for the first time, why Piero took the vanishing point of his perspective in the vertical line that he did. Also, the point is one-third of the height of Christ, and I have heard somewhere (at the 2006 Urbino conference) that the height of a man was sometimes taken as three *braccio*s (which sounds unreasonable, since 3 arms' length would be better). This would then explain the position of the vanishing point.

As we have seen in Interlude I, the concept of the divine proportion was surely in Regiomontanus' mind as he conceived his epigram. But, apparently, he deliberately did not put the *divus* Bessarion at the divine cuts of the epigram. (Or if he did, I cannot prove it.) The epigram has no boundaries, so it has no divine cuts. On the other hand, it was entirely appropriate that Piero should use the divine proportion in this way in his representation of Christ.

I find it curious that this feature of the painting does not appear to have been mentioned in the modern literature,[238] which concentrates on Piero's use of perspective and the quadrature of the circle on the floor around the column behind Christ. However, as is well known, Piero discussed the divine proportion himself in his treatise *Trattato dell'abaco* (before 1477-78), first in the exercises in plane geometry dealing with the pentagon, and again in his discussion of solid geometry, namely, the measurement of a regular dodecahedron, with twelve pentagonal sides, inscribed within a sphere.[239] Furthermore, Piero's knowledge of the mathematics of the divine proportion inspired the treatise *Divina proportione* published by his student, Fra Luca Paccioli, in 1509.[240] The practical solution to the perceived "problem" of representing incommensurability is "as good an approximation as possible". To use the words of Luca: "*non potalmente la verità : ma e molto presso*", "not exactly the real value, but it'spretty close".

There is a reticence of art historians nowadays to even mention the divine proportion in relation to Renaissance art. The same scholars who are happy to recognize that the diagonal of a unit square is incommensurable, and yet who will have Renaissance painters draw it anyway, have a strange reticence when it comes to the divine proportion. Yet a mathematician armed with a graduated ruler and compass can divide a given line in the divine proportion as quickly as he or she could construct a square and draw its diagonal. So we cannot accept the strange pronouncement of Rudolf Wittkower in 1960 (my emphasis):~[241]

> " ... as compared with the commensurable ratios of the small whole numbers, the incommensurable Golden Section played an insignificant part in Renaissance and post-Renaissance art, *precisely because of its incommensurability*."

Likewise, even historians of mathematics specialized in the art of Piero have denied the possibility of his use of the divine proportion. Thus we read in the *Oxford Dictionary of the Renaissance*:~[242]

---

239 Daly Davis, *Piero della Francesca's Mathematical Treatises*, especially p. 1, n. 1, on various editions, and pp. 30-38, on Piero's discussion of the divine proportion. See also Herz-Fischler, *Mathematical History of Division in Extreme and Mean Ratio*, pp. 144-149.

240 See Herz-Fischler, *op. cit.*, pp. 149-151.

241 Wittkower, "The Changing Concept of Proportion", p. 205.

242 Article "Golden Section".

> "There is no evidence that 'divine proportion' played any part in the work of painters or sculptors. What is now known about the Renaissance suggests very strongly indeed that the supposed importance of the 'golden section' is an artefact of nineteenth-century historiography."

Also, J. V. Field in 2005 categorically dismisses the possibility of the divine proportion being of any significance in any Renaissance work of art, let alone Piero's:~[243]

> "This proportion, which has some interesting mathematical properties, gained considerable mystical significance in the late nineteenth and early twentieth century. In the manner of occultists (the same habit may be observed in astrologers at all periods), enthusiasts for the 'golden section' laid claim to a respectably long ancestry for their opinions, in this case tracing them back to the Renaissance, and in particular to Renaissance art."

In this case, the exception perhaps disproves the rule.

A first look at the painting and the epigram

First, we look at the painting with the epigram enlarged to fit over it: see **Fig. III-10**. The reader should try to imagine the excitement of Berthold Holzschuh when he did this for the first time, and our astonishment when he showed us this in the Frankfurt Seminar in May 2005. The idea of looking at the painting backwards, as in **Fig. III-11**, was proposed by myself in the next seminar. Nobody liked it.

The epigram has been superposed here so that it fits roughly over the flagellation scene. We should keep in mind that the inscription is only 3.8 cm long, whereas the painting is 81.5 cm wide. We should also keep in mind that the painting has two principal axes, similar to, but not quite identical with those in the epigram.

The graphics in **Figs. III-9** and **10** may convince any reader who has got this far that the epigram and the painting are related.

For – lo and behold – where Basileus Ioannis is seated on a throne we find the vertical axis II with the letters BA, then IO, then ANNIS. Behind him we have the column of letters SVB CD ANNIS in axis I, which will provide a time scale. Next to him we find the Ottoman Sultan with the letters IDRI in a vertical column, then S to the right side, suggesting the Muslim name Idrîs. This is, of course, not the name of the Ottoman Sultan, but it will serve to identify a space for the Muslim ruler.

---

243 Field, *Piero della Francesca*, p. 4.

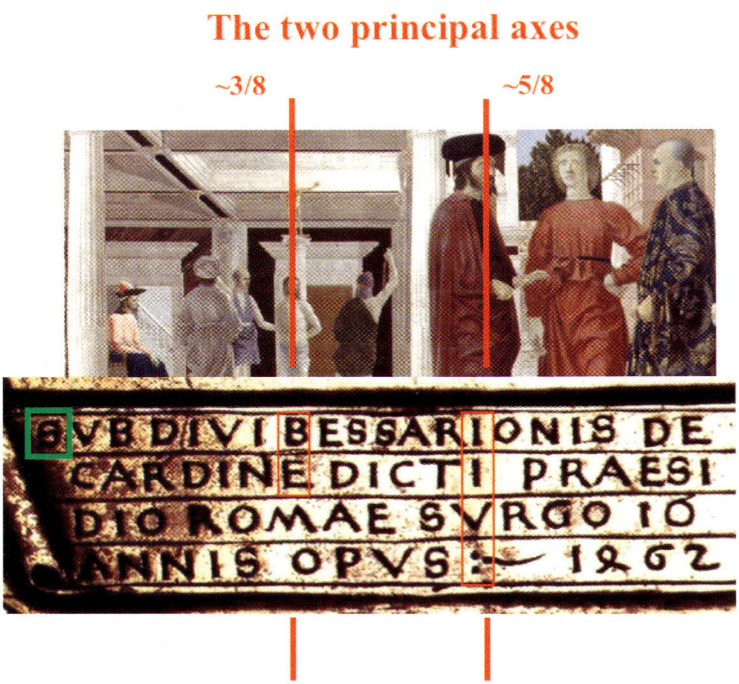

Fig. III-9:~ The epigram on the astrolabe enlarged to cover part of the painting. Notice the coincidence of the principal axes. The "B" in B-ESSARION defines the left principal axis that lies between the eyes of Christ. Also, the "I" in BESSAR-I-ON defines the right principal axis that lies between the eyes of the bearded man.

Then Christ is close to axis IV at the left principal axis, with the letters BE, then NE DICTI, then ROMAE, the expression BENEDICTI ROMAE hinting at the Church, the word BENEDICTI alone hinting at Christians in general.

BessarIon is at the right principal axis, with the axis VI through the I of his name, passing between his eyes.

The letters RGO with an I above and another below in axis VII identify IO(annes de) ReGIO (monte).

And perhaps the man on the right has the name Ioannes or Giovanni, identified by the standard abbreviation IO.

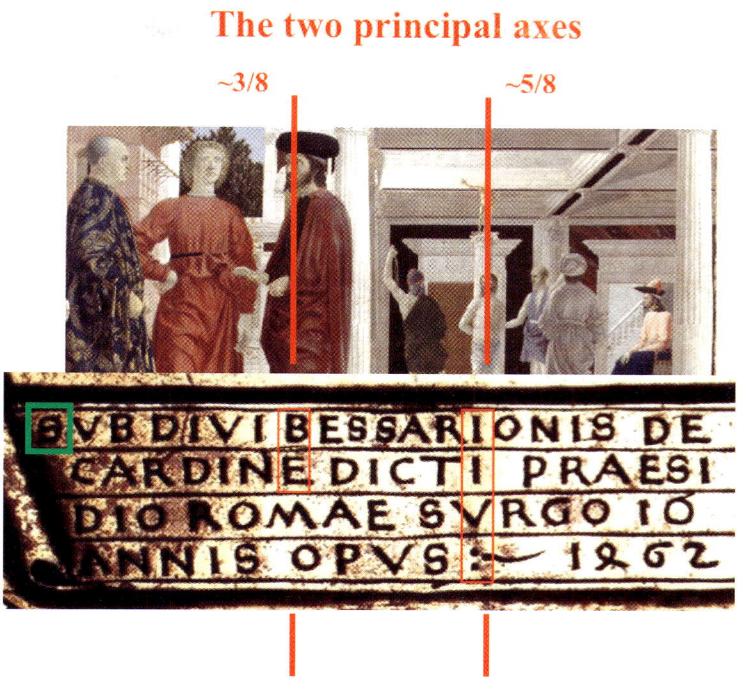

Fig. III-10:~ The epigram superposed on the painting viewed backwards. This is equivalent to reading the text backwards over the actual painting.

These, then, are seven axes (a-g) that we can identify in the painting, corresponding to six of the main axes of the epigram:~

a	b	c	d		e	f	g
I	**II**	**-**	**IV**		**VI**	**VII**	**VIII**
SVB	**B**	I	**B**E	BESSARION		**I**	
C	**A**	D	NE DICTI			**R**	
D	**IO**	R	ROMAE			GO	IO
ANNIS	ANNIS	IS				**I**	

So we could immediately assert that the painting features Ioannis VIII, the Ottoman Sultan, Christ, Bessarion, Regiomontanus, and a sponsor named Giovanni. (This scheme, with Giovanni Bacci on the right, was the proposal of Carlo Ginzburg.)

But let us have another look. In axes I+II we find **VB** and **CARD**, calling to mind the family of Ubaldini dalla Carda, as well as **IOANNI BACI**, calling to mind Giovanni Bacci. We were looking for another “Ioannes”.

In axis III we find a **DIV**vs figure, recalling the angelic St Bessarion.

In axis IV we find **B**essarion, with the B at the left principal axis.

In axis VI we discover **IOVIS**, the classical god Jove or Jupiter.

Finally, in axis VIII we find **SEDES** for a throne, the throne of **(H)ERODES**.

To summarize:~

I+II	**I+II**	**III**	**VI**	**VI**	**VIII**	
VB	**B**	DIV	**B**ESSARION	**IO**	S **DE**	**DE**
CARD	**CA**			**I**	**ES**	R **ES**
	IO			SV		**O**
	ANNI					

Now the fact that the epigram contains two symmetrical principal axes corresponding to those of the painting invites consideration of the same letters *backwards*. If we do this, we would have the following additional identifications for our original scene:~

b		**d**	**e**	**f**	**g**	**g**
VIII		**VI**	**IV**	**III**	**I+II**	**I+II**
S **DE**	**DE**	IO	**B**ESSARION	DIV	**VB**	**B**
ES	R **ES**	**I**			**CARD**	**CA O**
	O	**V**				**IO**
		S				**ANNI**

If we now combine these two schemes, we have:~

(a) a time scale between two different scenes, that are to be separated by 1400 years, not 400 years as before;

(b) the Byzantine Emperor Ioannis VIII, now on the throne of Herod;

(c) the Ottoman Sultan;

(d) Christ with a statue of Jove (!) on top of a pillar behind Him;

(e) Bessarion;

(f) a *divus* Regiomontanus; and

(g) a pair of potential sponsors, (Ottaviano) Ubaldini dalla Carda and Giovanni Bacci.

This, indeed, is not bad for a start. We could conceive a symbolic flagellation scene on the left:~

IOANNIS VIII	SULTAN	CHRIST
on SEDES of [Pontius Pilate]		IOVIS on column

And we could conceive three or more "moderns" on the right:~

BESSARION	REGIOMONTANUS	OTTAVIANO VBALDINI
	DIVUS FIGURE	GIOVANNI BACCI

Things are getting better.

However, the epigram was not intended as *the key to the painting*. Or to put it another way, Regiomontanus did not compose his epigram so that we in 2005-06 could use it as the key to identify the persons in Piero's "Flagellation". Rather, he composed it for Bessarion. When Bessarion and Regiomontanus looked at the epigram more carefully, finding in it all sorts of things that Regiomontanus could never have intended, there surely was no "Flagellation". They must have been very surprised and very delighted and very excited about what they found. The epigram was somehow magic.

We know they eventually found *eight* people not six. In the sequel I shall attempt to show that the existence of the two principal axes at roughly 3/8 and 5/8 of the width is related to the fact that there are eight principal figures in Piero's painting.

And the god on the column does not look like Jove. So who is it? Maybe we can find him in the epigram.

The epigram is, I maintain, *the key to the inspiration behind the painting.* We need to look at it more carefully, and we have one advantage over Bessarion and Regiomontanus, namely, we know what we are looking for. We have already seen that we can identify the principal figures in the painting simply by looking at the corresponding axes in the epigram. But there is still much more in the painting that needs to be explained. Why have earlier investigators thought that the young man in red might be dead? Why have they been able to identify, at least to their own satisfaction, different people in the images on either side of the young man? Maybe there are more people to be identified in the epigram who would be relevant to the painting?

We should keep in mind that both men were familiar with all sorts of abbreviations in Latin and Greek. Also a given person might have a Greek name and a Latin one, or a Latin name and an Italian one (usually differing slightly from the modern Italian one). If we find all but one letter in a name, we should not be afraid to suspect that they saw the name from this incomplete collection of letters. For both of them, the first two or three letters would have been adequate to identify a particular name. If we find GONS or even GONSAGA, this is more than adequate for identifying somebody from the Gonzaga family. However, we shall seek as much of the name as feasible, lest we end up with a completely unreasonable scenario full of irrelevant people. Of course, I cannot be certain that what I shall find is what they found. But I shall use six criteria,[244] and I invite readers to do the same:~

(1) *what we find should be relevant either to the Passion of Christ or to the Fall of Byzantium or to Bessarion's personal and political agenda;*

(2) *if we find nothing more than what we have found already, we shall abandon the operation forthwith;*

(3) *if something that we find fits in the resulting painting with its eight figures and one classical god, we shall not reject it out of hand;*

(4) *if we find letters relating to a classical god, an angel, a throne, a door, a tree, roses and thistles, all more or less in the places in which they occur in the painting, we shall be encouraged to look further;*

(5) *even if we can find double or multiple identities for eight persons and one classical god in roughly the right order across the epigram, we shall not insist that Bessarion and Regiomontanus must have found them, or that Piero needed to be aware of all of them; and*

---

244 Compare the assumptions made by Marilyn Aronberg Lavin in her study of the painting (*Piero's Flagellation*, p. 15):

> "first, that every element visible in the painting contributes to an overall message, and secondly that all the elements necessary for understanding the message are to be found within the painting itself."

Clearly, we have no problem with the first assumption. The second explains why Lavin, and others before and after her, have not been able to explain the painting in a convincing manner.

(6) *we should never lose sight of the fact that the interpretation of selected word-clusters in the epigram is a purely subjective affair; the precarious nature of this undertaking is fully evident to this author.*

Finally, we know that the title "The Flagellation of Christ" is a modern title for the painting. The flagellation of Christ is not the principal subject. Christ is not really being flagellated. It is the *Passion* of Christ that has been used in the painting:~ His betrayal and His flagellation come together in a symbolic flagellation. However, it is not only Christ who has been betrayed, and not only Christ who is being scourged. That scene from the distant part is intimately linked to scenes in the recent past (*ca.* 1439) and in the present (*ca.* 1462). In the painting, as we might expect from its earlier title, all of these scenes *CONVENERVNT IN VNVM.*

TRIBUTE II:~ Berthold Holzschuh's interpretation of the axes in the epigram and their relationship to the painting

In May 2005, when Berthold Holzschuh noticed the correspondence between the principal axes of the epigram and the principal axes of Piero's painting, he very quickly saw seven axes in the epigram. When I told him that, in addition, IDRIS is a Muslim name, he immediately saw the eight axes shown below. He views six of these as containing *deliberate messages concerning the six principal individuals in Piero's painting*:~

S V B D I V I B E S S A R I O N I S D E

I II III IV V VI VII VIII

axis	hidden text		BH's interpretation
I	SVB		400 years
	C		1062
	D	IO	→ Byzantine astrolabe
	ANNIS	62	→ these words are a kind of time-scale corresponding to the left column in the painting
II	B		**Basileus Ioannis VIII**
	A		→ Pilate
	IO		
	ANNIS		
III	I		Muslim name [according to DAK]
	D		→ **Ottoman sultan**
	R		
	IS		
IV	BE		Church, Rome → **Christ**

	NE DICTI ROMAE	at left major axis of both the painting and at the golden sector of the the Flagellation scene
V	AR T	axis, centre Bessarion, ARchimedes
VI	I I V :~ 14	Bessarion-axis **bearded man = Bessarion** at right principal axis
VII	I R GO I	Regiomontanus-axis **young man in red = Regiomontanus**
VIII	DE continuing on the next line with CARDINE	compare dalla Carda **man with thistles (It. *cardo*) on his gown = Ottaviano Ubaldini dalla Carda**
Boustrophedon		
	RGO IO 62	Regiomontanus age 26 in 1462
VIII		
	DE E **S** I O	*Carmen figuratum* around a central **S** → Dei / Deo

So Holzschuh sees the three principal figures on the left of the painting as:

BASILEUS IOANNIS VIII	SULTAN MEHMET II	CHRIST / CHURCH at left golden sector of the Flagellation scene

and the three figures on the right as:

BESSARION	REGIOMONTANUS	OTTAVIANO VBALDINI DALLA CARDA

Holzschuh's conclusion is that *either* the epigram and the painting evolved simultaneously, perhaps in discussions between Bessarion, Regiomontanus and Piero, *or* that Regiomontanus saw the painting and conceived the epigram.

Remarks

I agree with all of the individual components and most of Holzschuh's interpretations thereof. I agree that the Christ-axis in the painting is around the golden sector of the Flagellation scene, whose width is defined by the axes of the columns at either side.

I cannot accept Holzschuh's overall interpretation that *the epigram and the scheme for the painting were conceived together* or that *the epigram deliberately mirrors the painting*. I do not believe that Regiomontanus necessarily knew "in advance" about any Muslim names (other than those of a few Muslim scientists) like Idrîs or about Italians named dalla Carda. Regiomontanus surely had the idea for the epigram already in Vienna.

Just suppose that Regiomontanus *had* shown the "raw" epigram to Bessarion before or during an encounter with Piero. Then, if Holzschuh's interpretation is correct, we have to imagine Regiomontanus first fixing up the actual epigram that he engraved on the astrolabe so that the letters BAIO of BA IO ANNIS formed one axis, and the letters IDRI and S formed another one deliberately to accommodate the idea of a Muslim ruler. DE CARDINE as a reference to dalla Carda happily "comes out in the wash". Further, I think that by aligning BA IO ANNIS as he did, Regiomontanus originally had in mind *only* of BAsileIOs IOANNIS, that is, Bessarion. Of course, then, happily, Basileus Ioannis also "comes out in the wash". I cannot believe that the occurrence of the letters IDRI and S in a sort of vertical axis is anything more than a coincidence.

INTER(L)UDE III:~ On order and disorder

The letters SRVEDNTI do not convey much, but they are an anagram for INTERVDS, which does not convey much more. However, if we allow that we can use any given letter more than once, we could stretch the letters to INTERVDVS, and we are nearly where we want to be right now. If we had an L, and we don't, we could imagine the word INTER(L)VDVS, and that is where we are. We shall use this notation of letters in parentheses for letters that we would like to see, but which are not there.[245]

*****

In the Museum für Moderne Kunst in Frankfurt there is an English version of the brilliant work "Ordine e disordine" by the Italian artist Alighiero Boetti (b. Turin, 1940, d. Rome, 1994).[246] The work is based of two sets of panels of 4×4 letters.

The first set defines what Westerners might consider "order", for the text is written horizontally:~

O	R	D	E
R	A	N	D
D	I	S	O
R	D	E	R

There are 100 of these panels, each with a different colour on each cell. They are neatly arranged on the wall of a small room in 10 rows of 10.

The second set, with the letters arranged vertically, here defines "disorder":

---

245 The reader should keep in mind that the letters do not all need to be there anyway. Any medieval Latin scribe could have interpreted *intld*^{9} as *interludus*. (On the abbreviation "9" see n. 24 above.)

246 See Lauter, ed., *Alighiero Boetti*, pp. 217-265, on numerous of his works based on letters of the alphabet, especially pp. 220-223 on "Ordine e disordine".

O	R	D	R
R	A	I	D
D	N	S	E
E	D	O	R

These panels are arranged randomly on the opposite wall. There are only 99 because apparently Boetti gave one away. It doesn't matter.

These images for us reveal more than ORDER AND DISORDER, if only for "INSIDERS" who can SEE. Foreigners are allowed, and names are favoured. An Italian might recognize the original ORDINE E DISORDINE. A Greek might see ADONIS. Regiomontanus would have seen DEO and DEI, used as configured letters in his epigram.

If we allow that any letter can be used more than once, or if we can use both panels, we can see SENSE AND NONSENSE and DRIED RAISINS and a lot more. An Arab might see IDRIS. Regiomontanus and Bessarion would have seen IOANNES.

We could allow abbreviations. We could then find I(esus) N(azarenus) R(ex) I(udaeorum).

With an extra "joker" panel "x" that can serve any letter, a child might find DINOSA(U)RS. Regiomontanus and Bessarion would have seen (C)ARDO and RE(G)IO and ODDAN(T)ONIO. They could also have found the chronogram for 2001.

Now suppose Regiomontanus and Bessarion had a matrix of letters like this:~

I	C	T	M
R	A	V	P
D	O	G	E
N	S	B	"x"

with a built-in joker "x" that need serve only L or F or H. These letters appear at first sight to be a hopeless jumble, but, with the same rules – or concessions – as before, we can extract a set of players for a new game, namely:~ IOANNES, IDRIS and OSMAN, IVDAEVS, IEVSVS, IOVIS, IVPPITER, (H)ERODES, BESSARION TRAPESVNTIVS, IOANNES DE REGIO MONTE, GIOVANNI BACCI, OTTAVIANO VBA(L)DINI DA(LL)A CARDA, (L)VDOVIGO

GONSAGA and PETRVS DE BVRGO SANCTI SEPVLCRI. Bessarion and Regiomontanus would have also seen the chronogram for 1606.

Now suppose that these same letters were spread out on a canvas in such a way that, *mainly quite by chance*, the eight personal spaces for eight of the above names were more or less clearly defined and in an acceptable arrangement. Suppose that, in addition, one could read the text backwards, with the beginning now on the right. Then we look at the same letters again, now "displaced" from their original positions because we are beginning from the right rather than the left. Imagine if, just by luck, one could find another set of names corresponding to the same eight spaces that we had before. Then we would have double identities for each of eight persons.

I propose that this is precisely what Bessarion and Regiomontanus found when they looked at the epigram more carefully. The two of them determined the multiple identities of eight persons and a classical god from these letters alone, by reading the text forwards and backwards. The technique they used was not far removed from the idea of medieval Latin and Byzantine Greek monograms, in which a letter that is repeated in the name need appear only once in the monogram. Furthermore, in both traditions, there were no strict rules about where the letters in the name should appear in the monogram. Add to this the fact that sometimes the monograms were composed back-to-front, as on signet rings. In Part IV we shall look at these monograms in more detail.

All this explains why, in Piero's symbolic "Flagellation", we find, for a start, a Byzantine emperor, an Ottoman sultan, a man who betrayed Christ, Christ Himself with a classical god above Him, a man who ordered Christ to be flagellated, a man bereaved of a son or godson, the dead youth himself, and a potential sponsor. But each of these has another identity. And besides, it is not all about betrayal, torture and bereavement. Christ's Passion was for Bessarion and Regiomontanus a message of hope. And there was also hope for the ageing Bessarion's immediate future in the young Regiomontanus, whom, of course, we shall also find in the painting.

PART IV:~ A NEW HYPOTHESIS TO EXPLAIN PIERO'S PAINTING

> "Merely uttering the name of Piero della Francesca is often used as a substitute for all argument, as though it were a magic spell." John North, *The Ambassadors' Secret* (2004), p. 96.

> "Like all kinds of historical detective work the solution of iconographic puzzles needs luck as well as a certain amount of background knowledge. But given this luck the results of iconography can sometimes meet the exacting standards of proof. If a complex illustration can be matched by a text which accounts for all its principal features the iconographer can be said to have made his case. If there is a whole sequence of such illustrations which fits a similar sequence in a text the possibility of the fit being due to accident is very remote indeed." Ernst H. Gombrich (1909-2001) in his "Aims and Limits of Iconography" (1972/1985), p. 6.

> "Scienza o magia? Se capisco è scienza, se non capisco è magia." / "Science or Magic? If I understand, it's science; if I don't, it's magic." Title of an exhibition in the Palazzo Re Enzo, Bologna, in 1991.

> "One of the perpetual delights, as well as one of the continual hazards, of following wherever the books lead, is the fact that we are forever being dropped without warning into new topics where we are total innocents, and we must attempt to become Instant Experts ... . The repeated scramble to master a historical framework new to us is at the same time exhilarating and terrifying." M. A. Rouse and R. H. Rouse, *Authentic Witnesses – Approaches to Medieval Texts and Manuscripts* (1991), p. 10, cited in King, *The Ciphers of the Monks*, p. 15.

The epigram is almost magic, once one recognizes its potential.

---

247 Clearly, the number of letters in a hexameter followed by a pentameter will be of this order.

There are, by coincidence, 62 letters.[247]

There are 15 different letters with the following frequency of occurrence:~[248]

A: 5 E: 5 I: 11 O: 6 V: 4

B: 2 C: 2 D: 5 G: 1 M: 1 N: 4 P: 2 R: 5 S: 8 T: 1

If there were one more "B", these letters could be used to form the name BESSARION three times. It is a pity that there is no "L" and no "F", but we shall learn to manage without these letters. We do not expect to find Bessarion's acquaintance, Francesco Filelfo. (The fact there is no "H" is not a problem, for this letter is infrequent in Latin, and in Latinized Greek words with, say, X (*chi*) and Θ (*theta*), these letters can be represented by "C" and "T".)

It is not difficult to find PETRVS DE BVRGO SANCTI SEPV(L)CRI by jumping about inside the epigram. Of course, once one starts this kind of exercise, one can also find in the inscription much more, such as IN VINO VERITAS, or, more or less straight across the board, OMNI SANCTI ORATE PRO NOBIS. In fact, one can find almost anything one wants except words or names involving F, H, L, Q and X, or Italian Z. We should beware of claiming to find too much. But we can indeed find a lot more than we have found already.

A closer look at the epigram

We shall never know when Regiomontanus composed his dedication and when he engraved it on the astrolabe, or when Bessarion first saw the epigram or the astrolabe with the epigram engraved on it. Perhaps he was privy to a preview of the epigram on paper? What precisely did Regiomontanus mean by the expression SVB PRAESIDIO when referring to the Cardinal? Between late September and late November, 1461, the two men journeyed together from Vienna to Rome. At some time, either on this journey or at their final destination, one or other of them noticed something about the epigram that Regiomontanus cannot possibly originally have intended in it. For the epigram contained – *mainly quite by chance* – the inspiration for something else. There is no need to suppose the intervention of any third party. If Regiomontanus had not already noticed, as soon as Bessarion saw the epigram he would have recognized BAS(ILEVS) IOANNIS and (G)IOVANNI BACCI and VBA(L)DINI and (L)ODOVICVS in the first four columns of letters on the left hand side. And there was more … … . Enough to fill a whole painting.

---

248 For a discussion of the frequency of letters in an oration of Cicero see Heinke, "Classical Cryptography", pp. 118-119. For more on the frequency of letters in the epigram see the section entitled "How was this possible?" at the end of Part IV.

250 * On abbreviations in Greek see the references cited in nn. 29 and 32 above.

First, Christ is there, for the IN and the RI are to be found aligned in two adjacent vertical axes near the left principal axis. Second, the Church is there, designated by the configuration of letters BE-NE-DICTI ROMAE, of which the "B" is at the left principal axis. Third, Bessarion is there. Fourth, Regiomontanus is there.

Next, the letter B alone (or the letters BA or BAS) and the letters IO-ANNIS on the left side immediately suggest the Byzantine Emperor, Basileus Ioannis VIII.[250] Then there is an IO on the far right, suggesting another or the same IO-annes.

However, the letters of the inscription can be used in reverse, by reading from the right hand side and interpreting as if from the left hand side. Immediately, (G)IOVANNI BACI jumps to the right hand side, which is where a sponsor might be found. And in the same way, Ioannis VIII ends up on the SEDES. Clearly, we are not going to complain that only one "C" is available for the actual name, BACCI; that would really be petty. But the letters on the left also spell out (L)ODOVICVS (Ludovigo Gonzaga) and O(TT)AVIANO VBA(L)DINI DA(LL)A CARDA, as well as VRBINO, so we have two more potential sponsors and their location. Also, the letters SEDES, for "seat" or "throne", are to be found on the extreme right. By reading the inscription in reverse, the SEDES ends up on the extreme left, ready for Ioannis VIII.

	Left principal axis	Right principal axis		
BA	CHRIST / CHURCH	BESSARION	REGIO	IO
IO		"IO"	IO	IO

Furthermore, the combination of an "I" with an "O" occurs four times across the epigram, and, with any luck, could serve:~ the Byzantine Emperor IO-annis VIII, "IO-annes" Bessar-IO-n, IO-annes de Reg-IO Monte, and (one of) the potential sponsor(s), G-IO-vanni Bacci, not to mention IOvis. And whoever noticed all this surely also saw that the letters RG and IO, for Regiomontanus, Bessarion's new favourite, can be juggled to represent GRGIO, suggesting George of Trebizond, who was anathema to Bessarion. From the letters SVRGO IO, the name GeORGIOS or GIORGIO is even more apparent. And we can also find names and words suggestive of a Muslim ruler and a deity of classical Antiquity, in precisely those positions where we find them in the resulting painting.

Mainly quite by chance, various groups of letters across this remarkable inscription suggest names of people other than the main ones we have mentioned. However, the "chance" is "controlled" by the interests of the two or more men who searched the epigram for such names. The geometry of the epigram, with its pair symmetrical principal axes, suggests two main scenes, one centred on Christ and one centred on Bessarion. Above all, however, it was Regiomontanus' idea of representing two people at different times, one dead and one alive, in a single "concept" – the word BESSARION referring to the Cardinal and the angelic saint (over 1000 years apart), and the word SVRGO embodying Sergios and Regiomontanus (400 years apart)[251] – that was to make the painting like none other before or after.

Almost miraculously, the short text of four lines that Regiomontanus had conceived, replete with letters, words, numbers, vertical axes, hints of places and times, is adequate to reveal names for images of several potential events:~ the betrayal and flagellation of Christ; the ineffectiveness of Ioannis VIII to implement the Union of the Churches and subsequent fall of Byzantium; Bessarion's summons to the Cardinalate; George of Trebizond's supposed treachery; the death of Bessarion's godson Buonconte; the arrival of Regiomontanus on the scene; and more besides. Some of the ideas to be gained from the epigram can only have been noted with the assistance of someone familiar, if only vaguely, with the Islamic world, most probably, Bessarion himself.

The "game" – looking for "monograms"

The playing-board

The left and right principal axes are at approximately three-eighths and five-eighths of the width of the epigram. This invites consideration of the eight divisions from left to right across its width, not least because there are eight principal figures in the painting. I label these divisions **1** to **8**, and their borders **1/2**, *etc.*, so that the principal axes are roughly at **3/4** and **5/6**. The borders are only for our convenience. The figures that correspond to the divisions in the painting do not divide it into eight equal sections; rather, they are controlled by the two principal axes. Also, the fifth figure from the left is not in space **5** anyway.

The scheme below shows the eight divisions across the inscription, with the two principal axes emphasized. We read the text forwards and backwards, so that, essentially, each main division contains two distinct groups of letters. We label the divisions from left to right **1a** to **8a**, and, in order to read them backwards, also **8b** to **1b**. See **Fig. IV-1**. As we shall see, each double division from **1** =

251 And it may be that the angel looked somewhat like the young Regiomontanus himself: see the text to nn. 107 and 518.

1a+1b to **8** = **8a+8b** roughly corresponds to one of the images in the painting, but the actual situation is far more complicated.

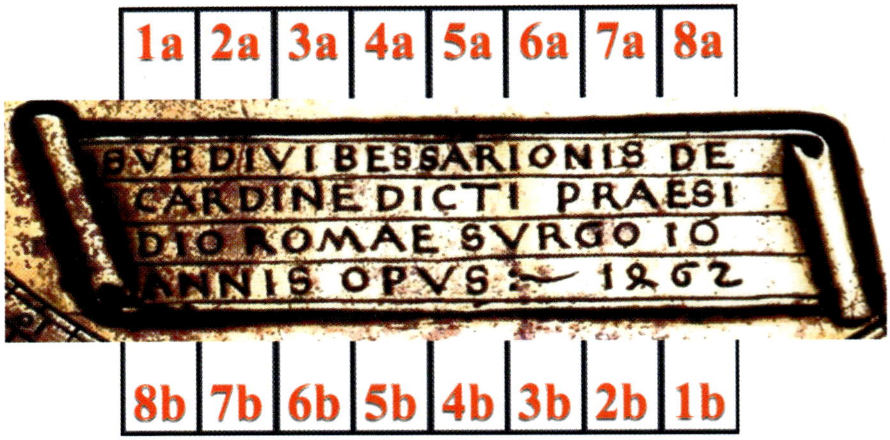

Fig. IV-1:~ The eight divisions from left to right in the epigram are labelled 1a-8a. As we have already shown, the letters in these provide hints of single identities for the eight persons and one classical god in the painting. The eight divisions from right to left, labelled 1b to 8b, provide further identities for these nine figures.

We are going to look for the identities of the eight persons in the painting as if they were each assigned the same "space" across the painting. In other words, we shall be working in two dimensions, as shown in **Fig. IV-2**.

Now the space **1a** is occupied by BAS IOANNIS, suggesting Basileus Ioannis. In space **8a** we find SEDES, "seat" or "throne", so we imagine it in **1b**, and BASILEVS IOANNIS VIII ends up sitting on a throne in place **1**.

Space **8a** is occupied by IO, suggesting someone named Ioannes in Latin. But IOVANNI BACI, suggesting Giovanni Bacci, is at **1a**, and we eventually want him at the right hand side, so we imagine him in **8b**. Now **8a** serves IO and **8b** serves IOVANNI BACI, and so place **8** can be occupied by Giovanni Bacci. But **8b** is also occupied by O(TT)AVIANO VBA(L)DINI DA(LL)A CARDA and (L)ODOVICVS (Gonzaga), suggesting two more candidates for place **8**. I label the three **8.1**, **8.2**, and **8.3**, and occasionally adopt this kind of label for other multiple identities.

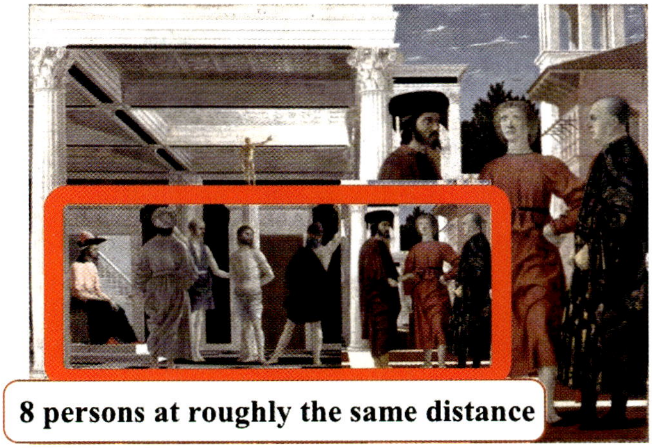

Fig. IV-2:~ The eight persons in the painting at roughly the same distance from the viewer. This is achieved by according 5/8 of the width to the five figures of the Flagellation scene and 3/8 to the three "moderns".

It would be absurd to expect to find all the letters for a given name in precisely the eight artificial divisions that we have defined only for our own convenience. We can rest content if we find most of the necessary letters in the appropriate space and the most of the rest in the space to the left or right.

The purpose of the game is to identify the multiple identities of each of eight persons and one classical god who could be relevant to Bessarion and/or Regiomontanus and whose images would correspond to those we eventually find in the painting "*Convenerunt in unum*".

The rules

> "The Byzantine monogram normally contains all letters of the name or word(s) in question (a repeated letter, however, is used only once). … … The arrangement of the letters seems to follow primarily aesthetic principles; attempts to discover underlying rules have failed. Hence the decipherment often proves difficult and in many cases remains ambiguous." From the article "Monogram" in *The Oxford Dictionary of Byzantium*.

In the sequel, we continue to capitalize letters that are found in or near the eight strips of the inscription; how near we mean by "near" is a matter of our discretion, because – I repeat – *the borders are only for our convenience*. We underline once letters that are in the adjacent space(s) to the left of the one under consideration and underline twice those that are in the adjacent space(s) to the right. We put letters in parentheses that are not near to the corresponding strip, or, as in the case of "L" and "F", not in the inscription at all.

A single letter may be used twice in the interpretation of a given group of letters. We do not expect to encounter names written out as linear text: we have to look around for the appropriate letters. Both of these "rules" would have been familiar to Bessarion from Byzantine monograms[252] and to Regiomontanus from Christian Latin monograms,[253] as in such medieval treatises as the *De inventione linguarum* attributed to the early-9th-century German scholar Hrabanus Maurus: see **Fig. IV-3** and **IV-4**.[254] These monograms were sometimes designed back to front, for example, on signet rings. Two examples must suffice:~

---

252 See the remarks in the article "Monogram" in *The Oxford Dictionary of Byzantium*, cited at the beginning of this section, to which one could add that sometimes the monograms are engraved backwards, as on signet rings and stamps (see below).

If the word has only five letters, all can be represented in a +-shaped configuration, even if, say, two are the same. The examples given in that source show ΠΑΥΛΟΥ, "of Paul", with the A written inside the Π, the sides of the A serving for the Λ, and a small O and Y on top of the P, and ΘΕΟΤΟΚΕ ΒΟΗΘΕΙ, "O Mother of God, help!", more or less as we have reproduced it here, in two +-shaped arrangements, the first with Θ and E at either end of the horizontal axis and O, T and K reading down the vertical axis, and the second with Θ and H on the horizontal axis and O and B on the vertical one.

Paderborn 2001 Exhibition Catalogue contains illustrations of several such monograms on pendant crosses (pp. 305-307), on weights (pp. 259-261), on two attachments for a polycandelon (pp. 218-219), as well as on a marble column (pp. 94-95), and some engraved backwards on signet rings and stamps (pp. 232-241).

253 See the splendid publication Gardthausen, *Das alte Monogramm*, with some 400 examples from Antiquity, and also the articles "Monogramme" in *Dictionnaire d'archéologie chrétienne et de liturgie*, and "Monogramm" in *Lexikon des Mittelalters*, also Ernst, *Carmen figuratum*, pp. 529-539.

254 On Hrabanus see the references given in n. 7 above.

A dozen of his monograms are reproduced in Ernst, *Carmen figuratum*, p. 532. Here again a given letter can serve more than once. Thus SANCTA is represented in a +-shaped configuration with SAT on the horizontal axis and TAC downwards on the vertical axis. In the same way, MARIA is written RAI horizontally and MAA vertically downwards, and PAVLVS is written with PAL horizontally and VAS vertically. Again, DOMINVS is written with D and S at the ends of the horizontal axis, and M and N at the ends of the vertical one, with a diamond-shaped O in the middle to provide the V, and with an I inside it.

For more complex Latin monograms collected by Du Cange (1610-1668) and published by him in 1678 see King, *The Ciphers of the Monks*, p. 334, from the 1883-87 edition of his Latin *Glossarium*.

O	O	T	M
Θ–T–E	Θ–|–H	S–A–N	R–A–I
K	B	C	A

ΘΕΟΤΟΚΕ ΒΟΗΘΕΙ
O Mother of God, help!

SANCTA MARIA

At least a novice can read these monograms, since all the letters are present, even though one has to figure out where to begin and then establish the order in which to take the letters in order to form the name or word; there were other monograms in circulation whose significance was known only to the initiated.[255] And, fortunately for us, we are not dealing with any secret alphabetic codes that were used in the Papal offices in the 15th century.[256]

I am going to claim the equivalent of the following: if we can find a disordered cluster of letters ACIMNRST, we can assert that these point to the words SANCTA MARIA.

Relevant names and words can be in Latin, Italian or Greek. Some Italians will be identified by their Latin names, others by their Italian names: the differences are minimal, but we should be careful to distinguish, for example, between Latinized Federicus and Italian Federigo (modern Italian Federico). In our procedure, for example, a cluster of letters AGONS points to the name GONZAGA.

We have already seen (1) that the letters REGIO are hidden in the name SERGIOS, and (2) that both SeRGIOS and ReGIO are hidden in the letters SVRGO IO. From the same letters SVRGO IO one can envisage GeORGIOS. Did Bessarion notice this? I strongly suspect that he did. The only question is whether Regiomontanus beat him to it.

---

255 We should keep in mind that the two men met in Vienna, where the Emperor Friedrich III had his own very special monogram, namely, "aeiou": see the article "AEIOU" in *Lexikon des Mittelalters* and the literature there cited.

These letters are engraved on two instruments from the Vienna workshop surely destined for the Kaiser: first, an ivory horary quadrant dated 1438, and second, a brass pocket sundial dated 1451. On these see Appendix D, items VI-3 and VII-1.

On a monogram associated with Friedrich IV also featuring the letters "aeiou" see King, *The Ciphers of the Monks*, pp. 331 and 333, but also n. 16 on p. 331.

256 See the references cited in n. 12.

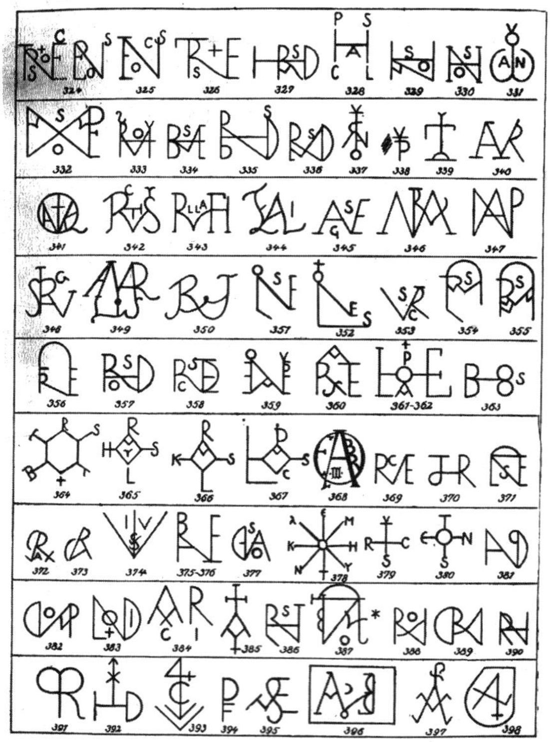

Fig IV-3:~ Several dozen classical monograms out of some 400 collected by Viktor Gardthausen. [From his *Das alte Monogramm* (1924).]

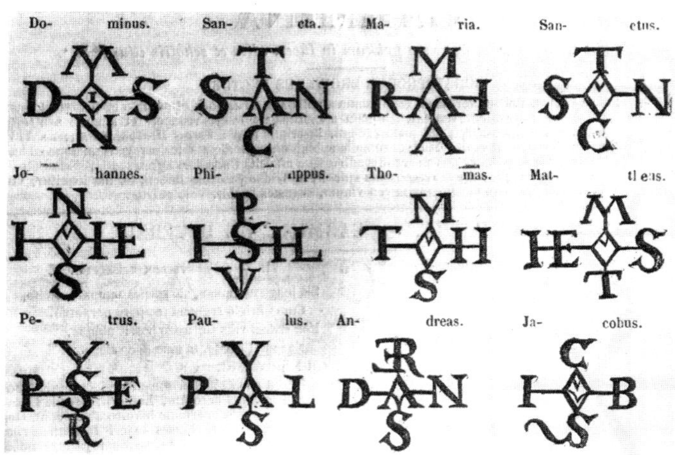

Fig. IV-4:~ A selection of monograms in Hrabanus Maurus' *De inventione linguarum*. [From Ernst, *Carmen figuratum*, p. 532.]

As we shall see below, a single group of letters, such as TRAPESVNTIVS, can apply to more than one person depending on which way – from left to right or from right to left – one is looking at the inscription.

We should not forget that this is a subjective exercise because, at least for some of the spaces, we know more or less what we are hoping to find, such as a Turkish ruler in space **2** and a classical god at **3/4**. Many persons can be identified in the letters of the epigram who are completely irrelevant to the painting.[259]

And we should keep in mind that, if this kind of "game" was really played by those who conceived the painting, there would surely have been several players, each with their own interests. We can be confident that Bessarion recognized some of the figures we shall identify, perhaps with the help of Regiomontanus. But other letter configurations may have been noticed later, perhaps by Piero, and here I am thinking particularly of the whole symbolic Passion scene with its furniture already in the Lateran College, and such items as the tree and the roses.

It does not seem worthwhile to attempt to separate what I have found into subgroups, somehow separated according to the "likelihood" of their relevance. So I shall simply present all of my findings for each of the eight main spaces.

---

259 ** For example, one can generate a list of participants at a hypothetical (and imaginary) session at Bessarion's academy. This would only be worthwhile if we had a Renaissance picture of such a gathering, featuring an assembly of 8 or 16 mainly unidentified participants.

The symbolic betrayal and flagellation scene

For each of the five spaces 1-5 we shall investigate groups of letters reading from the left and then from the right. We begin with the two most important images that we find on the left hand side of the painting.

The Christ figure

First we consider the Christ figure, cognisant that more is intended in his image. We find the following groups of letters:~

4a		
VI	I	BE
E	N	NE DICTI
	R	ROMAE
S	I	

4a IEVSVS and INRI, for Jesus Christ; and also BE-NE-DICTI ROMAE, which could signify the Papacy or the Lateran College, or simply BENEDICTI, for Christians in general.

4b					
RI	ARI	SA	N	R	ES
CTI	CTI	CT		T	E IC
SV	S	SV		E SV	AE
VS		VS		PVS	

4b RI, that is, R(ex) I(udaeorum), "King of the Jews", and CHRISTVS, the Messiah, deemed Son of God, who is CARITAS, as in the phrase "*Deus est caritas*", usually translated "God is love" (*First Epistle of John*, IV:8 and 16).

4b SANCTVS PETRVS, for St Peter, symbolizing the Church, also represented by ECC(L)ESIA.

We anticipate here:~

4 A Christ figure, not wounded by the threatening whips, also representing St Peter, the Church, and prehaps the Papacy, now to be situated at the divine cut of the symbolic betrayal and flagellation scene, defined by the central axes of the columns on either side.

This part of the painting shall deal with the PASSIO C(H)RISTI:~

4a	**4b**
SS	RI
I	CTI
A	S
OP	

The symbolic PASSIO C(H)RISTI in Jerusalem will take place with the throne, stairs, column and doors that tradition claimed had been brought to the Lateran College from Jerusalem via Constantinople.[260]

260 See the detailed account in Ginzburg, *The Enigma of Piero*, pp. 66-71. The booklet Camaparani & Amodei, *La Scala Santa*, casts some aspersions on the genuineness of some of these items.

The statue on top of the column behind Christ

Near to the letters for Christ and the Church, we find room for:~

3a+4a	4b
ES	SA
E	T
R M	SV
S	VS

3a+4a/4b a STATVA, statue, of the classical god (H)ERMES, or Ερμησ, Hermes, herald of the gods.

Again, counting spaces from right to left, we find in the complementary space:~

3b+4b		
RI IO	I N	A ION
TI P I	CTI	CTI
E VR SV	SV	SV
S		S

These suggest:~

3b+4b IVPPITER, Jupiter, or IOVIS, Jove, the chief god of the Romans and the god of Rome; the word INVICTVS, "invincible" (see below); and CONSTANTINIVS or CONSTANTINOS, for the Emperor Constantine I (Κωνσταντινοσ).

We also find in **3b+4b** INVICTVS, recalling the expression "*Sol invictus*" used in Late Roman religion, and the images thereof on the coinage of Constantine. A statue of Helios, intended to represent Constantine, had apparently been erected in Constantinople.[261] So we can be fairly sure that our predecessors wanted the Sun to be portrayed here in one way or another.[262] But we can also find:~

3a+4a	3b+4b	3a+4a
I	A ION	VI B
IN ICT	CTI P	N I T
O A	S	MA
S OP	S	S

suggesting the city of CONSTANTINOPO(L)IS (twice!) and even BISANTIVM. We recall a remarkable depiction of a personification of the city of Constantinople as an angel on a Byzantine gold goblet from the 7th or 8th century.[263]

I suggest that the statue will represent:~

3/4 Jupiter or Jove, the god of imperial Rome, dominating the flagellation scene, which was orchestrated by the Romans (though they were not held responsible), to be embodied in the Greek god Hermes, appropriately represented with a sphere and staff: he has no helmet with wings because he is more than Hermes. The column on which he stands / they stand had been brought to Rome from Jerusalem *ca*. 326 by Helena, mother of the Emperor Constantine, together with the *scala sancta*. It is not only Christ and the Church who are at the centre of the symbolic betrayal and flagellation scene, but the entirety of the Greek intellectual heritage, here perhaps symbolized by Hermes, "betrayed" and "flagellated" – in the mind of Bessarion – by George of Trebizond (see below). The naked young male figure is also intended to represent Helios, the sun. But it is also a statue of Constantine, and, deliberately illuminated by Piero, it also symbolizes the Byzantine heritage as well as the city that was named after him, Constantinopolis.

---

261 See, for example, Calvesi, *Piero*, p. 110. On the other hand, in the article "Helios" in the *Oxford Dictionary of Byzantium* it is stated that it is unclear whether Constantine supported the cult of Helios and had erected such a statue. In any case, the same source states that the image of the sun remained the focal point of Byzantine imperial propaganda.

262 Who knows what the young monk Bessarion had heard about the sun from his teacher Plethon at Mystra? See Walker, *Spiritual and Demonic Magic*, pp. 60-63, but also Medvedev, "Solar Cult in Plethon's Philosophy?".

263 See Evans *et al.*, *The Arts of Byzantium*, p. 33.

But there is more, and it is perhaps even more charming. We also find the letters:~

3b-4b
RI TI E SV G S

These suggest, at least to me, TRIS(M)EGISTVS, so that the Hermes mentioned above would be transformed into Hermes Trismegistus. This elusive sage and his gnostic writings were as popular in some academic circles in 15th-century Italy as Plato.[264] Bessarion himself acquired in June 1460 a Greek manuscript of the *Corpus Hermeticum*, and his library included several others containing Hermetic materials.[265]

If Bessarion found ERMES in the epigram, he would also have found TRIS, "thrice" in Greek, from *tria*, "three", which is all he would have needed for the identification of TRISmegistos. If he didn't find ERMES in the epigram, why is Hermes in Piero's painting? Certainly Regiomontanus would have heard about Hermes Trismegistus from Bessarion. In his inaugural lecture on al-Farghânî at Padua in 1464 he mentioned that some people claimed that Hermes had established the origin of the science of music.[266]

---

264 See the articles in *Dictionary of the Middle Ages*, *Oxford Dictionary of Byzantium* and *Oxford Dictionary of the Renaissance*. A particularly useful source for the influence of Hermeticism in magic, alchemy and science during the 15th-18th centuries is *Venice 2002 Exhibition Catalogue*. See also *Florence 1999-2000 Exhibition Catalogue* on the first translations in Italy by Ficino. (On the Islamic tradition see the sources cited in n. 270 below.)

265 *Venice 2002 Exhibition Catalogue*, II, pp. 14-26, 29-36, 52-57, *etc*. In one of these are some notes in Latin by Regiomontanus: see pp. 56-57.

266 This comment is recorded in Zinner, *Regiomontanus*, p. 113. It is translated in Swerdlow, *The Renaissance of Astronomy in the Age of Humanism* (see n. 113 above), p. 10 of the oration:

> "it is claimed that ... music was discovered by a certain Jubal, although some attest that Hermes established the origin of this science while walking on the shore when by chance a tortoiseshell resounded aginst his foot."

The man on the throne

In the first space we find the following suggestive letters:~

1a		1b			1a	
B SV**B** A **A** I	 **IO** ANNIS	 **IO**	S DE ES	DE R ES O	 C O ANN	S CA A

1a BA(SILEVS) or even BASI(LE)VS IOANNIS, or ΒΑΣΙ(ΛΕ)ΥΣ ΙΩΑΝΝΗΣ, that is, the Byzantine Emperor, Ioannis VIII; and

1b IO again, for Ioannis again, or Ioannes, and SEDES, a seat or throne, the throne of (H)ERODES, Herod.

Now that we have identified the throne, we find that, but for the ever-absent "L", we can also find:

1a CO(L)ONNA, Italian for column, and SCA(L)A, Latin and Italian for the staircase known as the *scala Pilati*.

So we identify:~

1 The Byzantine Emperor **IO**annis VIII, *ca.* 1438 (possibly copied from a medallion, of the kind that exist in Florence, Brescia and Vienna, or a bust, such as exists in the Vatican,[267] although Piero had seen him in Ferrara and Florence and had painted him elsewhere). He embodies Herod and Pontius Pilate, for he is to be seated on Herod's throne, in front of the *scala Pilati* or *scala sancta*, which had been brought from Jerusalem to Constantinople, and thence to the Lateran Palace in Rome.[268] The scene is set in a hypostyle hall. Ioannis VIII ignores what is happening to the Church, even though he is looking directly towards it, as Pontius Pilate washed his hands of what would happen to Christ.

---

267 For these see Gouma-Peterson, "Pietro's Flagellation", figs. 5-10, and *Venice 1994 Exhibition Catalogue*, p. 66 and 150, also Lavin, *Piero*, p. 25, and Ginzburg, *The Enigma of Piero*, fig. 31. Each side of six of Pisanello's medallions are illustrated in Zuffi, *La Renaissance*, pp. 108-109. On Ioannis VIII's distinctive hat, also depicted in Piero's Arezzo frescoes, see, for example, Gouma-Peterson, "Piero's Flagellation", p. 219 and figs. 5-8, and *Arezzo 1992 Exhibition Catalogue*, pp. 124-125.

268 See n. 260 above.

The man in Oriental dress

In the next space we notice the letters:~

2a		2b
I		NI
D	N	A
R	OMA	GO
IS	S	

2a IDRIS, for the Muslim personal name Idrîs, and also, nearby, OSMAN, calling to mind another Muslim personal name, 'Uthmân. (The less common Muslim name OSAMA, Arabic Usâma, can also be derived from this second set of letters.)

Only Bessarion or someone familiar with the Muslim world would have known that both Idrîs and Osmân are Muslim names. Idrîs is not a pure Arabic name: it was originally derived from Greek Ανδρεασ, referring either to the apostle Andrew or to the cook of Alexander the Great who achieved immortality in the Alexander legend, or from Εσδρασ, for Ezra. To medieval Muslim scholars, the name Idrîs would call to mind the prophet Idrîs mentioned in the *Qur'ân*, but also the sage Hirmis, sometimes called Idrîs, and sometimes called Dâniyâl, after two other Biblical characters with this name.[269] Bessarion might have known that the Arabic name Idrîs was used in Muslim sources for Hermes, "the prophet of the philosophers".[270] It is unlikely that he knew of any connection between the name Idrîs and Greek Andreas; certainly, St Andrew was on his mind because the Byzantine pretender Thomas Palaiologos had brought the head of the apostle to Italy in 1461, and Bessarion was present in April, 1462, when this precious relic was handed over to the Pope.[271]

---

269 On the names Idrîs and Dâniyâl see the respective articles in *The Encyclopaedia of Islam*.

270 See the articles "Hirmis" and "Idrîs" in *The Encyclopaedia of Islam*. For Arabic compilations mainly on alchemy and astrology attributed to Hirmis see Martin Plessner, "Hermes Trismegistus and Arabic Science", *Studia Islamica* 2 (1954), pp. 45-59; Sezgin, *Geschichte des arabischen Schrifttums*, IV, pp. 31-44, V, pp. 189-190, and VII, pp. 50-58; also King, *Survey of the Scientific Manuscripts in the Egyptian National Library*, pp. 27-28 (no. A22).

271 See, for example, Jöckle, *Encyclopedia of Saints*, pp. 31-34. On the arrival in Rome of St Andrew's head, see also Coffin, *The Villa in Renaissance Rome*, p. 167. (The rest of the apostle's body had been brought to Amalfi by Crusaders after the sack of Constantinople in 1204.)

The Turkish pronunciation of the Arabic name 'Uthmân is Osman, and it was the mysterious Osman I (d. 1324) who gave his name to the Ottoman dynasty.[273]

In any case, there was only one Muslim that Bessarion and Regiomontanus would have thought of to fill the space for somebody with a Muslim name, that is, the Ottoman Sultan Fâtih ("the Conqueror") Mehmet II.[274] Regiomontanus was well aware of the superiority of Muslim scholarship in the sciences even though he was familiar with relatively little of it.

We can find in the same space **2a** somebody who was in the background of Christ's Passion:

2a
D RD E O R S

2a King Herod. Several interpreters of Piero's painting have assumed that the man in the turban was Herod. There was a succession of rulers of Judaea who bore this family name:[278]

- ❖ Herod the Great (73-4 B.C.E.), first made Governor of Galilee by the Romans, and in 40 B.C.E King of Judaea. He ordered the massacre of the children of Bethlehem.
- ❖ His son, Archilaus, ruled over half of his father's kingdom, and was responsible for Jesus' family settling in Nazareth, outside his domains, when they returened from Egypt.
- ❖ Another son, Antipas, tetrarch of Galilee (4 B.C.E. - 39 C.E) put John the Baptist to death and was present at Jesus' trial.

---

273 * See the article "Othmân I" in *The Encyclopaedia of Islam*.

274 See the article "Mehemmed II" in *The Encyclopaedia of Islam*, and, on his scientific interests, also Bádenas, "Byzantine Intellectuals at the Ottoman Court"; Akasoy, "Mehmed II as a Patron of Greek Philosophy and Science"; and Heiderzadeh, "Turco-Persian Scholarly Exchanges", p. 422.

278 ** See, for example, Metford, *Dictionary of Christian Lore and Legend*, p. 119.

❖ Herod the Great's grandson, Agrippa I, His son, Agrippa II, gave the early Christians trouble, and the latter was titular king of Judaea when Jerusalem fell in 70 C.E.

In the complementary space counted from the right we find:~

2b
NI A GO

2b A special place for the Ottoman Sultan, who spells AGONIA, torment and suffering in both Greek and Latin, for Byzantium, with its capital Constantinople lost in 1453, Athens in 1456, Serbia in 1459, the Morea in 1460, and its last bastion Trebizond lost in 1461.

Here we recognize:~

2.1 The Ottoman Sultan Fâtih Mehmet II, who brought agony, death and destruction to Byzantium once again. (The first destruction was at the hands of the Crusaders in 1204.) He is to be portrayed facing Christ, but, in particular, confronting Byzantium and its heritage, symbolized by the statue of Constantine – see below.[279] Possibly, in Bessarion's view, it was the ineffectiveness of Ioannis VIII to implement the Union of the Churches that had led to the loss of Byzantium to the armies of Mehmet II. We note that Piero's image of the Sultan does not give the impression of the young conqueror in 1453 at age 21. Rather he portrays a rather heavily built and older man, even older than Mehmet II was in 1462.

2.2 King Herod again.[280]

---

[279] For other late-15th-century paintings and medallions of Mehmet II see, for example, *Barcelona 2004 Exhibition Catalogue*, pp. 20-21, 121, and 126-127. On the painting of Gentile Bellini see most recently the article "Capturing the Conqueror" by Orhan Pamuk in *The Guardian* (London), April 8, 2006 (available on the Internet).

[280] See the text to n. 278 above.

The man to the left of Christ

Here we find:~

3a		3b	
VI DI E A S	VI D O A S	IO IS I S RG S	NIS TI PRAE SVR

3a IVDAEVS, that is, a Jew, to which is related the name of IVDAS, ΙΟΥΔΑΣ, Greek for Hebrew Yehûdah, literally, "blessed", that is, Judas Iscariot;

3b GeORGIOS or ΓΕΟΡΓΙΩΣ or at least GIORGIO and then TRAPESVNTIVS, calling to mind George of Trebizond.[281]

My interpretation is:~

3.1 The historical Judas Iscariot, the disciple close enough to touch Jesus. We are dealing with the betrayal of Christ as well as the flagellation.

3.2 That figure should perhaps also represent George of Trebizond: see **Fig. I-6**. His image is to touch that of the Sultan, and he is to be associated not with Christ but with the Hermes figure (see below).

The introduction of George of Trebizond into the picture raises all sorts of problems and questions. But given a baker's dozen (13) of Bessarion's scholarly circle there was one who, in Bessarion's opinion, had betrayed "the cause", and that was George of Trebizond.[284] The lack of similarity between the portrait of George in **Fig. I-6** and the Judas-image in the painting perhaps confirms that in the latter we are not dealing with portraits.

---

281 It seems unlikely we should consider here George Amirutzes (see n. 52), also of Trebizond, not least because in 1461 he had consoled Bessarion about the fall of Trebizond. However, he did support Fâtih Mehmet's program to prepare Arabic and Turkish translations of Greek scientific works such as the *Almagest*. On his version of Ptolemy's *Geography* see also n. 52 above.

284 ** See the text to n. 49 above, and the reference to John Monfasani's discussion of the problems of the surviving images.

Personal reputations are based not on what people are, or what they say or write, but on what other people think of them or what it serves other people to think of them and to say about them. Ludwig Pastor, in his monumental 40-volume *History of the Popes from the Close of the Middle Ages* (1891),[285] could write:~

> "George of Trebizond is the most unpleasing of the Greeks of that day. Conceited, boastful and spiteful, he was universally hated."

This is, of course, not true. But Bessarion, who at one time entertained George in his own house, came to dislike and despise him as well as to criticize most of what he had written. Regiomontanus, who came from a milieu where some people did not like Greeks at all,[286] wrote against George even more violently.

And maybe even poor Judas, the disciple of Christ for years, was not totally bad either. The very name Judas means "blessed" in Hebrew, and Jesus had a brother with the same name (which conveniently gets changed to Jude in English translations of the *Gospels*). In a moment of weakness, Judas betrayed Christ. His fictitious biography in the *Legenda aurea* even makes him kill his own father and marry his own mother.[287] For 2000-odd years he has symbolized the archetype of the traitor. Yet in the 2nd-century Coptic *Gospel* attributed to him, now published from a 4th- or 5th-century copy, it is stated that Christ asked Judas to betray Him in order that God's plan for mankind be fullfilled.[288]

---

285 Quoted from Pastor, II, p. 202, note, in the article "George of Trebizond" from the *Catholic Encyclopedia* (found on the Internet).

286 An anonymous medieval writer from Württemberg (cited in Neugebauer, "Lesefrüchte", p. 1 (see n. 161) after Cerulli, *Etiopi in Palestina,* Rome, 1943, I, p. 341), had written of them:

> "Item die Indianer [!!] und die Jacobini und Suriani gleichen sich vast mit unserm glauben und unter allen glauben in dem tempel ist uns Kristen niemant widerwärtiger denn die Kriechen."

287 See, for example, n. 210 above.

288 See Robinson, *The Secrets of Judas*. For a translation of the *Gospel* and considerable information about the whole business of its discovery and subsequent adventures see Richard Pearse, "The Coptic Pseudo Gospel of Judas (Iscariot)", at:
http://www.tertullian.org/rpearse/manuscripts/gospel_of_judas/ .

The faceless soldier

Finally, we turn to the soldier threatening to whip Christ from the right. We find the letters:~

5a				5b	
RION	S R O	SAR	SAR	SAR	SAR
CTI	D	T	IC	T	IC
E SV	E	E S	E	E	E
VS		P	VS	P S	VS

5a CENTVRIONVS, which serves to identify a soldier.

5a (H)ERODES, that is, King Herod, again, and SATRAPES, "provincial governor", here referring to Pontius Pilate, also IV(L)IVS CAESAR.

5b SATRAPES and IV(L)IVS CAESAR again.

We identify for this position:~

5 A soldier with a distinctive (Jewish?) cap[290] perhaps embodying King Herod, but also Pontius Pilatus, Governor (*satrapes*) of Judaea, who represented Julius Caesar.

Between him and the Christ figure is to be a door of the Lateran Palace, for in **4a+5a** we find the letters PORTA:~

4a+5a
AR
T
A
OP

Some will argue that Pontius Pilate, Herod and Julius Caesar did not flagellate Christ. I hope they will realize that I am not trying to say that they did. The iconography of the hat worn by this figure should be investigated further.

[290] * On the subject see, for example, Sansy, "Chapeau juif".

Four symbolic betrayal and flagellation scenes

The scene for four symbolic betrayal or flagellation scenes was set, starting from the "time-scale" of the column on the extreme left (SVB *MCD* ANNIS), as in the akrostych in the Latin epigram (SVB CD ANNIS):~

	1	2	3	4	5
1	HEROD (PILATE)	HEROD	JUDAS	CHRIST JOVE / ROME	CENTURION HEROD / PILATE
2	BASILEVS IOANNIS			CHURCH	
3		SULTAN		BYZANTIUM	
4			GEORGE TR	HERMES / GREEK LEARNING	

The three men on the right

Emboldened by the fact that we can find five reasonable candidates with double or multiple identities for a quadruple betrayal and flagellation scene, plus a quadruple identity for the clasical god atop the column behind Christ, we continue looking for possible candidates for the three "moderns" on the right.

The bearded man

We find in the epigram:~

6a		**6b**	
BESSAR**IO**N	ARI N	BESSAR**IO**N	**I**
	TI P		NE
	SV		A
	S		S **O**

6a BESSAR-I-O-N or ΒΗΣΣΑΡ–Ι–ΩΝ, even with the appellation TRAP(E)SVNTIVS, from Trapezunt or Trebizond, with the "I" of BESSAR-I-ON at the right principal axis close to **5/6**.

6b "IOANNES" B-ESSARION again, with the "B" at the left principal axis near **5/6**.

With this information we identify:~

6 Basile**IO**s Bessar**IO**n of Trebizond at the right principal axis, reminiscent of how he had been in 1438 or 1439 before he became a cardinal (Piero had seen him in Ferrara and Florence), but also now as "**IO**annes" Bessar**IO**n in 1462, at a time when he was still pleading with Federigo for a crusade but was at least relieved by the fact that Regiomontanus was about to complete Peuerbach's *Epitome* of the *Almagest.* At the same time, he felt that the entire Greek intellectual heritage and the Byzantine Empire had been betrayed by George of Trebizond. Bessarion *ca.* 1462 is also to be depicted "by association", because he has now come back to Italy with Regiomontanus (see below). *Cardinal* Bessarion is to be deliberately hidden, as in the astrolabe dedication, but also because his image perhaps embodies someone else (see below).

The young man in red

In the next space we find:~

7a	7a	7a
NIS I AE R E **O** GO	A N N S TI A E SV G GO	N S NIS D AE A V GO O

7a the young man, IOANNES de REGIO monte, that is, Regiomontanus;

7a VANGE(L)ISTA GONSAGA, the dead son of Ludovigo Gonzaga; and

7a ANGE(L)VS or ΑΓΓΕ(Λ)ΟΣ, for a potential angel, with the face of an ADONIS or ΑΔΩΝΙΣ, bringing back memories of the angelic St Bessarion on the astrolabe.

In the complementary space we find:~

7b	7b	7b	7b	7b
B V C NE O O N S	B V ARDINE O R NIS	DI A D O NIS	DIV S	DI A O O NNI

7b BVONCON(T)E, or BONCON(T)ES,[293] the dead son of Federigo da Montefeltro, godson of Bessarion;

7b BERNARDINO or BERNARDINVS, the dead son of Ottaviano Ubaldini dalla Carda;

The red colour of his smock is surely to signify that Regiomontanus was the protégé of a cardinal, and Buonconte the godson of a cardinal.

The single angelic image for Regiomontanus and the three dead young men is to be reminiscent of:~

---

293 ** See Franceschini, "La morte di Buonconte", p. 497, for his signature on a letter of 1455: "Servus Boncontes Montis Feretrij", and Clough, "Cardinal Bessarion and Greek at the Court of Urbino", p. 161, for the text of a letter from Bessarion to Buonconte, beginning: "Bessarion Cardinalis Bonconti salutem".

7b an ADONIS or DIVVS, this last attribute reminding us of the DIVVS angel Bessarion.

We can also perceive the letters for:

7b ODDO AN(T)ONIO, the half-brother of Federigo da Montefeltro, who was murdered in 1444.

Now we can also find the letters:

7a		7b
S	I I	B
RAE	I RA	AR
O	VRGO	OR

suggesting two features that occur in the 7th space:

7a ROSAE, roses, symbols of the VIRGO (M)ARIA, and

7b ARBOR, a tree.

Thus we can identify:~

7.1 **IO**annes Reg**IO**montanus *ca.* 1462, who had been a mere child in 1439, and now eagerly accompanies the Cardinal, his smock coloured red, with a laurel tree behind him signifying his prowess as a poet and the fact that the painting was inspired by his poem. But the face of the young man is not that of Regiomontanus.

7.2-4 The young man embodies three dead youths, barefoot and in a simple smock, staring into space, and framed on one side with the same Marian five-petalled rosettes that are found on Regiomontanus' astrolabe. The three youths are:~

7.2 The dead Buonconte, Bessarion's godson.

7.3 The dead Vangelista Gonzaga.

7.4 The dead Bernardino Ubaldini dalla Carda.

7.5 The image is to be displayed with an angelic Adonis-like face as on the angel representing the *divus* St Bessarion on Regiomontanus' astrolabe. It is to be a typical "Piero angel".

7.6? Oddo Antonio or Odd'Antonio, half-brother of Federigo da Montefeltro, murdered in 1444, is perhaps also embodied here, which would confirm older hypotheses.[294] Also, if the angel looks like anybody, he looks like the portrait of Oddantonio now in the Kunsthistorisches Museum, Vienna, that was copied in 1578 from a now lost original in the Ducal Palace at Urbino.[295]

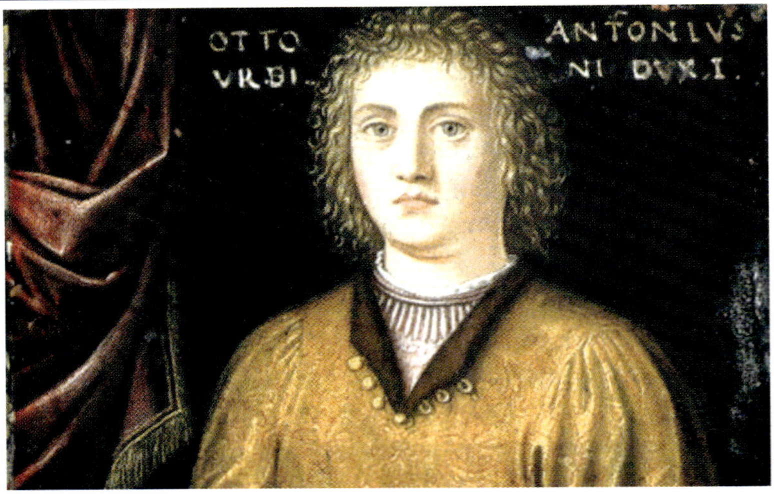

Fig. IV-5: A portrait of Oddantonio, copied in 1578 from one in the Ducal Palace, now lost. [Courtesy of the Kunsthistorisches Museum, Vienna.]

But there is more. In the same space we find the letters:~

7a

IS
R <u>E</u>
GO <u>I</u>

With these, or from the letters SvRGO IO, we identify:~

7.7 SERGIOS, the man who made the whole venture possible, and who should under no circumstances be forgotten here.

---

294 See the text to n. 197.

295 See Clough, "Federigo da Montefeltro's Artistic Patronage", fig. 13 on p. 12.

The man on the right

Now we turn to the last space, where we find:~

8a	**8b**	
E	B	V
AESI	CA	
IO	I	**IO**
		ANNI

8a room for someone else named IOA(nn)ES. (We have already mentioned finding IO in **1b**.)

In the complementary space we find:~

8b (G)IOVANNI BACCI.

Thus the first of the three "moderns" could be:~

8.1 G**IO**vanni Bacci, jurist and clerk to the Camera Apostolica. It is highly probable that he was one of the sponsors, if not the main sponsor, of the painting.

Also we find:

1a
S B
AR
O
A

from which we can read BORSA, late Latin for "purse", derived from the Greek. So we have Bacci as:

1a The commisioner of the painting, as suggested by Ginzburg. He does indeed have his hand on an object that could be taken for the opening of a purse.

The man on the right is shown with a red *becchetto* over his shoulder, and one end of it can be seen behind his feet. The *becchetto* is "a long scarf hanging loose from a kind of turban, much worn in Italy in the Quattrocento" (Ginzburg, 2000). Whatever the significance of the red colour of the *becchetto*, I am not going to claim anything here. It has already caused enough trouble, and will surely cause more now. An identical *becchetto* is also featured in Piero's "St

Jerome and a Devotee" (now in Galleria dell'Accademia in Venice).[296] Here the devotee, supposedly the Venetian patrician, Gerolamo Amadi, but more probably a member of the Bacci family, is kneeling down on the right, facing the saint, who is seated on the left. He is wearing a dark red tunic. The *becchetto*, in a slightly brighter "cardinal" red, is shown over his right shoulder, with one end falling on his upper back, parts in front of his chest and below his waist, and the other end terminating on the ground in front of him, as if it were a pool of blood.

8b
SV C I NN

It happens that in space **8b** we can also find "NVNCIVS", an epithet applied to Bacci – *nuncius* to Caesar" – without further explanation in a 17th-century biographical notice. This, or the Italian equivalent, Carlo Ginzburg in his first study of "The Flagellation" took as signifying Bacci's mission to the Byzantine Emperor to summon Bessarion to the Cardinalate.[297] I prefer not to argue that these letters are significant here. Ginzburg first interpreted the red shawl over the shoulder of the man on the right as a cardinal's shawl, and, as such, a symbol of Bacci's mission to Constantinople.[298] The evidence for the mission was the mysterious appellation "*nuncio* to Caesar". Ginzburg later retracted his earlier claims, identifying the "red shawl" as a *becchetto* and no longer insisting on Bacci as the *nuncio*. My interpretation does not depend on whether or not Bacci himself actually went to Constantinople. However, clearly someone delivered the invitation to Bessarion, and maybe the colour of the *becchetto* really was intended to symbolize that mission.

---

296 See, for example, Calvesi, *Piero della Francesca*, pp. 142-143 and n. 94 on p. 138.

297 See Ginzburg, *The Enigma of Piero*, pp. 76, 86, 120, and also 97, n. 69, for an English version of the 17th-century notice.

298 See Ginzburg, *The Enigma of Piero*, pp. 119, 120-122, and also 126. In this revised 2000 study, Ginzburg retracted his earlier claim that this was a cardinal's shawl, and proposed instead that it was simply a *becchetto*, compare his remarks on pp. 119 and 126. For the most recent discussion see Bridgemann, "Observations on Dress in the Work of Piero della Francesca", pp. 86-87.

Next we find:~

8b							8b
V	VB				SV	VB	SV
A	A	A	CAR	CAR	CAR	R	C
IO	DI	D	D	D O	D	IO	DIO
N	N	A	A			N	

8b O(TT)AVIANO VBA(L)DINI DA(LL)A CARDA, of VRBINO, also CARDO and CARDVVS, inviting a beautiful play on images (see below); and

8b (L)ODOVICVS for Ludovigo (Gonzaga), his family name as GONSAGA already in **7.4a**.

Thus the second person embodied in the figure of the man on the right can be:~

8.2 Ottaviano Ubaldini dalla Carda, father of the deceased Bernardino, perhaps another sponsor,[299] wearing a blue gown with gold brocade,[300] of which the most important feature is the embroidered thistles (Latin *carduus*, Italian *cardo*), a reminder of the several puns on *cardo* in Regiomontanus' astrolabe epigram.[301]

However, perhaps the facial profile of the man in place **8** is to be that of:~

8.3 Ludovigo Gonzaga, Marquis of Mantua, a man passionately interested in astrology, father of the deceased Vangelista, perhaps yet another sponsor.[302]

This man is to be standing in front of a villa, for which we find:~

---

299 On Ottaviano see Lavin, *Piero's Flagellation*, pp. 62-67, and also Appendix G below. Regiomontanus referred to him three times in his later correspondence: see Curtze, "Regiomontan's Briefwechsel", pp. 292 "*Octavianus*", 301 "*magnificus dominus Octavianus*", 309 "*dominus Octavianus*".

300 For the most recent discussion see Bridgemann, "Observations on Dress in the Work of Piero della Francesca", pp. 84-85.

301 See the text to n. 82 above.

302 On Ludovigo see Lavin, *Piero's Flagellation*, pp. 53-62.

8b
S CA A

for CASA.

Part of what is happening in this scene with the three "moderns" will be in Rome – ROMAE – and VRBINO can be found twice in **6b-8b** and VRBINVM once.

To summarize my findings for the three "moderns":~

6	7	8
BESSARION ~1439 BESSARION ~1462 (with REGIO)	REGIOMONTANUS BUONCONTE (dead) BERNARDINO UBALDINI (dead) VANGELISTA GONZAGA (dead) SERGIOS (dead)	GIOVANNI BACCI OTTAVIANO UBALDINI LUDOVIGO GONZAGA
	ANGELIC SAINT BESSARION (long dead)	
	ODDANTONIO (?)	

In addition we can find from the epigram the following items across the painting:

1	4	7	8
SCA(L)A SEDES	PORTA	ARBOR ROSAE	CASA BORSA CARDO

1	4	7	8
STAIRS THRONE	DOOR	TREE ROSES	HOUSE PURSE THISTLE

Coincidence? Maybe. I find the letters ROSAE in **7a** particularly astounding.

Table 5 presents an overview of all the persons identified using the epigram, as well as the various features of the hypostyle hall and the foreground, with an indication of whether they correspond to the space in the epigram in which they are found (x) or whether they correspond to the opposite space (9-x). In **Table 6**, the same list is presented in a different way, namely, according to the positions of the persons in the painting.

Table 5

INDIVIDUALS FOR THE PAINTING IDENTIFIED IN THE EPIGRAM

Note: Letters in parentheses are not found near the appropriate spaces, and, in the case of L, they are not found in the epigram at all.

EPIGRAM SPACE	PERSON	PAINTING SPACE
(The bold numbers 1-8 on the left refer to the spaces from left to right across the epigram; the numbers 1-8 on the right refer to the spaces from left to right of the painting)		
1	BASI(LE)VS IOANNIS	1
1	(G)IOVANNI BACCI	8
1	O(TT)AVIANO VBA(L)DINI DA(LL)A CARDA	8
1	(L)ODOVICVS (Gonzaga)	8
2	IDRIS, OSMAN, (H)ERODES	2
2	BVONCON(T)E, BERNARDINO	7
3	IVDAEVS, IVDAS	3
3	"IOANNES" BESSARION	6
3/4	ERMES, SO(L) INVICTVS	3/4
3/4	CONSTANTINOPO(L)IS, BISANTIVM	3/4
4	PASSIO C(H)RISTI	4
4	IEVSVS, INRI	4
4	BENEDICTI ROMAE	4
5	CENTVRIONVS	5
5	(H)ERODES, SATRAPES (Pilate), IV(L)IVS CA(E)SAR	5
5	C(H)RISTVS, RI (Rex Iudaeorum), CARITAS	4
5	SANCTVS PETRVS	4
5/6	IVPPITER, IOVIS, CONSTANTINIVS	3/4
5/6	TRIS(M)EGISTVS	3/4
5/6	CONSTANTINOPO(L)IS	3/4
6	BESSARION TRAPESVNTIVS	6

6	GIORGIOS TRAPESVNTIVS	3
7	IOANNES (de) REGIO (monte)	7
7	VANGE(L)ISTA GONSAGA	7
7	ANGE(L)VS, ADONIS, DIVVS	7
7	ODDO AN(T)ONIO (?)	7 (?)
7	AGONIA	2
7+8	(H)ERODES	1+2
8	IOA(NN)ES	8
8	(H)*f*ERODES	1

In addition we list the following features that will appear in the painting:~

1	SCA(L)A	1
1	CO(L)ONNA (It.)	1-5
1	CASA	8
1	BORSA	8
1	CARDO (It.), CARDVVS (Lat.)	8
2	ARBOR	7
4-5	PORTA	4-5
5	STATVA	4
7	ROSAE [VIRGO (M)ARIA]	7
8	SEDES	1

****

Table 6

IMAGES FOR THE PAINTING IDENTIFIED IN THE EPIGRAM

Note: Letters in parentheses are not found near the appropriate spaces, and, in the case of L, they are not found in the epigram at all.

(The bold numbers 1-8 on the left refer to the spaces from left to right across the painting; the numbers 1-8 on the right refer to the spaces from left to right of the epigram in which the name or title is found.)

PAINTING SPACE	IMAGES	EPIGRAM SPACE
1	BASI(LE)VS IOANNIS	1
1	SEDES, (H)ERODES	8
1	SCA(L)A	1
1-5	CO(L)ONNA (It.)	1
2	IDRIS, OSMAN	2
2	(H)ERODES	7
2	AGONIA	7

2	(H)ERODES	2
3	IVDAEVS, IVDAS	3
3	GIORGIOS TRAPESVNTIVS	6
3/4	ERMES, SO(L) INVICTVS	3/4
3/4	IVPPITER, IOVIS, CONSTANTINIVS	5/6
3/4	CONSTANTINOPO(L)IS, BISANTIVM	3/4
3/4	CONSTANTINOPO(L)IS	5/6
3/4	TRIS(M)EGISTVS	5/6
5	STATVA	4
4	IEVSVS, INRI	4
4	PASSIO C(H)RISTI	4
4	C(H)RISTVS, RI (Rex Iudaeorum), CARITAS	5
4	SANCTVS PETRVS	5
4	BENEDICTI ROMAE	4
4-5	PORTA	4-5
5	CENTVRIONVS	5
5	(H)ERODES, SATRAPES (Pilate), IV(L)IVS CA(E)SAR	5
6	"IOANNES" BESSARION	3
6	BESSARION TRAPESVNTIVS	6
7	IOANNES (de) REGIO (monte)	7
7	BVONCON(T)E	2
7	BERNARDINO	2
7	VANGE(L)ISTA GONSAGA	7
7	ANGE(L)VS, ADONIS, DIVVS	7
7	ODDO AN(T)ONIO (?)	7 (?)
7	ROSAE [VIRGO (M)ARIA]	7
7	ARBOR	2
8	IOA(NN)ES	8
8	(G)IOVANNI BACCI	1
8	BORSA	1
8	O(TT)AVIANO VBA(L)DINI DA(LL)A CARDA	1
8	CARDO (It.), CARDVVS (Lat.)	1
8	(L)ODOVICVS (Gonzaga)	1
8	CASA	1

From the two-dimensional epigram to the three-dimensional painting

Even with all these identifications, we are not quite finished. Somebody – not necessarily Piero – had the idea to assign the same space to the five figures on the left as to the three figures on the right. The former were moved backwards and the latter forwards. See **Fig. IV-6**.

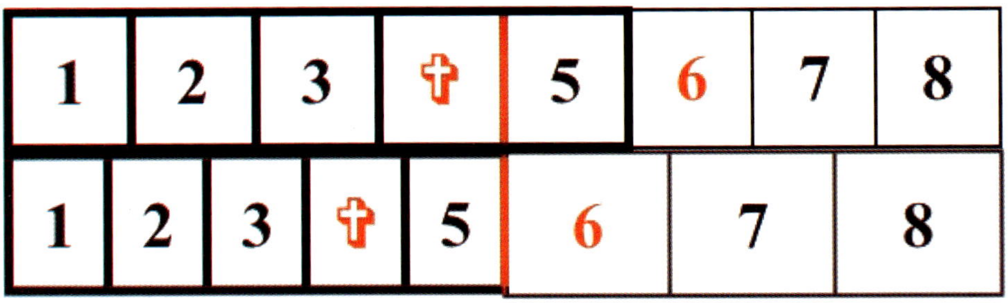

Fig. IV-6:~ The rearrangement of the people identified in the epigram – in two dimensions – so that the Flagellation scene in the background and the three moderns in the foreground occupy the same space.

Can we believe all this?

Is not all this too good to be true? The epigram appears to have revealed a surplus of information for the programme of the painting. Any historian of Quattrocento art would have been ecstatic about the discovery of a document identifying the three "moderns" alone, even as single historical individuals. Two identifications of the young man in red, say, as Regiomontanus *and* Buonconte, would have sufficed to provide a completely new interpretation of the entire painting that would have pleased nobody more than this author. Likewise, with Berthold Holzschuh, one could argue on the basis of a first glance at the epigram that the man on the right *must be* Ottaviano Ubaldini dalla Carda and then look no further. But he is not Ottaviano, or, rather, he is not *only* Ottaviano. Again, there is so much in the painting that we really do not need to have George of Trebizond in it at all. And perhaps he is not really there anyway, being simply a figment of my imagination.

Are all of these coincidences – *some 60-odd identifications of names and words in more or less the right positions* – not just simply that, the multiplicity of "findings" explicable as the result of the frequency of letters of the Latin alphabet? Have we not simply been misled by a circular argument claiming to have found just what we were looking for and having found only what fitted our own agenda?

Right in the middle of the epigram we can find the letters of the word **CAVEAT**. At least I should take heed for I cannot be certain that anybody other than myself ever saw most of, or even any of these letter-combinations in the epigram.

Some readers may wonder whether a scenario for a painting could not be derived from "any old epigram" by looking for "monograms" as I have done.[303] This I very much doubt.

I readily concede that much of what I have found is the result of coincidence; pure chance has played a large role in the utility of the epigram for identifying potential players. I am well aware that the letters of the Latin alphabet lend themselves readily to this kind of exercise.[304] I also know that other names can

---

303 Consider, for example, the epitaph composed by Regiomontanus' teacher, Georg Peuerbach, for his own tomb, and fortunately recorded in a manuscript, for his tombstone has disappeared (Zinner, *Regiomontanus*, p. 46). Let us take this as an example, even though it is over twice the length of Regiomontanus' epigram:

Extinctum dulces quidnam me fletis amici?
Fata vocant: Lachesis sic sua fila trahit.
Destituit terras animus: coelumque revisit,
Quae semper coluit: liber et astra colat.

At least we now have an F, H, L and Q, but now there is no G and no P, so that, for example, his own name is not lurking anywhere. Nothing at all comes to mind in the way of hidden vertical axes, let alone monograms.

Or consider an inscription to honour Bessarion accompanying his *cardinalia* above a portal in the facade of the legend-rich Rotonda of the Madonna del Monte in Bologna, placed there surely by papal decree in 1450 (taken from an 18th-century sketch illustrated in Lollini, "Bessarione e le arti figurative", p. 151). This is replete with standard abbreviations:~

DIVAE MARIAE VIRGINI PIENT^{I}SS
BISSAR^{I}ON C^{A}RD GR^{A}ECOR APOS SED L^{E}GAT
AEDEM HANC AEDIFICARI CVRAVIT
NICOL^{A}O V PONT MAX KL QVINT^{I}L DDCCCCL

Here we can certainly find a few names, but surely not a programme for a painting other than an Annunciation scene with an angelic Cardinal and a Pope as bystanders.

304 See n. 248 above. Notice also that the 15 letters A, B, C, D, E, G, I, L, N, O, P, R, S, T and V are present in Piero's inscription on his painting:~

OPVS PETRI DEBVRGO SCI (for SANCTI) SEPVLCRI

Allowing for the same letter to be used twice, we can "find" in the inscription the names IEVSVS, SANCTVS PETRVS, PONTIVS PILATVS, IVDAS, (H)ERODES, IVLIVS CAESAR, IOANNIS PALAIOLOGOS, BASILEIOS BESSARION, IOANNES

be found in the epigram, not least (F)EDERIGO at **8a**, TO(M)AS at **6a**, (T)OMAS at **6b**, and CORVINVS at **7b**.

On the subject of coincidence, I might point out that if the Byzantine astrolabe had not survived, Regiomontanus' epigram would probably never have been deciphered. It was a coincidence that Berthold Holzschuh happened to be familiar with the 1062 astrolabe when he was looking at the 1462 astrolobe in 2005. As Louis Pasteur (1822-1895) wrote, "Le hasard favorise l'esprit préparé." / "Chance favours the prepared mind."

And it was a coincidence that I just happened to have published the 1462 astrolabe several times and also to be familiar with the literature on hagiography and ciphers. And it was also a coincidence that these discoveries were made when the average age of Holzschuh and King was about 62.

But back to the *caveat*. Perhaps nobody *ca.* 1462 noticed any of the six main figures in the epigram beyond, say:~

BA IOANNIS / IDRIS / BENEDICTI ROMAE //

BESSARION / RGO IO / IO

These we could interpret as:~

IOANNIS VIII / MEHMET II / CHRIST - CHURCH //

BESSARION / REGIOMONTANUS / GIOVANNI BACCI

and this corresponds to what was proposed by Carlo Ginzburg, with our Regiomontanus replacing Ginzburg's Buonconte.[305]

Or suppose they never noticed any letter combinations beyond, say:~

BA IOANNIS / IDRIS / BENEDICTI ROMAE //

BESSARION / RGO IO / DE leading on the next line to CARDINE

Then we could certainly see here, as Berthold Holzschuh does:~

IOANNIS VIII / MEHMET II / CHRIST - CHURCH //

BESSARION / REGIOMONTANUS / OTTAVIANO DALLA CARDA

REGIO, BVONCONTE, LVDOVICO and VANGELISTA GONSAGA, OTTAVIANO and BERNARDINO VBALDINI DALLA CARDA, GIOVANNI BACCI, as well as CARDO and CARDVVS. To claim that the inscription was written in this way for this purpose would be absurd. See also n. 336 below.

305 See the text to n. 202.

This at least partially confirms Angelo Turchini's interpretation:~[306]

IOANNIS VIII / MEHMET II / CHRIST //

BESSARION / BUONCONTE / DALLA CARDA

with Regiomontanus now replacing, in every sense of the term, Buonconte.

However, I firmly believe that the programme behind the painting is far more complicated than Berthold Holzschuh is prepared to accept or any of our predecessors ever guessed.

It would certainly be unreasonable to claim that Bessarion and Regiomontanus and whoever else participated in this venture *must* have seen all of the letter combinations that I have found and documented above. **Mine is a purely speculative interpretation.** Therefore, **I cannot insist on all of these identifications**, even though, at least in my opinion, most of them do fit together rather nicely. On the other hand, maybe some identifications have been overlooked?

Perhaps somebody with better control over the historical milieu of Bessarion and Piero can come up with fresh insights and propose a compromise solution, a veritable and credible medial solution. That person will first have to come to terms with the appearance of Regiomontanus as Bessarion's new *protégé* on the Humanist scene of Renaissance Italy and will then, I think, have to study his epigram for a very long time.

Was Federigo da Montefeltro somehow involved in the painting?

> "In keeping with his international stature, Cardinal Bessarion maintained contacts with a good number of secular courts and governments. But his relations with the court of Urbino under Federico of Montefeltro stand apart because of their remarkable personal nature. For anyone interested in Bessarion's role in Quattrocento Italy or in the culture of Urbino during the Renaissance, the significance of their relationship cannot be gainsaid. In the course of more than twenty years Bessarion stood as the godfather to all three of Federigo's sons, took a genuine interest in their education, especially their Greek studies, and at the end of his life entrusted his magnificent library to the safekeeping of Federico." John Monfasani, "An Epilogue to Bessarion's Relationship with the Court of Urbino" (1984), pp. 95-96.

> "*Cui quero plus favet divus ille Bessarion, dominus meus Reverendissimus, quam Umbrorum principibus Federico*

[306] See the text to n. 205.

> *et Octaviano* ?" / "To whom, I ask you, is *divus* Bessarion, my much respected master, more devoted than to the Dukes of the Umbrians, Federicus and Octavianus?" Regiomontanus in a letter to Jakob von Speyer, court-astronomer to Federigo da Montefeltro in Urbino, dated 15th February, 1465. From Curtze, "Regiomontan's Briefwechsel", p. 292.

Where is the father of Buonconte?[307] Federigo da Montefeltro was well acquainted with Piero and also with Bessarion, who was one of the 28 leading intellectuals of all time illustrated above the intarsias of his studiolo.

The letters DE in space **8a** and the I and RGO in space **7a** could certainly be taken as pointing to (F)EDERIGO. Add to this the fact that one can find the letters DVCA VRBINA(TE) in space **8b**. I cannot believe that Bessarion would not have noticed at least the latter.[308] Did somebody say words to the effect: "If we had an "F", we could put Federigo here as well?"

But is Federigo in the painting? I can find no hint of a symbol for him in the image of the man on the right of the painting, unless it is the way the man is standing, with only the left side of his face visible. (Federigo was usually shown in this way after he lost his right eye in 1451.) Nevertheless, we seem to be dealing either with an intensely personal painting made either for Bessarion, but from which Federigo is the only person omitted, or for Federigo, in which he is not himself depicted. What would make some sense would be if the figure of Bessarion embodied Federigo, sharing the grief over the lost Buonconte, but, somehow, Federigo should not be disguised so much that he looks like a Greek ambassador. And would the purpose of the painting be to embarrass Federigo into mounting a crusade by reminding him what had happened to Byzantium when its sole hope for salvation had been lost because the Emperor had been ineffective in implementing the Union of the Churches?

We cannot dismiss the possibility that the painting was commissioned by Bessarion, with Bacci providing the financial backing, as a gift for Federigo. His

---

307 See the text to n. 196.

308 The potential word-plays on Federigo's name did not go unnoticed amongst his contemporaries, so that, for example, the philosopher Marsilio Ficino wrote of "*Federico [Fide-rigo] da fede regia e dal comando sull'Orbe [che gli sarebbe stato attributo per volere divino], duca Orbinate.*" Elsewhere, in his biography of the Duke, he knew the Germanic derivation of the name: "*Germanorum est nomen, quod 'pacis dives' significatur. Nam fred ... pax est. Ric significat ... divitem.*" Bruscolini, *Federico da Montefeltro*, pp. 43-44, also Bianca, "L'Accademia del Bessarione tra Roma e Urbino", pp. 78-79.

dead son Buonconte and his nephew – some say his brother[309] – Ottaviano are, if my hypothesis is correct, depicted in the painting. Perhaps he would not have required that he himself be depicted with all the others. Certainly, Bessarion's pleas for him to undertake a crusade were still valid in 1462, for by then, even though most of Byzantium was lost, *Europe* was at risk.

Extract from the family tree of the Montrefeltros

… …

|

Antonio (1348-1404)

|

Guidantonio (1378-1443)

|

Oddantonio (1427-1444)   Federigo (1427-1482)

|

Buonconte (1442-1458) / Guidobaldo (1472-1508) / several daughters

---

So I cannot see Federigo in the painting except, by association, in his beloved son Buonconte and in his nephew/brother and second in command in Urbino, Ottaviano. In any case, for better or for worse, and in spite of Bessarion's pleas, Federigo did not mount a crusade. And the planned crusade of Pope Pius II in 1464 did not leave the port of Ancona.

A second scenario? Thomas Palaiologos and Matthias Corvinus?

I have concentrated on Bessarion and Regiomontanus because it was surely as a result of their collaboration that the scene for "The Flagellation" was set. However, we should recall Maurizio Calvesi's hypothesis, cited above, on the *Conventus* of Mantua, and the possibility that the two "moderns" on the left might also have been intended to represent the Byzantine pretender Thomas Palaiologos and the young King of Hungary, Matthias Corvinus.[310] Likewise we recall Jean-Pierre de Rycke's proposal that the bearded man is Ioannis

---

309 I shall not pursue this here. In 2006, Duccio Alessandri alerted me to a recent interview with Luisa Ubaldini, the last surviving member of the family, in which she repeated the family tradition that Federigo and Ottaviano were brothers, both sons of Bernardino Ubaldini, after whom was named Ottaviano's son, who died shortly after Buonconte.

310 See the text to n. 206.

Argiropoulos. The letters ARGIROPOV(L)OS are right there in and around **6a**.[311]

Perhaps we should have paid more attention to the letters CORVINVS near **7b**? And we can find (T)OMAS at **6b**, and, with yet more stretching of the imagination, TO(M)AS at **6a**. This would mean that the person combined with Bessarion could indeed be Thomas Palaiologos (b. Constantinople, 1409, so two years younger than Bessarion), brother of the now deceased Ioannis VIII and pretender to the throne, who had come to Rome in 1461. But would Thomas be portrayed in such a positive fashion in a painting associating his brother with Herod and/or Pontius Pilate? Nevertheless, one of Thomas' last hopes for a crusade against the Turks was indeed the young king of Hungary, Matthias Corvinus, whose domains stretched as far as the Adriatic. Also, Corvinus was a vigorous sponsor of Humanistic scholarship.[312] Regiomontanus left Italy in 1467 to work at his court, and we have descriptions of several spectacular instruments that he made for Matthias during his stay in Buda that lasted until 1471, when he moved to Nuremberg.[313]

There was something else in which Bessarion and Thomas Palaiologos had a common interest. Thomas, fleeing the Morea in 1459, first collected the head of St Andrew from Patras, where it had apparently been housed since the martyrdom of the saint *ca.* 60 C.E.[314] Once in Italy, Thomas agreed to trade the relic for the permanent hospitality in Rome offered him by Pope Pius II, who organized an imposing ceremony to receive the relic.[315] Bessarion delivered a speech on April 26, 1462, on behalf of the physically present, but mute, St Andrew, and he clearly associated Pius II with St Peter. Andrew, through his spokesman, Bessarion, implored his brother Peter, represented by the Pope, to dedicate himself to the destruction of the Ottoman Turks. Since Peter, like Andrew, also needed a spokesman, it was left to his representative, the Pope, to pledge to uphold his brother's cause, which Pius, of course, did. This part of the ceremony is no surprise considering the lukewarm response Pius II received in Mantua in 1459 when he pleaded his cause – the defeat of the Ottoman Turks and the subsequent protection of Christians in those lands appropriated by their

---

311 See the text to n. 212.

312 See Gründler *et al.*, eds., *Matthias Corvinus und die Renaissance in Ungarn.*

313 Zinner, *Regiomontanus*, pp. 145-162. See also the essays on Regiomontanus in Hungary in Hamann, ed., *Regiomontan-Studien.*

314 The saint's headless body had been stolen by the Crusaders from Constantinople in 1204 and was housed in the Duomo Sant'Andrea in Amalfi. On St Andrew see, for example, Jöckle, *Encyclopedia of Saints*, pp. 31-34.

315 The following account relies heavily on the abstract of a recent lecture by Karen Frank (see the bibliography), which underscores the political importance of the relic.

foe – to secular dignitaries and princely ambassadors. Though his eloquence at the *Conventus* of Mantua failed to arouse the response that he had wished, Pius II must now have hoped that St Andrew's forced flight would make some impression on Christian secular leaders. This ceremonial opportunity, which did indeed attract attention from all over Europe, must have been irresistible to Pius II and his supporters within the curia. However, the crusade for which Pius II had hoped never left the port of Ancona, and he himself died in Ancona before it might even have otherwise set sail.

Next we should consider Silvia Ronchey's hypothesis that, whilst the bearded man is Bessarion, the young man in red is the 30-year-old Thomas Palaiologos as he looked at the Council of Ferrara-Florence in 1438-39. He too was a good-looking young man.[316] I do not find his name in position **7**, but who knows what the commissioner of the painting had in mind? Maybe he is there in position **6** or **7**?

In addition, we recall the interpretation of Stephen Hoffman that the scene on the right represents a debate between a Jew and a Christian (images **6** and **8**).[317] We can find IVDAEVS in **6b**, as in **3a** (as well as – less helpfully – BENEDICTI, which can be taken to stand for Christians, in **5b+6b**, as in **3a+4a**).

Also, we should keep in mind the very recent proposal of Jean-Pierre de Rycke, who hypothesized that the bearded man was Giorgios Argiropoulos, the young man in red the Psalmist David and Oddantonio, and the man on the right Franciscus Accoltus. Of course we can find GIORGIOS ARGIROPOV(L)OS more or less in **6a**, DAVIDVS in **7b**, as well as (F)RANCISCVS in **8b**, and we have already noted ODDOAN(T)ONIVS in **7b**.

In spite of all this, I am not ready to take that extra step to dare to suggest that there is a double or even triple message to the entire painting, namely, in addition to all that I have claimed, to represent the hopes of Thomas Palaiologos in the young Hungarian king or a debate between a Jew and a Christian or a scene involving Argiropoulus. (Nevertheless, I would not dismiss the possibility yet.) But would it have been acceptable to render a monarch in a single image together with dead nobility and a young astronomer? The suggestion certainly has the virtue that it would provide a double identity for the Bessarion figure, who does not necessarily look like Bessarion. I leave this for future discussions, for I make no claim to have exhausted the decoding of the painting.

---

316 See the text to n. 215.

317 See the text to n. 201.

Which came first – the epigram or the painting?

This is an extremely important but very delicate question. We can assume that Regiomontanus conceived the text of the epigram himself, but did he conceive its layout himself? I am of the opinion that he did, the idea of the vertical axes near the would-be divine cuts perhaps inspired by the fact that he was dealing with two dates, both involving the number 62.[318] On the other hand, we do not need to suppose that Bessarion and Regiomontanus found any of the identifications that we have proposed by actually looking at the epigram (width 3.8 cm!) engraved on the back of the astrolabe. The layout for the epigram could have been captured on, say, paper. Regiomontanus may also have engraved the epigram on the astrolabe *after* some identifications had been made and the painting conceived. We simply do not know.

The epigram and the painting *conveniunt in unum*

> "Each side of the composition (of *The Flagellation*) is an echo of the other." Bruce Cole, *Piero della Francesca* (1991), p. 67.

The identifications that we have made for Piero's eight images derived from a consideration of the epigram forwards and backwards. Now we consider the painting forwards and backwards.

Until now we have looked at a hypothetical painting in which the Flagellation scene occupied five-eighths and the three "moderns" the remaining three-eigths. We now confront the fact that the two scenes actually each occupy one-half of the actual painting.

We look at the two halves of the painting together with their mirror-images with respect to the central axis of the painting: see **Fig. IV-7**. We see the Flagellation scene forwards and backwards, and the three "moderns" backwards and forwards, each shown in relation to the epigram.

---

318 See the text to n. 98.

324 ** Kemp, *The Science of Art* (1997), p. 30.

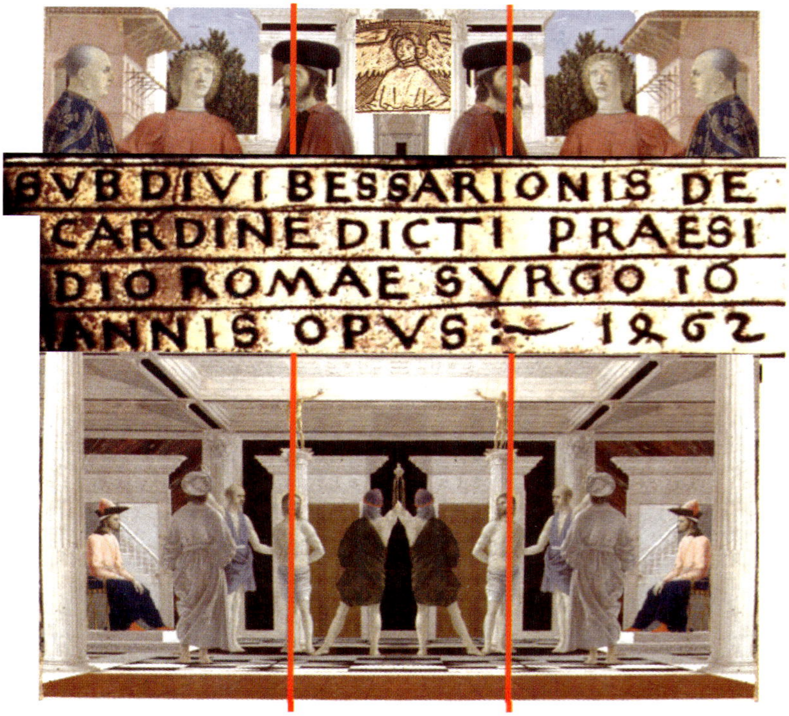

Fig. IV-7:~ The two halves of the painting rearranged together with the epigram. It is not only different persons who "come together in one" in each of the various images, but also two scenes, a symbolic flagellation of Christ and the three "moderns".

Common features of the astrolabe and the painting

Figs. IV-8a-b show schematic representations of both the masterpieces of Regiomontanus and Piero.

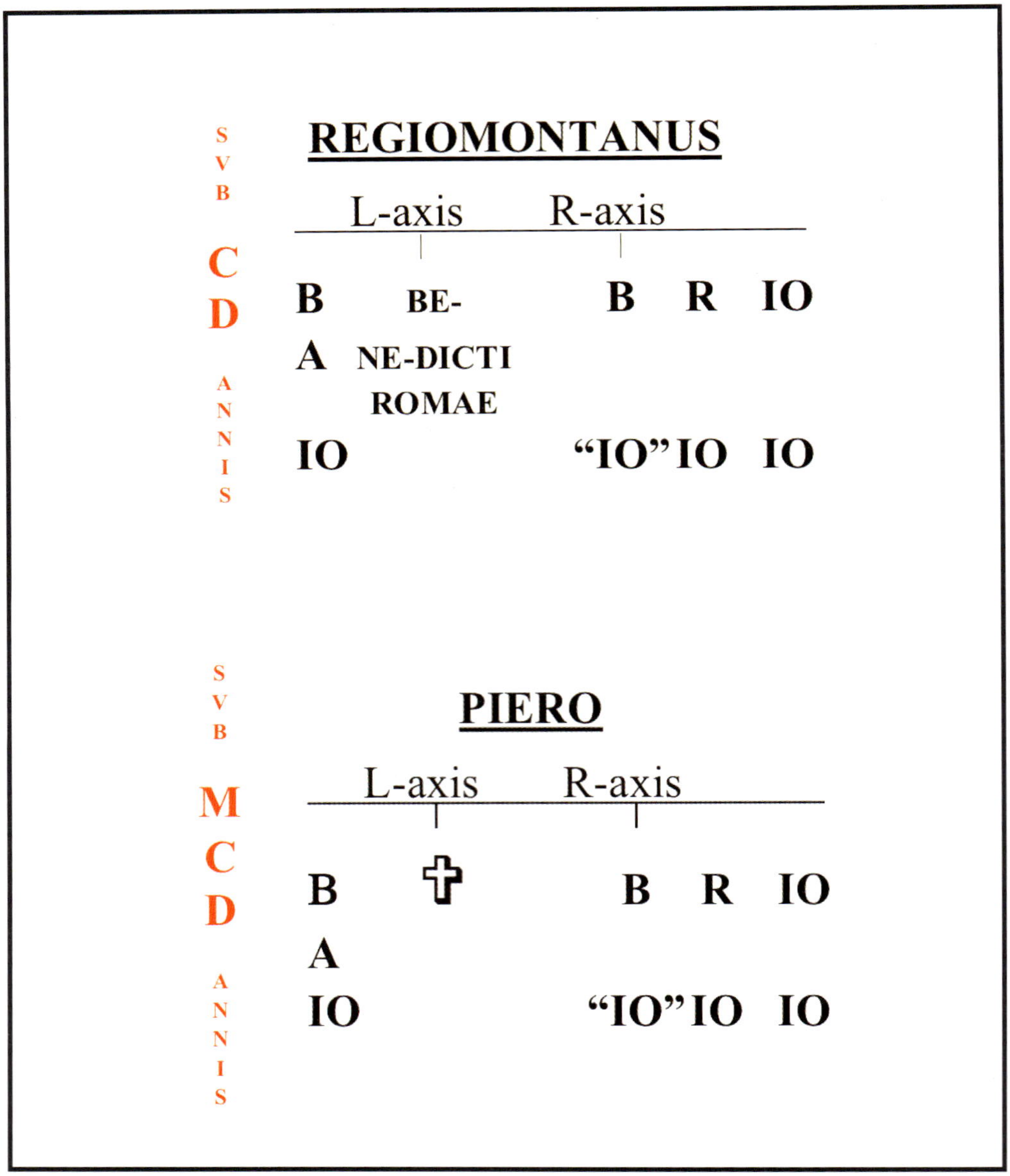

Now we recapitulate the correspondences between the angel and the epigram on the astrolabe and the painting and its angel. Their common features are shown in **Table 7**.

Table 7

COMMON FEATURES OF THE ASTROLABE AND THE PAINTING

Astrolabe	Shared	Painting
Stereographic projection		Perspective
	Eight axes in epigram and eight images in painting	
Eight "hidden" vertical axes in epigram		Eight "hidden" persons depicted
Divine proportion behind the conception		Divine proportion used for Christ
	Two principal axes, one on the left and one on the right	
	Bessarion not explicitly identified as a cardinal	
	Regiomontanus not explicitly identified as inspiration	
Two astrolabes in one		Three principal scenes in one
Time scale at left (SVB CD ANNIS)		Left column as time scale (SVB MCD ANNIS)
Two concepts in one		Two or more people in one image
Two concepts in one		Three classical gods in one
Benedicti (Romae) / INRI at left		INRI at focus of left scene
	Bessarion and Regiomontanus at right	
Two Bessarions in one		Bessarion and who else?
Two Ioannes in one epigram		Four Ioannes in one painting
Sergios and Regiomontanus in one		Dead youths and Regiomontanus in one
	Jerusalem, Constantinople and Rome in one	
	Three dates in one	
IN + RI	Jerusalem	INRI
1062	Constantinople	1440
1462	Rome	~1462
One person from Trebizond		Two persons associated with Trebizond
Fall of Trebizond in 1461		Ottoman Sultan
	Divus angel image	
Bessarion and *divus* angel in one		Regiomontanus and *divus* angel in one
	Regiomontanus and *divus* angel in one	
Word-play on *cardo* (Latin)		Image-play on *cardo* and *carda* (Italian)
Marian rosettes on throne		Marian rosettes around smock
Elegaic couplet		Laurel tree
Angel and epigram in one		"Flagellation" and three "moderns" in one
	The common theme in both is ***Convenerunt in unum***	

Static and emotionless

> "Piero's devotion to Neoplatonism, mathematics, and his art is nowhere more apparent than in his enigmatic masterpiece, *Flagellation*. Its Albertian vanishing point is rigidly certain, but where is the centre of interest? Is it the three men to the right in the foreground who stand together but seem to be ignoring one another? Or is it the party of men in the background centering on Christ (Christ in a background?), who is being whipped in **a scene as bereft of direct emotional expression as a still life of a bowl of fruit**? The *Flagellation* is not a modern painting. It does not exemplify patriotic, class, ethnic, or even painterly values so much as piety. **It is full of symbols of Platonized and personal Christianity**, and we do not and probably never will understand most of them, but (and this is the painting's special significance for us) they are **almost entirely quantitative and geometrical**. In their meanings, whatever they may be, they urge the viewer toward mysticism. In the nature of their language, they hustle the viewer toward a mathematical perception of reality." Alfred W. Crosby, *The Measure of Reality* (1997), pp. 194-196 (my emphasis), illustrated with the painting and Wittkower & Carter's 1953 reconstruction of its plan and elevation.

There is one more feature of the painting that distinguishes it from all of Piero's other works and which, I believe, supports my hypothesis of its programme stemming from the epigram. The painting is, in a sense, static, even perhaps sterile. There is no movement; nobody is doing anything; there is no communication between the eight persons depicted; there is no life. I repeat a remark by Martin Kemp:~ [324]

> "No picture could exude a more pronounced air of geometric control and no painting was ever more sumptuously planned."

Beyond the geometric control and the sumptuosity there is also the impression of an unnatural scene:~ people who should not be where they are. Christ should not be in the background. Christ should be alone, but he shares the divine centre with a classical god. Ioannis VIII and the Ottoman Sultan should not be witnesses to the Passion of Christ. Bessarion did not look like this in 1462. Regiomontanus was alive and well in 1462, but here he is as an angel or a dead person.

I wish to state categorically that I propose the epigram as the inspiration for the progamme of the painting *not* because I think that the three men – Bessarion, Regiomontanus and Piero – could never have come up with such a scenario on their own. They were all richly endowed with imagination. But to come up with this particular scenario, they perhaps needed an equally particular catalyst, and that was to be found in the quintessential Renaissance game of teasing meaning from the letters of the epigram. The inspiration for the entire scenario, with its highly geometrical structure, and also its static, emotionless images, comes, I maintain, from the epigram.

An overview of the persons featured in the painting

I propose the following hypothetical identifications of the main figures in the painting:

1	2	3	4	5
IOANNIS VIII HEROD (PILATE)	SULTAN HEROD	JUDAS GEORGE OF TREBIZOND*	**CHRIST** ST PETER (CHURCH)	CENTURION HEROD PILATE CAESAR

STATUE

JOVE / ROME
HERMES / GREEK LEARNING
SUN / CONSTANTINE / BYZANTIUM
HERMES TRISMEGISTOS / GNOSIS

6	7	8
BESSARION ~1439 BESSARION ~1462 (with REGIO)	REGIOMONTANUS BUONCONTE (dead) BERNARDINO UBALDINI (dead) VANGELISTA GONZAGA (dead) SERGIOS (dead) ANGELIC ST BESSARION (long dead)	GIOVANNI BACCI OTTAVIANO UBALDINI dalla CARDA LUDOVIGO GONZAGA

ODDANTONIO (?)

THOMAS PALAIOLOGOS (?) MATTHIAS CORVINVS (?)

* It is important to keep in mind that George of Trebizond, if he was intended at all, is to be associated with the domain of the god Hermes, not with Christ.

How was it possible?

Table 8 shows the letters needed to form a "monogram" for the individual names and titles that I claim to have "identified" in the epigram. Remember that there is no "F" and no "L" in the epigram, so these letters are not included in the table.

Table 8

Letters in the epigram needed for the names of the various persons identified

62 LETTERS	**A**	**E**	**I**	**O**	**V**	**B**	**C**	**D**	**G**	**M**	**N**	**P**	**R**	**S**	**T**
FREQUENCY	**5**	**5**	**11**	**6**	**4**	**2**	**2**	**5**	**1**	**1**	**4**	**2**	**5**	**8**	**1**
NAME/TITLE															
BASI(L)EVS	A	E	I		V	B								S	
IOANNES	A	E	I	O							N			S	
IDRIS			I					D					R	S	
OSMAN	A			O						M	N			S	
IVDAEVS	A	E	I		V			D						S	
GIORGIOS			I	O					G				R	S	
IEVSVS		E	I		V									S	
C(H)RISTVS			I		V		C						R	S	T
SANCTVS	A				V		C				N			S	T
PETRVS		E			V							P	R	S	T
IVPPITER		E	I									P	R		T
IOVIS			I	O	V									S	
INVICTVS			I		V		C							S	T
(H)ERMES		E								M			R	S	
TRISMEGISTVS		E	I		V				G	M			R	S	T
CONSTANTINVS	A		I	O	V		C				N			S	T
(H)ERODES		E		O				D					R	S	
BESSARION	A	E	I	O		B					N		R	S	
REGIO		E	I	O					G				R		
BVONCONTE		E		O	V	B	C				N				T
BERNARDINO	A	E	I	O		B		D			N		R		
VANGE(L)ISTA	A	E	I		V				G		N			S	T
GONSAGA	A			O					G		N			S	
ODDANTONIO	A		I	O				D			N				T
GIOVANNI	A		I	O					G		N				
BACCI	A		I			B	C								
OTTAVIANO	A		I	O							N				T
VBA(L)DINI	A		I		V	B		D			N				
CARDA	A						C	D					R		
(L)ODOVICVS			I	O	V		C	D						S	
TOTALS	**16**	**14**	**22**	**15**	**14**	**6**	**8**	**8**	**6**	**3**	**13**	**2**	**12**	**20**	**11**
FREQUENCY DOUBLED	**10**	**10**	**22**	**12**	**8**	**4**	**4**	**10**	**2**	**2**	**8**	**4**	**10**	**16**	**2**

The individual "totals" may usefully be compared with the individual "double frequencies", which represent the letters in the epigram viewed from left to right and again from right to left.

The distribution is also important, so we note it in **Table 9**:~

Table 9

Letters in the epigram and their frequency

SPACE	LETTERS	VOWELS					CONSONANTS									
1a/8b	SVBCADIAN	AA		I		V	B	C	D			N			S	
2a/7b	DIVRDIORNI			III	O	V			DD			N		RR		
3a/6b	IBNEOMSO		E	I	OO		B				M	N			S	
4a/5b	ESSDIAEP	A	EE	I					D				P		SS	
5a/4b	ARCTSVS	A				V		C						R	SS	T
6a/3b	IONIPVR			II	O	V						N	P	R		
7a/2b	ISRAGO	A		I	O					G				R	S	
8a/1b	DEESIIO		EE	II	O				D						S	
SUBTOTALS		**5**	**5**	**11**	**6**	**4**	**2**	**2**	**5**	**1**	**1**	**4**	**2**	**5**	**8**	**1**
				31							**31**		**TOTAL: 62**			

The vowels are liberally sprinkled throughout the epigram, and, anyway, it is the consonants that are generally more important for finding a clue to a name. Notice that we need more Ss than any other consonant, and they are indeed available and spread around. Besides the F and L that are missing altogether, one could say that it would be nice to have a few more Gs. Likewise, there are not enough Ts to go around.

What these numbers can never portray is the character of the arrangement of the letters. One has to *see* the VB / CA(RD) and B/CA/I in 1a/8b to even think of Ubaldini dalla Carda and Bacci. One has to see I/N in 3a and R/I in 3b to think of INRI.

What is significant – and what should be considered by any who would claim that the epigram is unrelated to the painting – is not that we have found all these names in the epigram, but rather that we have found these names more or less in the places where they correspond rather appropriately to each and every one of the nine images in the painting.

The purpose of the polysemous painting

In the author's opinion, the full scenario now begins to make sense. The purpose of the painting was sixfold:~

1) It was to call to mind the betrayal and Passion of Christ, and his eventual Crucifixion, which was a symbol of God's love for humankind and of hope for Christians.

2) It was to display Bessarion's feeling that the Church had been essentially betrayed through the ineffectivenes of the Byzantine Emperor, with the result that Byzantium, and even its last bastion, Trebizond, had been conquered by the Ottoman Sultan.

3) It was to display his feeling that the Greek intellectual heritage was under threat from George of Trebizond and that the latter's scientific collaboration with the Ottoman Sultan amounted to treason.

4) It was to serve as consolation for three men who had been bereaved of a teenage son, namely, Bessarion of his godson Buonconte, Ottaviano of his adopted son Bernardino, and Ludovigo of his son Vangelista.

5) It was to celebrate the arrival of the eager young Regiomontanus on Bessarion's scene. The young German, with his exceptional linguistic and mathematical abilities, embodied for Bessarion the talents of all three lost youths.

Last, but not least:~

6) It was to graphically illustrate just how far one could go with the concept *convenerunt in unum*, by means of an epigram based on the same concept.

The painting would have been purely for personal use, surely for either Bessarion himself or Federigo. Neither would not have wished to draw public attention to what Bessarion might have thought of Ioannis VIII and his depiction by Piero in the pose of Pilate. Also, if my hunch is correct, the depiction of George of Trebizond and Judas in the same image was not for the public.

Perhaps others will find still more in the painting, and so we indicate here the two additional scenarios that we have considered above:~

6	7	8
		FEDERIGO DA MONTEFELTRO (??)
THOMAS PALAIOLOGOS (??)	MATTAEUS CORVINUS (??)	

Piero's painting is both an image and a mirror image of Regiomontanus' astrolabe dedication and his angel, as well as of the universe of Bessarion. It is set on a stage to be viewed from left to right or from right to left, forwards or backwards, with the perspective of the mathematician to view three-dimensional space and the liberty of the artist to represent time. In both the epigram and the painting, the Cardinal is deliberately hidden:~ in the former, there is a brilliant play on the Latin word *cardo*, and in the latter, there are brilliant plays on the meanings of *cardo* in Latin and Italian.

In some of its aspects, the picture achieves essentially the same purpose as Regiomontanus' astrolabe and its dedication:~ a three-dimensional world is portrayed in two dimensions. Also, the astrolabe is generated by a projection known as stereographic, in which the rays emanate from a point, in this case, the southern celestial pole. The analogy of the lines of perspective converging *in unum locum* should not be overlooked. A different mathematical projection, called orthographic or orthogonal, underlies the markings on the back of the 1462 astrolabe, which achieve the same purpose, namely, to represent the universe in two dimensions, this time, by means of a projection from infinity. The four scales on the markings are to be read forwards, backwards, upwards and downwards.

In the painting there is also a play on time, since several different scenarios are portrayed:~ part of the Passion of Christ, the scene in Ferrara in 1438 and then in Constantinople in 1439, the deaths of Buonconte and Bernardino in 1458, and the arrival back in Italy of Bessarion and Regiomontanus in late 1461, and more. In parallel with the astrolabe dedication, a single image now represents two, even three or more, different people or the same person at different times. However, Piero and his sponsor have been as playful as Regiomontanus:~ for example, Bessarion could not have been portrayed as himself or as a cardinal, for this would have been, again, a "dead giveaway". In fact, he should not be portrayed as Bessarion.

There is also a play on certain events. We are witness to four betrayals and flagellations. It is not just Christ who is being flagellated. It is the Church that had been "flagellated" as a result of the inneffectiveness of the Byzantine Emperor. He is held responsible for its continued disunity, so that half of it has sustained a deadly blow. It is Byzantium that has been destroyed by the Turks. And it is the Greek heritage that is in danger from renegade scholars. Then there were three deaths of talented young men from Urbino close to Bessarion in the period 1456-60: these are recalled in a single image that also calls to mind Bessarion's encounter with Regiomontanus in Vienna in 1460.

There is also a play on geographical locations. As with the astrolabe dedication – in which Rome, Vienna (VI), Constantinople (Sergios), Trebizond (identified by Bessarion's year of birth), and Jerusalem (INRI) are explicitly or implicitly mentioned – the painting depicts or evokes scenes in Jerusalem (betrayal and flagellation of Christ), Rome (Roman column and Roman god, *scala sancta*, St Peter), Constantinople (Ioannis VIII and possibly Bacci's mission (?)), Urbino (Ottaviano, Buonconte and others), Constantinople and Trebizond (George of Trebizond and the Ottoman Sultan), Ferrara and Florence (Bessarion's feeling that the Church had been "betrayed" after its Union had been ratified there), and Vienna (Bessarion's encounter with Regiomontanus).

There is also a play on faces and iconography. The main figure is not necessarily the "real" Bessarion. It is certainly not the Metropolitan of Nicaea in 1438 or the Cardinal in 1462. Is this Bessarion's face as Piero might have remembered it from Ferrara or Florence during 1438-39? Or is there somebody else there too? George of Trebizond was deemed to have misrepresented the Greek intellectual heritage and to have sold out to the Ottoman Sultan. How appropriate to depict him confronting the Greek heritage symbolized by Hermes and Constantine as Helios, as Judas is shown confronting Christ. But this is not necessarily a portrait of George. The young man in red is not necessarily the "real" Regiomontanus in 1462. Buonconte was renowned for his beauty. But we are dealing with a composite angelic figure. The man on the right appears to be three men in one, differentiated by a profile, a gown, and a red *becchetto*, its colour perhaps symbolizing a mission. But why have we lost the mourning father Federigo? Or is he there too?

The painting thus reflects people at different times and places, each in different roles:~ it is a four-dimensional scene, with human experiences, beliefs and emotions, including obliviousness, as one additional dimension that has been charted with the genius of the Quattrocento. Bessarion is speaking, but not to the man on the right, who is looking straight at him or through him. Bessarion is perhaps saying the words *Convenerunt in unum*. Otherwise there is complete silence. The young man in red is not listening. He is at once the dead Buonconte, Bernardino and Vangelista, and at the same time the very much alive Regiomontanus, full of enthusiasm for the intellectual chase that Bessarion has proposed to him. He looks like Oddantonio da Montefeltro. Bessarion has his back to the flagellation scene because some of what it portrays is steeped in the distant past – a staged betrayal and flagellation of Christ, who is here depicted without wounds, and some of what it portrays is in the more recent past – the Byzantine Emperor, the Ottoman Sultan – and that nasty man who was particularly odious to him, George of Trebizond. The theme of the painting is most appropriately expressed by the words *convenerunt in unum*, and this refers to persons, times, and places, and the images themselves. We should not forget that the angel and the epigram on Regiomontanus' astrolabe had already achieved the same, albeit on a more modest scale.

Speculations on the commission

Unfortunately for some, the document we have identified as the inspiration for the programme of the painting is not a letter in a recognizable hand relating to a commission or a payment. Rather, it is a "document" prepared by one of the persons in the painting for another person also represented in the painting. The latter was familiar with the painter and with the other persons whose names we can find in the "document", not least, the most likely sponsor, who is also featured in the painting.

I propose that the epigram is not so much a key to the figures in the painting as it is a key to the programme for a painting featuring those figures in single images, a key to a hypothetical commission. It is an open question, which – if any – of the above identifications were proposed by Bessarion, with or without Regiomontanus. *Certainly, some of my identifications would have been outlined in the instructions to Piero from his sponsor(s), for we have the resulting painting.* One innovation in the painting is that Christ is now firmly at the right divine cut of the flagellation scene defined by the axes of the two columns on either side. The columns were introduced not least to allow Piero to control the geometry of the flagellation scene horizontally. Vertically, the column on the left guides us back in time:~ the scene with the three moderns is 1400 years after the historical Passion of Christ. One could say SVB MCD ANNIS.

Whether Bessarion and Regiomontanus together met with Piero or any potential sponsors – there seem to be three possibilities – is not known. If they did meet, we do not know where or when. If they would have met, say, in Urbino in late 1461, we do have other evidence that astrolabes were of interest there and then.[325] But we do not even have to suppose that they did meet for this purpose:~ a very private letter from Bessarion to a sponsor or even to Piero would have sufficed. Certainly, Piero would have been honoured to collaborate in a project involving the leading mathematician-astronomer of Europe. It has been noted that his mathematical abilities seem to have taken a step forward around this time; an encounter with Regiomontanus would explain this.

Piero would not need to have seen the epigram or known all that it had revealed to Bessarion and Regiomontanus (or some third party as yet unidentified).[326]

---

325 By a remarkable coincidence, the astrolabe featured in the intarsia of the *studiolo* of the Archduke Federigo has the same design as an astrolabe of Italian design, made by a German, and bearing two inscriptions indicating a connection with Urbino in the year 1462. See further King, "The Astrolabe in the Urbino Intarsia", and Appendix G.

326 For example, Niccolò Perotti (1429-1480), sometime secretary to Bessarion. He was present with Bessarion and Regiomontanus during their stay in Venice from July, 1463,

Piero did what he was told; the idea for the painting did not come from his initiative. Somebody either told him what to do, or wrote to him. Examples of paintings for which a commission order survives are rather few.

Martin Kemp, writing in 1997, stated that:~[327]

> "We cannot be certain that the scarcity of *invenzione* [written programmes by humanist advisors] in the earlier period [prior to 1500] necessarily reflects their rarity in practice, but the impression conveyed by contracts and related evidence is that content was generally a matter for verbal interchange between the artist and patron (or the patron's nominee) rather than written instruction. Even where the verbal instructions were quite precise, they inevitably fell far short of telling the painter what every aspect of the painting should look like, and a patron employing a reputable master would expect elegant and appropriate inventions in the main or subsidiary parts of the commission. The great majority of religious paintings … … concerned time-honoured or familiar subjects, and the artist would be expected to be visually literate in the way to tell the stories or characterise the figures. When something exceptional was planned, the burden of responsibility is likely to have shifted away from the artist, but not necessarily in all cases."

Let us suppose that the commission was transmitted by word of mouth. Then we have to imagine Bessarion and Regiomontanus sitting around a table with Piero and a potential sponsor, unless this was the penniless Bessarion himself. They would have explained to Piero, surely over a bottle of wine,[328] what the epigram had revealed to them, but only in essence. There would have been no need to even mention people who were not physically represented in the painting. However, we cannot prove that the two men met Piero. It is not even certain that they passed through Urbino or Sansepolcro on their journey from Vienna to Rome in late 1461, although it has been suggested that they did pass hrough

---

to July, 1464. Later, whilst professor at Bologna, he published a commentary to the epigrams of Martial.

327 Kemp, *Behind the Picture*, pp. 187-188.

328 On February 15, 1465, Regiomontanus suggested in a letter to Jakob von Speyer that they meet for a meal because they had so many scientific matters to discuss. Although he could not himself order pheasant or partridge for a banquet, he seems to imply that Bessarion could and would, and he promised that stimulating wines would not fail. See Curtze, "Regiomontan's Briefwechsel", p. 293:

> "Ceterum si convivio gaudeas, fasianos, perdices calamo nostro non ex cella epularie depromemus. Vegetes demum vini haud longe aberunt."

Urbino.[329] And then, once in Rome, Bessarion departed in early 1462 for a cure in Viterbo.

Let us now suppose that there *was* a written commission. As far as the persons to be depicted are concerned, that would only need to convey the following:~

On the left, a symbolic flagellation scene with five persons

1	**2**	**3**	**4**	**5**
Ioannis VIII in 1438, known to Piero	Ottoman sultan	man with face	Christ	soldier with back
		... ... traditional Flagellation scene ... ... also, classical god on column		

and, on the right, three moderns

6	**7**	**8**
Bessarion in 1438, known to Piero	angelic-looking young man in red facial features of Regiomontanus ?? did Piero meet Regiomontanus ?? facial features of Oddantonio ??	sponsor perhaps with identifying characteristics of three men known to Piero

If Piero did not see the epigram and was not privy to all that had could have been found in it, we could imagine a hypothetical commission for which we would dearly like to see a Quattrocento equivalent. We could take as a model the commission of the very remarkable painting *Le couronnement de la Vierge* by Enguerrand Quarton, 1453-54, now in the Musée de la Chartreuse at Villeneuve-lès-Avignon.[330] Here the commission document stipulates that God and Christ be represented identically on either side of the Virgin and specifies precisely the colours to be used for her garments. Scenes of Jerusalem and Rome are to be separated by a minuscule Christ on the Cross. An attempt at a hypothetical commission for the painting "*Convenerunt in unum*" would be

---

329 See, for example, Monfasani, "An Epilogue to Bessarion's Relationship with the Court of Urbino", p. 98, n. 26.

330 On this painting see Chevalier *et al.*, *Le Couronnement de la Vierge*, and the literature there cited, as well as, for example, Zuffi, *Il Quattrocento*, with a French translation *L'art au XV*^{e} *siècle*, pp. 168-169. A more recent study by Dominique Thiébaut, beautifully illustrated, is in *Paris ML 2004 Catalogue*, pp. 110-112, also p. 58, and a new book is Girard, *Le Couronnement de la Vierge*. For the text of the commission in a northern dialect of medieval French ("*langue romane*") see Chobaut, "Documents sur les peintres d'Avignon", pp. 142-145, with a modern French translation by Yves Grava in Chevalier *et al.*, *op. cit.*, pp. 55-56. I am grateful to my good friends Louisa Jones and Bernard Dupont, and Josiane Poquet, for providing me with a copy of the valuable publication of Chevalier *et al.*, and for lending me a copy of the 2004 Louvre catalogue.

worthwhile insofar as it could show that Piero need not even have seen the epigram.

Speculations on the date of the commission

We do not have to suppose that the painting was commissioned or achieved immediately after the presentation of the astrolabe in 1462. In Rome in early 1462 Regiomontanus had the opportunity to meet Leon Battista Alberti. Then Bessarion, as papal legate to Venice, and Regiomontanus spent a whole year together at the Benedictine monastery of San Giorgio, on the island of San Giorgio Maggiore, between July, 1463, and July, 1464. Bessarion's secretary, Niccolò Perotti, was with them in Venice. (Perotti in the 1470s published a book on the *Epigrams* of Martial, which he dedicated to Federigo da Montefeltro.[331]) It is certain that the three men spent some time in Venice plotting a campaign against George of Trebizond. Thereafter, Regiomontanus was in Rome for this first sojourn of his until mid 1467.

We should keep in mind that Bessarion was intimately involved in the Arezzo frescoes. The reason he was involved in "The Legend of the True Cross" is simple to explain. It was apparently Bessarion himself who brought to Italy from Constantinople (in 1440?) a reliquary containing pieces of the "True Cross", which he presented to a religious foundation, the Scuola Grande della Carità in Venice, on August 29, 1463: see **Fig. I-4**. In 1459 or thereabouts, Piero was in Rome, working in the Pope's personal chambers, and so it is highly probable that he would have seen the Cardinal, then in his early fifties, dusring this assignment. This would account for the nine partly-disguised images of a 50-year-old Bessarion-type-figure that appear in the Arezzo frescoes.[332]

Piero seems to have been oscillating between Sansepolcro and Arezzo during 1461-63 and even 1464 and 1465; indeed, there does not seem to be any record of his being anywhere else until 1469-70, when he had a contract in Urbino.[333] If Bessarion and Regiomontanus met Piero during this time it seems that they would have had to have passed through Sansepolcro and/or Arezzo, this either on their way to Rome in November-December, 1461; or on their way from Rome to Venice in July, 1463; or, less likely, on their way from Ancona to Rome in August, 1464.

Given Bessarion's ill-health it seems unlikely that in late 1461 they passed through Urbino, which is not on any direct road from Ravenna to Rome, not

---

331 *Washington LC 1993 Exhibition Catalogue*, pp. 56-57. See also Mercati, *Niccolò Perotti*, p. 167, *etc*.

332 On the depiction of Bessarion nine times in the Arezzo frescoes see n. 183 above.

333 See, for example, Calvesi, *Piero della Francesca*, pp. 182-185; and Centauro & Settesoldi, *Piero della Francesca – committenza e pittura*, p. 169.

least because of its difficulty of access. (It has been maintained that Bessarion may have passed through Urbino on this occasion, but there is apparently no evidence.[334]) It is more likely that after Ravenna they would have continued by way of Rimini and along the coast to Fano and then Fossombrone, and thence up and over to Sansepolcro and Arezzo:~ see **Fig. IV-9**. Otherwise, they might have met him on their way from Rome to Venice in July, 1463.

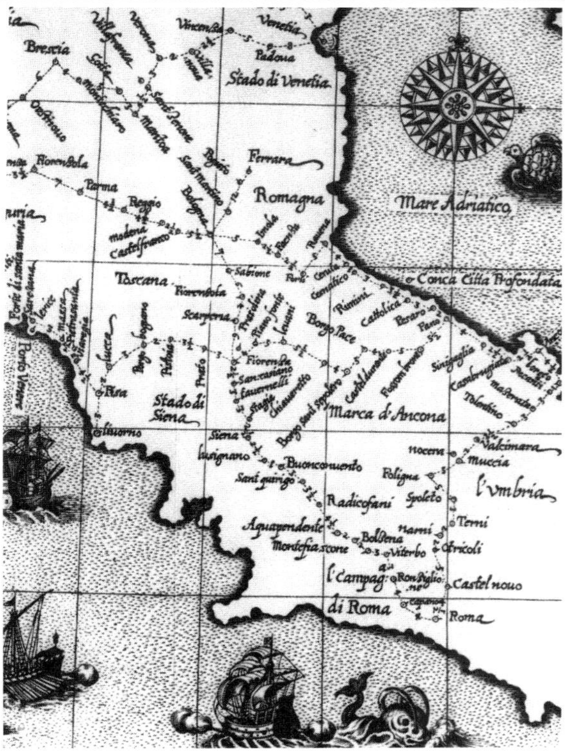

Fig. IV-9:~ The map of the routes between northern and central Italy by J. Furttenbach, giving distances in hours between coach-stops, from his *Carta itinerario dell'Italia*, in *Novum / Newes Itinerarium Italiae*, Ulm, 1627.

We need also to seek a possible encounter between Giovanni Bacci, who seems to be the sponsor of the painting, and Bessarion, surely accompanied by Regiomontanus. We know that Bacci – whose whole family was based in Arezzo – was in touch with Piero on account of the Arezzo frescoes, but where

334 See the sources cited in n. 329 above.

and when did Bacci meet up with the Cardinal and his *protégé* in late 1461, 1462 or thereafter? The answer may be Sansepolcro or Arezzo, and the date may be July, 1463.

The one consolation for not finding any convincing evidence for an encounter that could have served the commission of the painting is that others have come up with dates between *ca.* 1450 and *ca.* 1475, all mostly without a shred of documentary evidence.[335] Some of those who have proposed a date *ca.* 1450 have done so because the painting of Sigismondo Malatesta and San Sigismondo has a similar inscription and is dated 1451. Others recognize that Piero's "Flagellation" is his most remarkable work, and that it must have been painted later on in his career.

But can we really believe that Piero painted "The Flagellation" in his spare time, or during his occasional breaks, during his involvement with the Arezzo frescoes from 1457 to 1466, which are the dates according to the most recent research of Giuseppe Centauro and Enzo Settesoldi? Some months' respite would have been available to him because of the winter cold preventing work on the frecoes, as well as excessively hot summers. Perhaps one of these breaks, of several months' duration, was necessary precisely because he had an important new assignment. At least "The Flagellation" was a work that he could paint in the privacy of his own home.

What is more certain is that the personal themes of the painting beyond the Flagellation of Christ scene that were relevant to Bessarion – the ineffectiveness of Ioannis VIII, the threat of the Ottoman Sultan, the supposed betrayal by George of Trebizond of the Greek intellectual heritage, the need for consolation for lost sons and godsons, and the hope for the future in a talented young German astronomer – remained topical and relevant to Bessarion for a few years after 1461-62. But by 1467, Regiomontanus had disappeared from his scene, and when Regiomontanus returned to Rome in 1475, his former patron had already died.

So, if my hypothesis that the epigram provided the program for the painting is correct, the commission could have been given at some time starting late 1461 (assuming that Bessarion had already seen the epigram in some form or other), or 1462, when presumably Regiomontanus actually gave him the astrolabe (Jan., 2? June 6?), and ending, say, *ca.* 1465, after the Ancona crusade had come to nought, and after Regiomontanus had more than proven his scholarly worth.

---

335 To mention only some of the most recent proposals: *ca.* 1450, proposed in de Rycke, "La 'Flagellation' d'Urbino", and Ronchey, *L'enigma di Piero*, and *ca.* 1470 or thereafter, proposed in Laskowski, *Piero della Francesca*, p. 79. On the problems of dating Piero's works see Clough's 1970 study cited in n. 188 above.

Ideally, it would be nice to know as well the occasion of the presentation of the finished painting to its sponsor, but that would involve yet more speculation. We may never know when Piero finished this extraordinary painting with its intensely personal message. It must have dazzled Bessarion if and when he eventually saw it. Also, it would have pleased Federigo, if he saw it, not least to witness this tribute to his dead son (and perhaps even his long dead brother).

Piero's signature

If readers who have reached this point are of the opinion that the painting is related to, nay, was inspired by the epigram, they will then have to re-evaluate the painting and rethink its details and its environment, the latter task rendered particularly difficult because the principal *dramatis personae* were seemingly forever on the move. Also, they were up to all sorts of tricks ... .

What precisely did Piero intend from his inscription on the painting:~

OPVS PETRI DEBVRGO SCI (for SANCTI) SEPVLCRI ?

Clearly, it means:~

"The work of Piero of the Borgo of San Sepolcro."

but why did he combine the two words DE and BVRGO in this signature but not in two others?[336]

The influence of Piero's painting?

The "Flagellation of Christ" was, of course, a popular theme amongst Italian artists of the 14th, 15th and 16th centuries.[337] Some works that have been thought to have been influenced by Piero's "Flagellation" are the following:~[338]

---

336 Only three other works by Piero are signed by him. The inscription on his "Sigismondo Malatesta kneeling before San Sigismund" reads PETRI DE BVRGO OPVS MCCCCLI, and the words DE BVRGO are definitely written as two separate words. (Most available images cut off the inscription because it is not so nice as the painting. Cole, *Piero della Francesca*, p. 124, has the inscription correctly but omits the date; in Calvesi, *Piero*, p. 193, the two words are incorrectly written as DEBVRGO.) The inscriptions on his "St Jerome Penitent" and "Saint Jerome and a Devotee" are, respectively, PETRI DE BVRGO OPVS MCCCCL and PETRI DE BV[R]GO S[AN]C[T]I SEPVLCRI OPVS, respectively (see Calvesi, *op. cit.*, pp. 193 and 196).

337 For several examples see Lavin, *Piero della Francesca: The Flagellation*, pp. 23-30, and Ginzburg, *The Enigma of Piero*, figs. 37-38, 39 (Spanish), and 55-56.

338 Cecil Clough, who dated "The Flagellation" to Urbino, 1474-76, claimed that a miniature of *The Martyrdom of St Agatha* in a 15th-century antiphonal belonging to the Duomo in Urbino was inspired by the painting. See his "Federigo da Montefeltro's Artistic Patronage", fig. 14 on p. 13 and n. 76 on p. 17.

- ❖ A flagellation scene (h. 84 cm w. 60 cm) with a statue of a classical god [!] all in black atop the column behind Christ, with a fop on a raised seat [!] wearing red socks [!], by Luca Signorelli (after 1480).[339]
- ❖ A flagellation scene in the background of three figures on either side of the opening of a splendid portico, by Jacobo Bellini (Venice, *ca.* 1440-1460).[340]
- ❖ A minute flagellation scene at the far end of an imposing arcade, attributed to Vincenzo Foppa (b. Bagnolo Mella, *ca.* 1430, d. Brescia, 1515/16).[341]

A more drastic distinction between two separate scenes is found in:~

- ❖ A woodcut with the priest Melchisedek in front of a tower, brought into the foreground on the left, with *The Last Supper* on a slightly higher level on the right, the two scenes of slightly different height separated by the middle of three columns in the foreground of both scenes. This is found in the *Biblia pauperum* dated *ca.* 1470 that is now preserved in the British Museum.[342]

Particularly the first of the above-mentioned examples may merit further consideration in the light of what is now known about Piero's painting.

The struts for sun-shades

One feature of the architecture behind the man on the right we have not mentioned. Outside the four windows on the villa is a series of five metal struts supporting a wooden beam to which shades could be attached to protect the windows from the sun. The struts are known in Italian as *tenditoio*, pl. *tenditoi*,[343] a term still used for clothes-hanger in the sense of clothes-horse, the words *tendina*, *tenda* and *tendone* being used for curtains from small (blinds) to medium to large (awnings). Piero painted them on several houses in the Arezzo frescoes.[344] Their presence in this painting of his may have been intended to

---

339 See Binaghi Olivari, *La Pinacoteca di Brera*, p. 58.

340 See, for example, Laskowski, *Piero della Francesca*, p. 79, fig. 71, also Gilbert, "On Subject and Not-Subject in Renaissance Italian Pictures", p. 208 and fig. 8.

341 Gilbert, *op. cit.*, p. 208 and fig. 7.

342 See Calvesi, *Piero della Francesca*, pp. 108-109, where it argued that Piero and the "culturally backward anonymous carver" were drawing on "a common source".

343 Such things one learns when one visits the Arezzo frescoes with Duccio Alessandri.

344 Notably, on the house on the right of *The Rediscovery and Verification of the True Cross* (see, for example, Maetzke, *PdF*, 9, 135, 141, 220-221; Laskowski, *PdF*, p. 45; Renzi, *PdF*, p. 46; Field, *PdF*, fig. 6.18 on p. 212), and on the house "next to" God the Father above *The Annunciation* (see, for example, Laskowski, *PdF*, p. 34; Brilli, *Alla ricerca di PdF*, p. 70; and Centauro & Settesoldi, *PdF*, p. 145; Renzi, *PdF*, p. 58; Field, *PdF*, fig. 6.8 on p. 198). On the same kind of struts, now combined with a different kind

indicate that the wall is facing south, towards "*Sol invictus*". In any case, we may now need the *tenditoi* for the pigeons now that the cat is out of the bag.

The mysterious death of Regiomontanus in Rome

Niccolò Perotti apparently thought that the comet of late December, 1471, portended the death of Bessarion in November, 1472.[345] In any case, when Bessarion did die in 1472, his painting – if it was his – would have passed – if it had not already – into someone else's hands. As noted already, it was not mentioned in the list of his personal effects.[346]

Regiomontanus left Nuremberg for Rome around the end of July, 1475, at the invitation of Pope Sixtus IV to address – and help solve – problems of the calendar.[347] In 1470 Bessarion had made his predecessor, Pope Paul II, aware of the severe problems associated with the determination of Easter. Regiomontanus went unwillingly to Rome, fearing his own demise. And within a year he was dead.

One may wonder whether Regiomontanus really did die in Rome on July 8, 1476, as a result of the Plague, as is generally maintained. There was a rumour, circulated already in a poem by Jacobus Sentinus of Alderney that was printed in the first edition of Regiomontanus' *Ephemerides*, dated 1482, to the effect that he did not die of natural causes. By 1512 this had been elaborated to a claim that he was poisoned by the two sons of George of Trebizond.[348]

Let us suppose that the main thrust of my hypothesis about the letters of Regiomontanus' epigram providing the programme for the painting is correct, and that George of Trebizond (d. 1472 or 73) really did feature in that programme. What Bessarion and Regiomontanus between them said, wrote and

---

of sun-shade affixed within the window-frame, see the anonymous 1473 work *La guarigione del cieco*, *The Healing of the Blind Man*, illustrated in *Urbino 2005 Exhibition Catalogue*, p. 44. I do not know of any studies of Quattrocento domestic architecture discussing such struts.

Some years ago I investigated the literary and architectural materials relating to the imposing astronomically-aligned ventilators that graced the roofs of religious and secular buildings in medieval Cairo (only one fine example survives to this day): see King, *In Synchrony with the Heavens*, VIIc.

345 Monfasani, "Platina, Capranica, and Perotti: Bessarion's Latin eulogists", p. 98, n. 5.

346 See the text to n. 157.

347 See North, "The Western Calendar", especially pp. 94-97: "From John of Gmunden and Regiomontanus to Paul of Middelburg and the Fifth Lateran Council".

348 On Regiomontanus' second and final trip to Rome and his death there see Zinner, *Regiomontanus*, pp. 237-239; Mett, *Regiomontanus*, pp. 141-143; and also the article "Regiomontanus" by Edward Rosen in *Dictionary of Scientific Biography*, XI, col. 351b.

vowed to write against George,[349] was public knowledge, and maybe the sons found out about the association of George (*versus* the Greek heritage) with Judas (*versus* Christ) in Piero's painting. If the two brothers knew anything about Piero's painting, its inspiration and its implications, with their father essentially portrayed as Judas, in addition to what they certainly heard about Regiomontanus' plan to publish a total demolition of their father's work on the *Almagest*, they certainly had good reason to extinguish their father's detractor. Poisoning Regiomontanus would have been their sweet revenge. In any case, the greatest European astronomer of the 15th century was just 40 years old when he died.[350]

What happened to the painting?

We have not a shred of evidence of any reaction to the painting, not even a bill, unless it cost Regiomontanus his life. I very much doubt that Federigo had it hanging in his *studiolo* in the Ducal Palace, as was maintained by Cecil Clough.[351] Perhaps Bessarion gave it to Federigo in 1472, when he entrusted his library to his friend.[352] This would account for its presence in Urbino. Perhaps he gave it to somebody else in Urbino, for perhaps it was not housed in the Ducal Palace. We know that Bessarion also met with Ottaviano Ubaldini dalla Carda in Urbino in 1472,[353] and, as I hope to have shown, the latter was depicted in the painting. In any case, Piero's spectacular painting remained unknown

---

349 Michael Shank, "Regiomontanus against George of Trebizond", p. 189, characterizes Regiomontanus' *Defensio Theonis contra Trapezuntium* or, in an earlier version, *Corruptio Theonis per Trapezuntium*, apparently compiled between 1470 and 1474, as "an angry work" mixing geometrical proofs and arguments with "invectives, insults and asides aimed at George of Trebizond's equally massive and equally polemical "Commentary on the Almagest"". Shank points out that the two works between them comprise roughly 1,200 pages of unstudied astronomical polemics.

350 It is not without interest to compare his achievements with those of Jamshîd al-Kâshî, who died in 1429, and was arguably the greatest Muslim astronomer of his time. The article "al-Kâshî" by A. P. Youschkevitch and B. A. Rosenfeld in the *Dictionary of Scientific Biography* concentrates on his mathematical achievements. On his spherical astronomy see Kennedy, "Spherical Astronomy in al-Kâshî's *Khâqânî Zîj*". More information is to be anticipated in the *Biographical Encyclopedia of Astronomers* (forthcoming).

351 Clough, "Federigo da Montefeltro's Artistic Patronage", p. 11.

352 Labowsky, *Bessarion's Library*, pp. 36-37, and Monfasani, "An Epilogue to Bessarion's Relationship with the Court of Urbino", pp. 96 and 98-99.

353 Monfasani, *op. cit.*, p. 106.

until 1744, when it featured in an inventory of the sacristy of the Cathedral, just nextdoor to the Palace.[354]

There is a possible explanation of first reappearance of the painting in the 1744 inventory, rather than in the Palace next door, or anywhere else in Urbino, namely, that it was taken to the sacristy for safekeeping on the occasion of a natural catastrophe. The restorations of the cathedral following severe damage to it by the devastating local earthquake of 1741 could have prompted the move and a new inventory of the contents of the sacristy.[355] (Treasures from San Marco in Venice were housed there during World War II.) But this is sheer speculation.

The painting is now housed in the Galleria Nazionale delle Marche in Urbino. In February 1975, it was stolen but fortunately recovered in Switzerland just over a year later, after a handsome ransom had been paid.[356]

The three objects – the 1062 astrolabe, the 1462 astrolabe, and Piero's painting – have been waiting for hundreds of years to speak to us. The 1062 astrolabe has been much neglected or abused, mainly by historians of science, if not by historians of Byzantine art. The 1462 astrolabe, although properly described by Derek Price (1959) and Gerard Turner (1989), has been dubbed suspicious or even a fake (if only by certain scholars in one country rather far from Vienna). If Piero's painting would have been discovered in 2004, instead of in 1744, it

---

354 See, for example, Ginzburg, *The Enigma of Piero*, pp. 48 and 52, with references to the sources.

355 In *The American Weekly Mercury* (Philadelphia) from Thursday July 16, to Thursday July 23, 1741 (no. 1125) we read from the Court at St James's, the 6th Day of May, 1741:

> "The King's most Excellent Majesty in Council London, May 9. They write from Rome, that on the 23d of last Month N. S. they felt a little Shock of an Earthquake, which did no Damage there, but it did a great deal in Ancona and Urbino; all the fine Porcelain in the Palace of the Albany was broke in Pieces, and the Building very much damaged; the Territory of Fabriano and the Town of Cugo have suffered very much; the Church and Dome of Loretto were split by the Shock, and a great many People crush'd by that Accident."

And also in an internet guide to Urbino and surrounding villages, in this case, Fabriano, we read that the Church of San Domenico, originally dedicated to Santa Lucia Novella, was rebuilt after the 1741 earthquake, although the sacristy, magnificently decorated by Allegretto Nuzi and his school in the second half of the 14th century, apparently survived unscathed.

356 Inv. no. 1990 DE 229. See Dal Poggetto, *La Galleria Nazionale delle Marche ... nel Palazzo Ducale di Urbino*, pp. 119-120 and fig. 116.

would surely have been declared a fake by some because it is so completely different from all of his other known works.[357]

[357] Compare Carlo Ginzburg, *The Enigma of Piero* (2000), p. 48:

> "Every aspect of this famous small panel ... has been the subject of debate, except its authenticity."

CONCLUSION

> "In the 'Cloisters' of the Metropolitan Museum in New York there hangs a magnificent tapestry which tells the tale of the Unicorn. At the end we see the miraculous animal captured, gracefully resigned to its fate, standing in an enclosure surrounded by a neat little fence. This picture may serve as a simile for what we have attempted here. We have artfully erected from small bits of evidence the fence inside which we hope to have enclosed what may appear as a possible, living creature. Reality, however, may be vastly different from the product of our imagination; perhaps it is vain to hope for anything more than a picture which is pleasing to the constructive mind when we attempt to restore the past." Otto Neugebauer, *The Exact Sciences in Antiquity* (1969), p. 177.

> "Virtually nobody accepted my interpretation. This surprised me not in the least. ... ... I have always thought that an inclination towards audacious hypotheses is and should be perfectly consistent with rigorous research into evidence." Carlo Ginzburg, *The Enigma of Piero* (2000), pp. 118 and 120.

The universality of the painting

Further reflections on the painting on the non-scholarly level will surely also focus on the potential universality of its message as a vehicle for illustrating

---

373 ** See www.frankfurt-ballett.de/press/clouds.html (2005).

feelings about and reactions to treachery, betrayal, obliviousness, and torture, as well as bereavement and hope. As this study was being completed, I was privileged to witness the spectacle "Clouds after Cranach" produced by Bill Forsythe. This had as its silent theme Lucas Cranach's "Klage unter dem Kreuz" (1503), showing Mary's lamentation beneath the Cross. The various axes of the painting served as a backdrop to a mother's bereavement and helplessness before officialdom in the midst of violence in the Middle East.[373]

Piero's "*Convenerunt in unum*" has even more potential, for its message can be essentially expressed in a globalized scenario:~

1	**2**	**3**	**4**	**5**
RULER / OFFICIAL	ENEMY	TRAITOR	SAVIOUR	OPPRESSOR
(OBLIVIOUS)	RULER (CRUEL)	CALUMNIATOR	VICTIM	(MILITARY)

STATUE

DEFUNCT RELIGIOUS IDEAS FROM ANTIQUITY
FOREIGN DOMINATION
CULTURAL HERITAGE
INTELLECTUAL HERITAGE

6	**7**	**8**
AMBASSADOR / TEACHER	LOST HOPE / FUTURE HOPE	SPONSOR
FOREIGN REFUGEE	TALENTED YOUTH	(BENEVOLENT)
BEREAVED GODFATHER/TEACHER	DEAD YOUTHS	BEREAVED FATHER
	HISTORICAL ROLE-MODELS	

The eight images of people alone call to mind a palette of human situations:

1: Ineffective leadership, indifference;

2: Aggression;

3: Treachery; calumny

4: Self-sacrifice; salvation; hope

5: Brutality

6: Scholarship; bereavement; refugee status

7: Hope for the youth of the future; memories of lost youth

8: Benevolence; bereavement

One could argue that Piero's "*Convenerunt in unum*" is a painting about life.

Appendixes

Appendix A – What is an astrolabe?

"The greatness of the Astrolabe is due to the fact that, like other great scientific apparatus, it is the result of a combination of many discoveries and inventions. It is the product of the human mind acting simultaneously in several directions, which, aided by the mechanical skill of generations of instrument makers, has created a new organ by means of which previously laborious operations have been simplified and new paths have been cleared for future progress." Robert T. Gunther, *The Astrolabes of the World* (1932), I, p. v.

---

[381] ** On these notions see Aaboe, *Episodes in the Early History of Astronomy*, pp. 1-23: "What every young person should know about naked-eye astronomy", and for more medieval mathematical detail, see King, *In Synchrony with the Heavens*, I-1.1-2.

> "Like the modern electronic computer, the astrolabe in the Middle Ages was a source of astonishment and amusement, of annoyance and incomprehension. Imprecise as the astrolabe may have been in practice, it was undoubtedly useful, above all in judging the time. The instrument might have been used, more often than not, in the dark, but 'dark' is hardly the word to describe the age in which it was so widely known and so well understood." John D. North, "The Astrolabe" (1974/1989), p. 220 (reprint).

> "Weshalb interessieren Sie sich für Astrolabien? Sie sind doch Professor!" / "Why (on earth) are you interested in astrolabes? You're (already) a professor." Question put to the author by a (junior) German museum curator (of scientific instruments!) in 1991.

The possibility that this study may fall into the hands of persons who have not previously encountered an astrolabe encourages me to present some more information on those mentioned in the main text. But before I do that, I should present some basic information on the standard astrolabe. Such information is, alas, not so readily available.

There is something magic about the astrolabe. The world is divided between people who know what an astrolabe is, and those who have no idea. Therefore, much nonsense has been written about the astrolabe by people in the second category, some of which is recorded in encyclopedias and exhibition catalogues. The astrolabe calls for a certain basic understanding of astronomy, which nowadays alas cannot be taken for granted. The astrolabe tends to confuse people and they mispronounce it as "astrolobe" and worse. In the early 1990s at a cocktail party in Frankfurt, a lady asked me how my work on Skylab was progressing. (The equivalents of the name astrolabe also caused problems in the Middle Ages; one Arabic folk etymology derived the Arabic *asturlâb* from *astur Lâb*, "the lines of Lâb", where Lâb was, inevitably, a son of Idrîs.) So it may be useful here to try to answer the question: What is an astrolabe? In Appendixes B-F I present more information on various astrolabes relevant to the present study.

For the purposes of ancient and medieval astronomy, and for many other purposes, it is useful to imagine the heavens on a sphere of arbitrary radius about the observer.[381] That celestial sphere appears to rotate with respect to the horizon and meridian of the observer, which together can be thought of as providing a fixed frame enclosing the sphere. The apparent rotation is about a celestial axis, the altitude of whose pole above the horizon is a measure of the latitude of the locality of the observer. With such an arrangement, one can simulate the apparent daily rotation of the heavens about the observer and his horizon. We further conceive the celestial equator and the ecliptic on the sphere, and with the rotation of the sphere about the celestial axis, all celestial bodies appear to move in small circles parallel to the celestial equator. The day-circles of the sun, which are limited by its day-circles at the solstices, depend on the position of the sun on the ecliptic. This model also enables us to control the position of the ecliptic relative to the horizon and meridian. The astrolabe – the most popular instrument of the Middle Ages – achieves all of the above in two dimensions.[383]

The astrolabe is a two-dimensional model of the three-dimensional celestial sphere, reduced to a plane by a mathematical projection known as stereographic. The theory of stereographic projection dates back to Hipparchus (about 150 B.C.E.). The projection is effected from the south celestial pole, for it is the northern sky that is of interest to an observer in the northern hemisphere. The projection is achieved in the plane of the celestial equator, so that the centre represents the celestial pole and clockwise rotations correspond to rotations of the celestial sphere. On the standard astrolabe, the projection is bounded by the circle corresponding to the outermost reaches of the sun from the celestial pole, that is, by the circle corresponding to the winter solstice, the celestial Tropic of Capricorn.

The basic astrolabe consists of two main parts. The celestial part, called the rete, is in the form of a perforated frame with pointers representing various fixed stars and an excentric circular ring representing the ecliptic. The terrestrial part, called the plate, bears markings for a specific latitude, including the horizon and

---

383 * The best introductions to the astrolabe in English are still North, "The Astrolabe", and *Greenwich Astrolabe Booklet*. For introductions in other languages see Michel, *Traité de l'astrolabe*, and Poulle, "L'astrolabe"; King, "Die Astrolabiensammlung in Nürnberg", I, pp. 101-114, and Stautz, *Die Astrolabiensammlungen in München*, pp. 9-122; García Franco, *Astrolabios en España*, pp. 49-129; Chabàs & Bosch, *L'astrolabi pla*; and Trento, *Astrolabio*. The last is a most useful addition to the literature, with many illustrations of historical interest. Information about a reworked English translation of Michel's work completed by Jim Morrison is available at www.astrolabes.org .

It should be mentioned that, contrary to many claims in the tertiary literature, the astrolabe was not used in navigation. The so-called "mariner's astrolabe" is nothing more than a device for measuring celestial altitudes.

the meridian, as well as altitudes above the horizon. The celestial part can rotate over the terrestrial part, very much as it appears to do in reality. A different plate is needed for each terrestrial latitude, and the set of plates for different latitudes with the rete on top fits inside a frame known as the mater, which also provides a circumferential scale for the ensemble: see **Fig. A1**.

The rete features a horizontal (equinoctial) axis and sometimes a vertical (solstitial) axis, and it is bounded by a circular frame (corresponding to the Tropic of Capricorn): see **Fig. A2**. In the upper part of the central space is the circular frame corresponding to the ecliptic, which is divided counter-clockwise according to the signs of the zodiac. The vernal equinox, or beginning of Aries, is on the left of the ecliptic scale, at the intersection with the equinoctial axis, and the autumnal equinox, or beginning of Libra, is at the corresponding point on the right. The summer solstice, or beginning of Cancer, is at the lower limit of the ecliptic, and the winter solstice, or beginning of Capricorn, is at the uppermost limit. The outer frame is usually open at the top so as not to interfere with this part of the ecliptic. There is usually an additional frame below the ecliptic, corresponding roughly or exactly to the circle of the equinoxes. This serves to strengthen the rete (if it is connected to the ecliptic as well as the outer frame) and to provide a base for star-pointers.

On each plate, the local meridian is the vertical diameter, and the perpendicular diameter is also drawn. There are three base circles corresponding to the Tropic of Cancer on the inside, the celestial equator in the middle, and the Tropic of Capricorn on the outside. There are two other main sets of markings, corresponding roughly to the upper and lower parts of the plate. The upper markings show the horizon as an arc of a circle passing through the intersections of the east-west line, that is, the horizontal diameter, with the celestial equator. Above this is a series of arcs of circles decreasing in radius up to a point that corresponds to the zenith, or point directly overhead. These are the altitude circles, and they may be engraved for each 1°, 2°, 3°, 4°, 5° or 6° of argument.

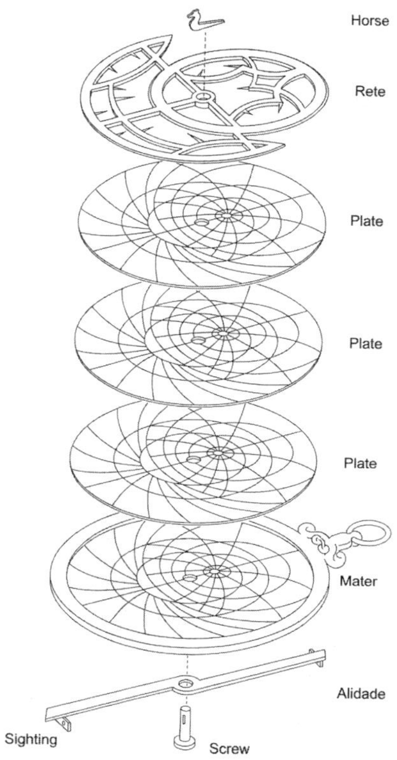

Fig. A1:~ The components of the standard astrolabe. The rete fits over the plates, which fit into the mater. The plates are held in position by pegs that fit into a hole inside the rim of the mater, and the rete is free to be rotated over the appropriate plate. Three plates are shown here, whose six sides, together with the mater, can be engraved with markings for the seven climates. This was a feature of (a) the earliest Greek astrolabes, (b) the earliest Eastern Islamic astrolabes, (c) the earliest Western Islamic astrolabes, and (d) some of the earliest European astrolabes. The ensemble is held together at the centre by a pin, which is fitted with a wedge, often in the form of a horse's head. [From King, *The Ciphers of the Monks*, p. 361, originally adapted, with permission, from North, "The Astrolabe".]

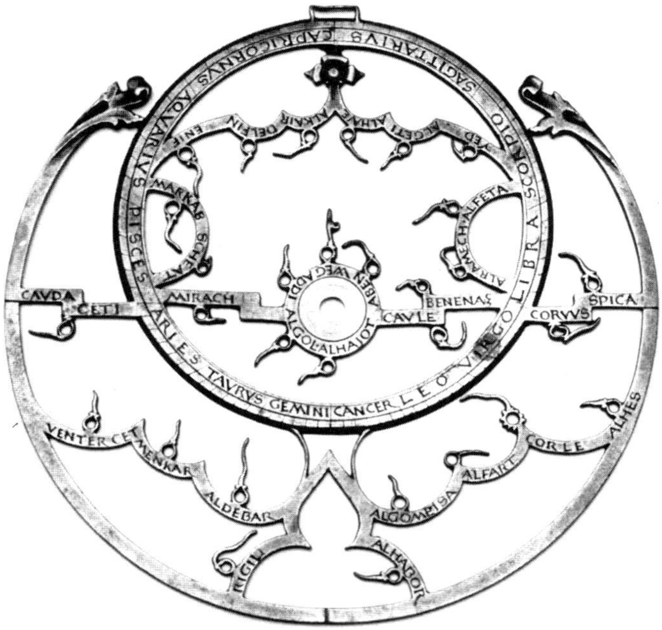

Fig. A2:~ The rete shows the apparent relative positions of various fixed stars, indicated by named pointers, and the apparent path of the sun, known as the ecliptic, represented by an excentric ring divided for the signs of the zodiac. This particular one is from another astrolabe made by Regiomontanus (#452), but unsigned and undated. It was probably completed before 1462 because (1) several old Latinized Arabic names are used for the stars rather than the new Latin ones, and (2) the design is less elegant than that on the 1462 piece. Note again the Marian lily and the crown on the pointer for COR LE(onis). [Photo by Gerard Turner, courtesy of the owner in Paris.]

Below the horizon on the plates, there is a set of arcs of circles representing the seasonal hours. These curves divide the arcs of the three base circles below the horizon into 12 equal parts. The seasonal hours, one-twelfth divisions of the length of daylight throughout the year, were in standard use in Antiquity and the Middle Ages and into the Renaissance. They vary according to the local latitude and the time of the year. Equinoctial hours, or 24th divisions of the length of day and night, as we use today, were used by astronomers in the Middle Ages, and are represented on the astrolabe by a rotation of the rete over a plate by an amount of 15° (= 360°/24). They are sometimes marked on the rim of the mater of the astrolabe.

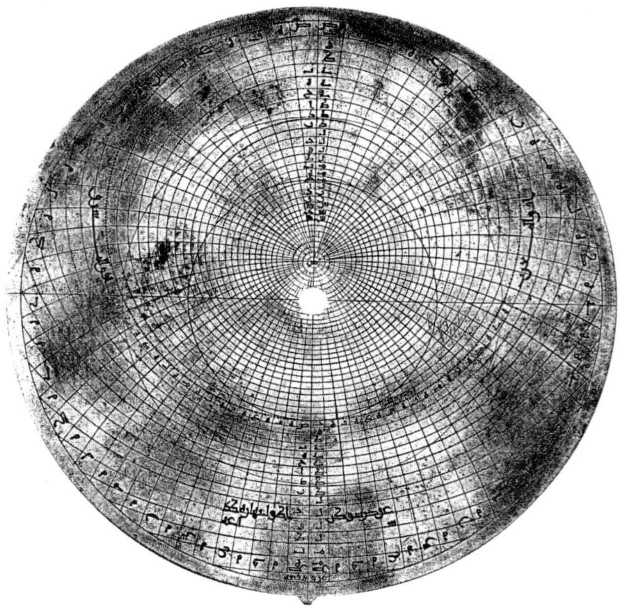

Fig. A3:~ This plate is from the astrolabe by al-Khujandî dated 984/85 (#111): see **Fig. B2**. It serves the latitude of the Arctic Circle and has markings for the azimuth orthogonal to the altitude circles. This kind of plate serves the conversion of ecliptic and equatorial coordinates. [Photo courtesy of a former owner.]

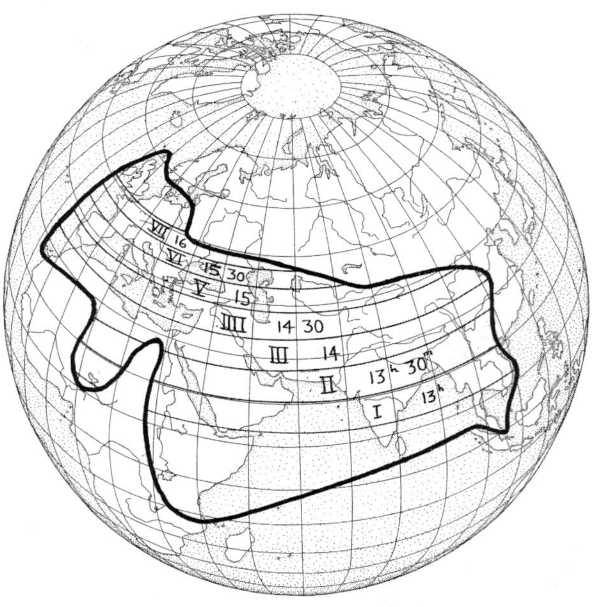

Fig. A4:~ The seven climates of Antiquity with a boundary showing the world known to Ptolemy. The centres of the climates, as well as their boundaries, were defined by the longest length of daylight. For the centres these are I: 13 hours, II: 13½ hours, … , VII: 16 hours. The boundaries are associated with the ¼ hours on either side of the centres. The climates were of prime importance in ancient and medieval geography and astronomy, particularly in instrumentation. [Courtesy of a student at Frankfurt University in the early 1990s, who left me with this but forgot to give me his name.]

Astrolabic plates for different latitudes developed from the ancient Greek notion of providing a separate plate for each of the seven climates of classical geography, defined in terms of the length of longest daylight – see **Fig. A4**. This was a means of ensuring that the instrument remained 'universal' in its application.[384] The earliest surviving Eastern Islamic astrolabe, and also the earliest known Western Islamic astrolabe, have markings for the seven climates, as do some of the earliest European astrolabes. But there was also a tendency, attested already on the only surviving astrolabe with Greek inscriptions, dated 1062 (see below), to choose latitudes of important localities for the plates. Thus, the plates on an Islamic astrolabe might serve a string of latitudes corresponding to certain local urban centres, as well as, perhaps, Mecca and Medina. In the same way, medieval European astrolabes might serve a string of latitudes for localities in the cultural region of the maker, as well as, perhaps, Jerusalem. Alongside the latitude, the length of longest daylight may be stated, underlying which will be a value of the obliquity of the ecliptic.[385]

The back of the mater on the standard astrolabe – see **Fig. A5** – bears one or two altitude scales on its upper rim and is fitted with a sighting device called an alidade attached at the centre. With these, one can measure solar or stellar altitudes. For timekeeping at night, one can then use the appropriate star-pointer on the rete, but for the sun, one first needs to know the position of the sun on the ecliptic.

[384] On the geography of astrolabes see now King, *In Synchrony with the Heavens*, XVI. On the use of the climates on astrolabe plates see *ibid.*, XVI-2-4. On the notion of universality in medieval astronomy see *ibid.*, VIa-b.

[385] On the possible values of the obliquity of the ecliptic underlying the associated lengths of longest daylight see the articles "Mayl" [= declination and obliquity] by D. A. King and "Mintakat al-burûdj" [= ecliptic] by Paul Kunitzsch in *The Encyclopaedia of Islam*. On the lengths of daylight for specific latitudes, see also King, *In Synchrony with the Heavens*, II, p. 931.

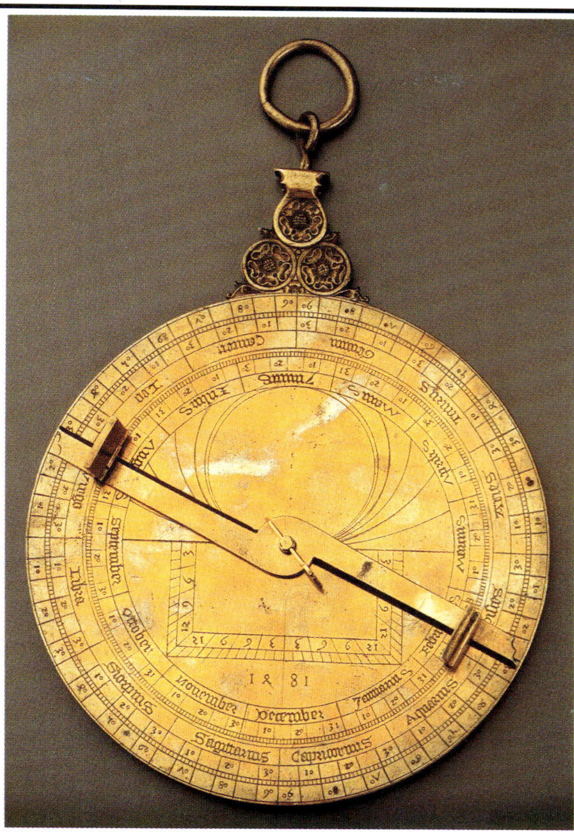

Fig. A5:~ The back of an astrolabe of the "Regiomontanus-type" dated 1481 (#530), that is, after the death of our astronomer. See **Fig. D4** for the front of this instrument. The instrument was made either in Vienna or in Nuremberg. Inside the standard altitude, solar and calendrical scales, we find a double horary quadrant, of the kind that Regiomontanus himself clearly did not favour, for it is inaccurate in European latitudes. Below this is a double shadow square to base 12. Note also the Marian rosettes on the throne. [From Debeauvais & Befort, *Astrolabes de Strasbourg*, p. 163.]

The basic astrolabe is a Greek invention, probably to be associated with Theon of Alexandria in the 4th century. The Muslims encountered the instrument in its standard form in Harran (now in south-eastern Turkey) in the 8th century, and in the 9th and 10th centuries, Muslims scholars devised all manner of additional features, some of which were incorporated into the kind of astrolabe that became standard, but many others of which are known to us only from textual sources.[386] Also, in the 11th century, Muslim scholars in al-Andalus developed the universal astrolabe.

Europeans apparently first encountered the astrolabe in Catalonia in the 10th century; at least, the earliest surviving European astrolabe comes from that milieu (#3042). Very few innovations of consequence were made to the astrolabe in Europe before the 16th century that had not been previously conceived by Muslim scholars, but their achievements were mainly lost in time.[387] In this overview, I have deliberately separated the description of the basic astrolabe from the description of Muslim contributions to the standard astrolabe.

Muslim and European craftsmen developed the rete into an object of beauty. Rete-design depended on which individual stars needed to be portrayed, and special features were introduced to reduce the lengths of various star-pointers. Popular on Western Islamic astrolabes were frames in the form of prayer-niches along the lower rim. Very popular on medieval European astrolabes were half-quatrefoil designs at the ends of the horizontal axis and full quatrefoils on the upper part of the vertical axis; these provide useful means both of strengthening the rete and of making areas of the heavens accessible with shorter star-pointers than would be needed if the pointers were attached to the basic frames of the rete. This very Gothic quatrefoil seems to have been influenced by early Islamic astrolabes, in turn perhaps copied from Byzantine retes.[388]

On the plates Muslim craftsmen also marked curves for the azimuth in addition to those for the altitude. Special curves were included for the times of Muslim prayer, defined in terms of shadow lengths and twilight phenomena.

On the back of the astrolabe they inserted calendrical and zodiacal scales to show the solar longitude in terms of the date in any of several solar calendars.[389]

---

[386] These are surveyed *ibid.*, X-4 and XIIIa-5. Additional information is in Charette, *Mathematical Instrumentation in 14th-Century Egypt and Syria*, and *idem* & Schmidl, "al-Khwârizmî on the Astrolabe", and Lorch, *al-Farghânî on the Astrolabe*.

[387] See King, Review of Turner, *Elizabethan Instrument Makers*, pp. 149-150.

[388] On the quatrefoil as decoration on astrolabe retes see now King, *In Synchrony with the Heavens*, XVII.

[389] On the early history of such scales, which is not at all clear, see *ibid.*, X-4.7.

For various reasons it is not a good idea to try to date any instrument by the correspondence of the solar and calendrical scales: see G. Turner, "Dating Astrolabes". Even the most innovative instrument-maker of the Islamic world, Ibn al-Sarrâj of Aleppo (see nn. 154 above and 423 below) had problems with his solar / calendrical scale: see King, *op. cit.*, II, figs. X-4.7.1 and XIVb-5.1b. See also the text to n. 510 below.

Once one has found the solar longitude for the day in question, one can then find the corresponding point on the ecliptic on the rete.

Additional markings inside the upper area of the back bounded by these two scales often include a set of arcs of circles that enable the user to calculate the time of day in seasonal hours for any terrestrial latitude. These horary markings were devised in Baghdad in the 9th century to solve graphically an approximate formula for timekeeping that came to Muslim astronomers from Indian sources.[390] They are easy to engrave and were still being put on the backs of astrolabes in Europe in the 17th century, long after their origin (and, in some cases, also their original purpose) had been forgotten. They serve to determine the time of day quite reasonably in the latitudes of the Mediterranean but not in those of Northern Europe.

Below the horizontal diameter on the back is usually a single or a double shadow square, consisting of a set of horizontal and vertical scales for measuring shadow lengths corresponding to solar altitudes measured on the upper altitude scale diagonally opposite.[391] The former measure horizontal shadows of a vertical object and the latter vertical shadows of a horizontal object. At solar latitude 45° both shadows are equal to the length of the object casting the shadow, invariably taken as having length 12 "digits", sometimes also, on a different scale, 7 "feet".[392] These shadow scales were also devised in Baghdad in the 9th century and were standard on astrolabes through to the Renaissance.

The uses of the astrolabe are manifold. The astronomer 'Abd al-Rahmân al-Sûfî (Shiraz, *ca.* 965) identified 1,760 operations that one could perform with it, many of course superfluous.[393] Having just made a measurement of the altitude of the sun or a specific star, one places the point of the ecliptic corresponding to the position of the sun or the appropriate star-pointer on top of the altitude circle corresponding to the observed altitude on the plate serving the observer's latitude. Care must be taken whether one is dealing with altitudes in the eastern or western sky (left and right side of the meridian, respectively). Then the

---

390 The first history of the universal horary quadrant, three forms of which, with or without a movable or fixed declination cursor, were invented in 9th-century Baghdad, is in King, *op. cit.*, XIIa. On the underlying approximate formula for timekeeping by the sun see *ibid.*, XI. See also n. 439 below for a newly-discovered Arabic text on the construction and use of the universal Graeco-Roman sundial based on this formula.

391 On the development of shadow-scales see *ibid.*, XIIa-A.

392 On these bases see *ibid.*, I, pp. 25 and 211, and also p. 469 and the references there cited.

393 See Destombes & Kennedy, "Al-Sûfî on the Astrolabe", for a list of the 386 chapter-headings of a shorter version.

configuration of the rete on top of the plate mirrors the instantaneous configuration of the heavens with respect to the local horizon and meridian. Two successive configurations can be used to measure the passage of time. One can, for example, rotate the rete until, say, the sun is on the western horizon: the amount of rotation indicates the time remaining until sunset (by day) or the time elapsed since sunset (by night). Or one can move the sun from the horizon to the twilight curve and investigate the duration of twilight. Or one can tell at a glance for any celestial circumstance which point of the ecliptic is simultaneously rising over the horizon and inspect the configuration of the ecliptic with respect to the horizon and meridian. Since this configuration was widely thought to be of significance in the affairs of men, this use of the astrolabe was of prime importance in astrology. The horary markings on the back enable the user to find the time of day approximately with facility, and without any kind of solar scale, but the result is unsatisfactory for latitudes in Northern Europe.[394] The shadow-squares can be used for surveying, although here a large instrument is preferable.[395]

While some astrolabes are large and bear mathematically accurate markings (the largest known is about 2.15 m in diameter, no longer suitable for putting in one's pocket), others may be quite small (the smallest known is 49 mm in diameter).[396] However, the vast majority of these instruments were not used for practical purposes; rather, they became collectors' items already in the Middle Ages.

In the Islamic world, at least in Iran and India, as well as the Maghrib, the production of astrolabes continued until the 19th century. Elsewhere, particularly in Egypt, Syria and Turkey, the simpler astrolabic quadrant for a single latitude was generally preferred. All this activity attests to a continued interest in traditional astronomy and astrology in those regions. In Europe, on the other hand, the 16th century witnessed the production of elegant and often sumptuous

---

394 See the text to n. 390.

395 Only one astrolabe is known that has four shadow-squares on the back, namely, the Byzantine astrolabe of 1062, on which see Appendix C below. This thus corresponds to the so-called "plane table" of the surveyors, when the instrument is held front down parallel to the ground. Indeed, it was from the Arabic term *al-'idâda* (whence our "alidade") that, by a gross distortion, the mysterious (and otherwise unexplained and inexplicable) word "theodolite" was coined. See the remarks in G. Turner, *Elizabethan Instruments*, p. 73, which merit pursuing. Turner dismisses any connection with the Arabic, but he did not consult any Arabist friends of his.

396 On the diameters of astrolabes see Price *et al.*, *Astrolabe Checklist* (first cited in n. 62 above), pp. 69-84.

astrolabes,[397] but, thereafter, the astrolabe was generally replaced by other instruments. The instrument was occasionally featured in works of art.[398]

The most important collections of astrolabes in the world – Oxford, Florence, London, Washington, Chicago, Nuremberg, Munich, Cracow, Cambridge, Madrid, Milan, Rome and Paris (Institut du Monde Arabe) – are well worth a visit. Many pieces in Florence, London and Oxford are illustrated and described on the Internet.[399]

---

397 See three recent catalogues of 16th-century English, Flemish and Spanish instruments, mainly astrolabes: G. Turner, *Elizabethan Instruments*; van Cleempoel, *Louvain Instruments*; and *Madrid FCA 1997 Exhibition Catalogue*.

398 See the new studies Sarma, "Astronomical Instruments in Moghal Miniatures", and Koenraad van Cleempoel, "Representations of Astrolabes in Western Art".

399 See EPACT in the bibliography.

Appendix B – The landmark astrolabes

> "It is a pity that so many modern historians and philosophers of modern science seem to be "born-again" theoreticians. They miss that sense of the craft of experimental science that would have come to them from experience at the bench. Too often they suppose that science is a department of intellectual history and that the only role of experimentation is to confirm ... a succession of theories. Worse still, the devices used for this testing are awarded even less significance and are assumed to have evolved over the centuries from simple instruments of measure." Derek J. de Solla Price, "Instruments of Reason" (1981), p. 15.

> "A general Catalogue of the Astrolabes of the World has long been overdue. Many writers on the subject, being impressed by the scientific or unusual artistic excellence of some particular astrolabe or especially by the antiquity of the instrument have been inspired to learn its use and to publish an account of their single example: but it can have fallen to the lot of a few persons to have had the advantage of being able to examine as many astrolabes as the present author has had the good fortune of handling." Robert T. Gunther, *The Astrolabes of the World* (1932), I, p. v.

> "The astrolabe, the most important astronomical instrument of the Middle Ages, was designed to measure the altitude of the stars, moon or sun without any mathematical calculation [!]. It was used in much the same way as the astronomical quadrant [!] or sextant [!!], but in addition it bore various diagrams [?] or scales [?] which made it possible to determine immediately the positions of the sun, moon and planets [!!] – most significant, the earth [!!!] – in relation to the fixed stars [!]." From *Washington NGA 1992 Exhibition Catalogue:* Circa *1492* (co-published by Yale University Press!!), p. 215.

> "The days are gone when every account of an astrolabe had to be prefixed with an essay on the history of the instrument and its uses ... ." John D. North, "The Astrolabe and the Imagination" (2000), p. 3.

Readers unfamiliar with the literature on medieval astrolabes and other instruments will soon realize that the topic is rather controversial. One reason is that astronomical instruments before the age of the telescope have seldom been taken seriously by historians of astronomy, and their study has often been undertaken by scholars unfamiliar with the history of astronomy. Very few astrolabes have been published by professional historians of science. Since epigraphers and even art historians as well as historians of science have seldom showed any interest in these beautiful objects and their inscriptions, the field has been wide open to amateur enthusiasts. This is a particularly treacherous field for art historians, even if they know what an astrolabe is and how it works.[401] However, greater dangers to the instruments themselves are posed by scholars who work only on historical texts, especially texts on instruments in general and on astrolabes in particular; such scholars are often interested mainly in "transmission", invariably, of course, to Europe. ("Texts are sacred; instruments are not.") It is a kind of unwritten rule that such colleagues would never think of looking at an actual instrument, but on the rare occasion that they do, they are happy to pontificate on a single instrument out of context, or on a single aspect of an instrument, such as the stars on the rete, if only to counter a colleague who actually knows something about instruments. It is another unwritten rule that most would not risk publishing such pronouncements. One only has to look at a few instruments to realize that in the main they do not always correspond to what we would expect from the textual tradition on instruments; in fact, the instruments are themselves as worthy of study as the texts, and, in many cases, they are more historically important than any texts. And, of course, the greatest danger posed to the instruments themselves comes from armchair specialists who do not understand them at all and declare them fake. A severe problem that we shall confront is that once an instrument has been labelled in any way suspicious by those who do not understand it, it is extremely difficult to reinstate the piece to its proper place in the history of human endeavour.

---

401 * See Hall, "More about the Detroit van Eyck: The Astrolabe … .". In this case, the author at least went to a lot of trouble to learn what an astrolabe is, and how it works, although the conclusions that he drew based on the astrolabe featured by Jan van Eyck's copyist are incorrect (it was most inappropriate to derive a date from a calendrical scale that is not on the instrument depicted). The author was certainly very poorly advised by the museum with the largest collection of astrolabes in the world, which happens to possess an astrolabe (#169) from the very same medieval Italian workshop as the one illustrated by van Eyck. See further nn. 60 above and 542 below.

403 * These are (1) #292: the 1342 English astrolabe signed by Blakeney (Gunther, *Astrolabes*, II, pp. 468-469); (2) #3053: the 1375 astrolabe made in Barcelona by Petrus Raimundus (King, "Astrolabis de la Catalunya medieval", pp. 198-200); and (3) #4523: the 1420 astrolabe signed by Antonio de Pacento (?) in Lanciano (Stautz, "Ein Astrolab aus dem Jahr 1420").

A particular problem with medieval European astrolabes is that they are invariably unsigned; indeed, only three signed pieces are known from before 1462![403] Another problem is that before the instruments of the 15th-century Vienna school (see Appendix D), European astrolabes are generally not dated; in fact, only the three signed pieces just mentioned are properly dated[404] and only one more securely datable.[405]

What does it mean when somebody "signs" an instrument? In the entire history of medieval instrumentation we have only one text that describes various astrolabes that were "made" and "signed" by a certain person. This is a document signed by two teachers of a Rasulid Yemeni Sultan *ca.* 1295 mentioning six astrolabes made by the Sultan, one of which survives (#109). And in this remarkable case, the Sultan wrote a book on astrolabe construction, and the statements by his teachers are appropriately appended to his own copy of his treatise.[406]

In the instrument trade, signed and dated pieces are much more valued than unsigned und undated ones, so there is not much serious interest in medieval European astrolabes. Also, instruments with more than one layer of inscriptions or composite pieces are readily dismissed, although in fact they can be of more historical significance than unified pieces.

Robert T. Gunther in 1932 published an imposing pair of volumes on the astrolabes at Oxford, the largest collection in the world, incorporating as many astrolabes in other collections as he could.[407] His volume on Islamic instruments reflects the facts that he knew but little Arabic and was poorly advised by the Arabists at Oxford. His volume on European astrolabes is much more reliable, although the medieval instruments relied heavily on the Islamic tradition. Alas, his volumes have never been corrected or updated. Reliable catalogues of the

---

404 A Munich astrolabe (#623=#3208) in the tradition of Jean Fusoris (*ca.* 1365-1436) of Paris is dated. However, the date can be read 1447 or 1487: see Stautz, *Die Astrolabiensammlungen in München*, pp. 191-199 (no. 4), esp. pp. 191 and the detail on p. 199. Stautz favoured 1487, which does seem preferable. In any case, none of the other 20-odd astrolabes associated with the Fusoris school bears any date, let alone the name of Fusoris. The date on this one shows that astrolabes of the same design were made in Paris for close to a century. On Fusoris-type astrolabes see King, *The Ciphers of the Monks*, pp. 397-398 and 400-401, and the references there cited, also Section 6.7 of the website www.uni-frankfurt.de/fb13/ign/instrument-catalogue.html .

405 This is a German astrolabe (#2027) dedicated to Ludolfus de Scicte, treasurer of the Stiftskirche in Einbeck. Two documents state that he held this post in 1322 and in 1342. See item III-1 in Appendix D.

406 See King, *In Synchrony with the Heavens*, XIVa.

407 Gunther, *Astrolabes of the World.* See Owen Gingerich's comments in a review of the 1976 reprint.

extensive collections in Spain (Islamic and European); the Museum for the History of Science at Oxford (only a rather small selection); the British Museum, London (only European); the National Museum of American History, Washington, D.C. (both Islamic and European); the Germanisches Nationalmuseum, Nuremberg (both); the Deutsches Museum and the Bayerisches Nationalmuseum, Munich (both); the Adler Planetarium, Chicago (both); and the National Maritime Museum, Greenwich (both), have been published.[408] Also catalogues are available of Flemish, Spanish and English instruments from the 16th century.[409] However, there are many more minor collections and private collections that merit catalogues.

Since there are no published descriptions of the majority of instruments, the whole subject of medieval European instruments can be said to rest on rather weak foundations. I have prepared descriptions of many of the surviving Islamic and European instruments from before *ca.* 1500,[410] but only relatively small extracts from these have been published. In a recent essay, I have tried to illustrate the use of astrolabes as historical sources.[411] Each of what I call *the landmark astrolabes* has had something of a rough time in the modern literature. These pieces are:~

---

408 See, respectively, García Franco, *Astrolabios en España*; Maddison, *Billmeir Supplement*; Ward, *Instruments in the British Museum* (based on Price); Gibbs & Saliba, *Astrolabes in Washington*; King, "Die Astrolabiensammlung in Nürnberg"; Stautz, *Die Astrolabiensammlungen in München*; Webster & Webster, *Western Astrolabes in Chicago* (a volume on the Islamic and Indian astrolabes prepared by David Pingree is currently in final page proof).

409 See van Cleempoel *et al.*, *Astrolabes at Greenwich*; *Madrid FCA 1997 Exhibition Catalogue*; van Cleempoel, *Catalogue of Louvain Instruments*; and G. Turner, *Elizabethan Instruments*.

410 This was part of a project generously funded over several years by the German Research Organization (DFG). However, our original intention to publish a catalogue of the entire corpus is temporarily on hold because of the need for further funding and additional personnel, and not least the impending closure of the Frankfurt Institute. A website on medieval instruments advertised to medievalists, Islamicists and instrument specialists is at:

www.uni-frankfurt.de/fb13/ign/instrument-catalogue.html .

This website may be suppressed when the Frankfurt Institute is closed. The section of the catalogue dealing with the earliest astrolabes from Iraq and Iran (8th-11th centuries) is now published: see King, *In Synchrony with the Heavens*, XIIIb-c.

On the rich archive of microfilms, photographs and papers relating to instruments that were collected by Ernst Zinner, which is currently housed at the Institute, see Schaldach, "Das Ernst Zinner-Archiv", and Ackermann, "Dormant Treasures".

411 King, *In Synchrony with the Heavens*, XIIIa: "The Neglected Astrolabe – A supplement to the standard literature on the favourite astronomical instrument of the Middle Ages".

1) The sole surviving Byzantine astrolabe (#2), recognized in 1926 for what it is by an eminent British classicist, and subjected to various distortions since the late 1940s. See Appendix C.

2) The earliest known Islamic astrolabe (#3702), from late-8th-century Baghdad, until recently preserved in the Archaeological Museum in that city. It was described in 1957 by two Iraqi scholars who did not recognize what they had before them because they had nothing with which to compare it: see **Fig. B1**. Also most, but not all, of the original (*kûfî*) engraving had been obliterated and replaced with early modern (*naskhî*) engraving. This piece was essentially forgotten until the 1990s, and is now published in detail, although its present whereabouts are uncertain.[412] I shall mention it again in Appendix C.

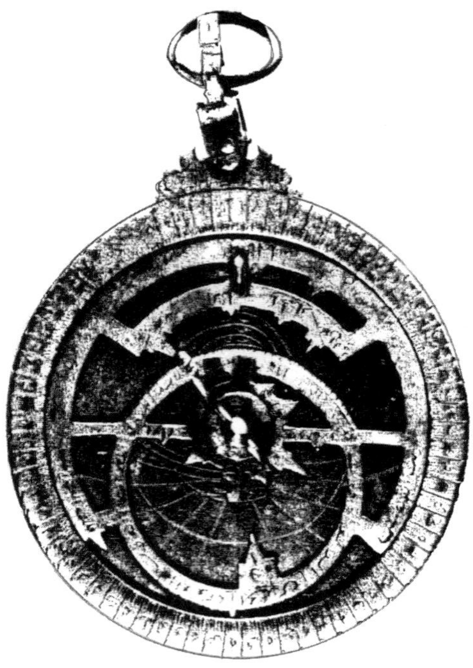

Fig. B1:~ The front of the oldest known Islamic astrolabe (#3702), a mere 8.5 cm in diameter. The star-positions correspond to *ca.* 500, which suggests that they were computed (by someone in the Hellenistic tradition) for *ca.* 700 using Ptolemy's incorrect value for the precession of the equinoxes, In other words, they date from before the much improved measurement of precession conducted

412 *Ibid.*, XIIIb.

in Baghdad in the early 9th century. The design of the rete is copied from a Byzantine one, for it is identical with that on the 1062 Byzantine astrolabe (see **Fig. C1**). The mater and three plates bear markings for each of the seven climates, and there are no markings on the back beyond an altitude scale. The original inscriptions have been replaced by more modern ones, perhaps in the 18th or 19th century. [Photos courtesy of the Archaeological Museum, Baghdad.]

3) The magnificent astrolabe of the astronomer al-Khujandî, constructed in 984/85, probably in Baghdad (#111): see **Fig. B2**. This has zoomorphic decoration on the rete as well as a quatrefoil, and it incorporates several of the improvements to the standard astrolabe that were made in the 9th and 10th centuries, including an horary quadrant for latitude 33° (Baghdad). It was offered to the Museum of the History of Science at Oxford in 1929 but was rejected because it was thought to be more recent than the seller claimed: Gunther, on the advice of a distinguished Arabist at Oxford, dated it to 1376.[413] In 1990 its last owner escaped from Kuwait with it in his pocket, and it recently passed to a museum in Qatar. It is now published in detail.[414]

Fig. B2:~ The front of the spectacular astrolabe (#111) of al-Khujandî, made in the year 984/85. This is by far the most remarkable and most carefully-constructed astrolabe known from the early Islamic period (to, say, 1200). The zoomorphic features on the throne and rete transform the piece into a veritable work of art. The distinctive quatrefoil was perhaps originally adopted from a Byzantine astrolabe. The plates of the standard variety serve latitudes 21° (Mecca), 27°, 30°, 33° (Baghdad), 36° (Rayy), 39° and 42°. There are also special plates for latitudes 0° and 90°, for horizons of all latitudes, for conversion between the ecliptic and equatorial coordinate systems (see **Fig. A3**), and for astrological purposes. On the back we find a sexagesimal trigonometric quadrant, an horary quadrant for latitude 33° (Baghdad), and astrological information arranged in concentric rings. [Photos courtesy of a former owner.]

---

413 Gunther, *Astrolabes*, I, p. 245 (no. 111).

414 King, *In Synchrony with the Heavens*, XIIIc-9, also XVII on the quatrefoil and its later influence.

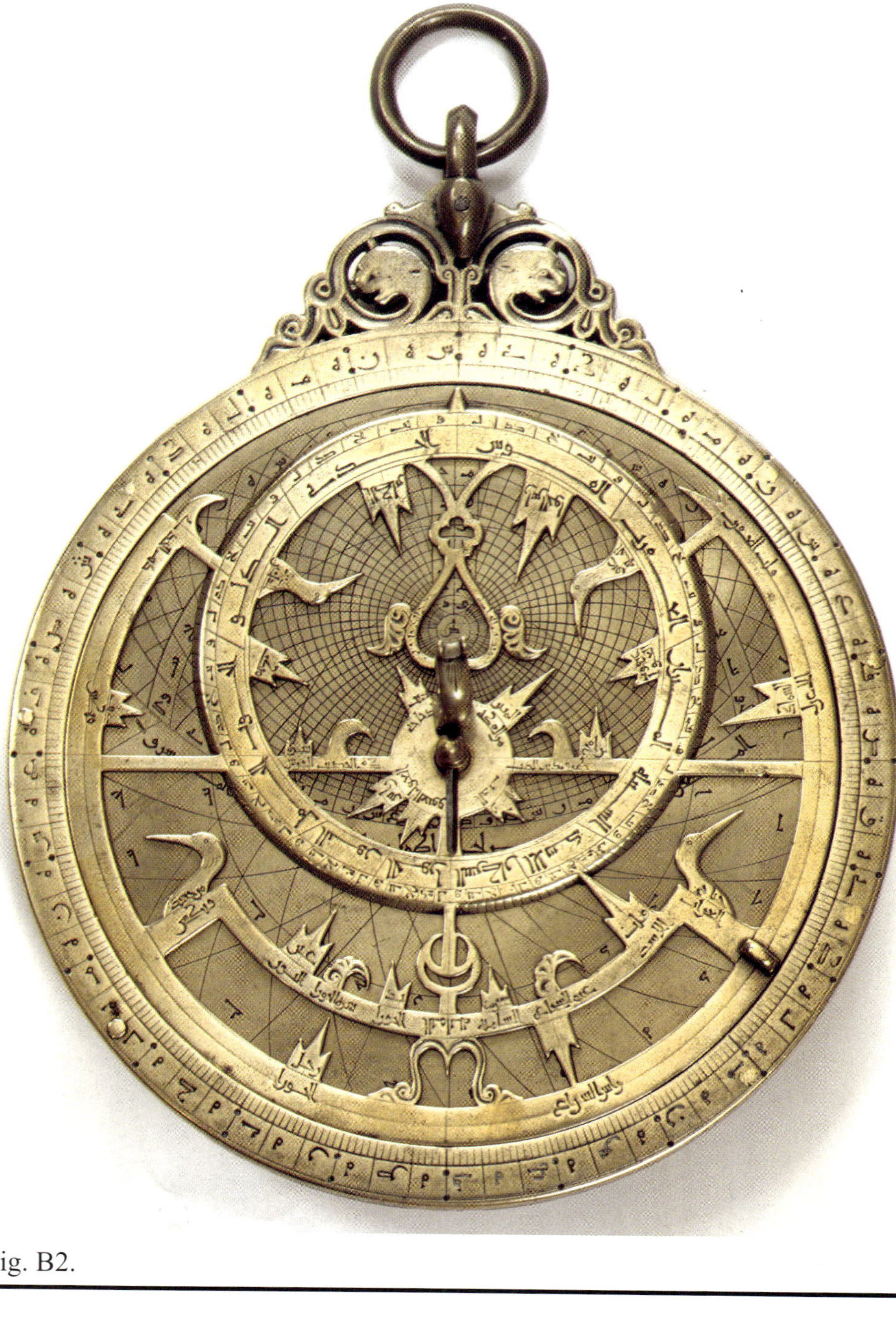

Fig. B2.

4) The earliest known European astrolabe (#3042) is from 10th-century Catalonia: see **Figs. B3a-b**. It was recognized in 1962 for what it is by the French instrument specialist Marcel Destombes, and then pronounced suspicious by scholars versed in the medieval manuscript tradition of astrolabe texts but not with the instruments themselves. The early Catalan provenance of this piece was confirmed in 1995, but the damage done to this piece by modern "scholarship" is irreversible.[415]

---

[415] The first description was in Destombes, "Un astrolabe carolingien", a fine piece of scholarship, in which the author also considered all of the earliest astrolabes known to him. After its appearance, rumours were rife in Paris that the astrolabe was a modern fake, although, inevitably, no critique of Destombes' description was ever published. Destombes bequeathed the piece to the Institut du Monde Arabe in Paris rather than let it pass to any French national museum. It was reinstated in 1995 when the distinctive engraving was shown to be attested in 10th-century Catalan inscriptions, the Latin on the later engraving was shown to display Catalan influence, and the astrolabe had been compared with several dozen earlier Islamic pieces and other early European ones. See various studies in Stevens *et al.*, eds., *The Oldest Latin Astrolabe*, especially Mundó, "Analyse paléographique", on 10th-century Catalan inscriptions, and King, "The Earliest Known European Astrolabe", p. 379, on the Catalan influence in the later Latin engraving (recognized by Kurt Maier), also *ibid.*, pp. 371-372, and n. 151 above on the alphanumerical notation. The star-pointers are not labelled but the stars represented can be identified. Their positions do not correspond very well with the actual ones: see Kunitzsch & Dekker, "The Star-Positions on the 'Carolingian' Astrolabe".

The latest affront to this piece, in *Paris IMA 2000-01 Exhibition Catalogue*, p. 242 *ad* no. 245, ignores all of the most significant research on it and gives the provenance as "Catalogne, Italie, Sicile ou France (du Nord) (?)" and the date as "X^{e}-XIe siècles (?)". The damage done to the piece, and the impossibility of restoring it to its rightful place in history, is summarized in the words "parfois envisagée, l'hypothèse d'un faux moderne est de moins en moins retenue".

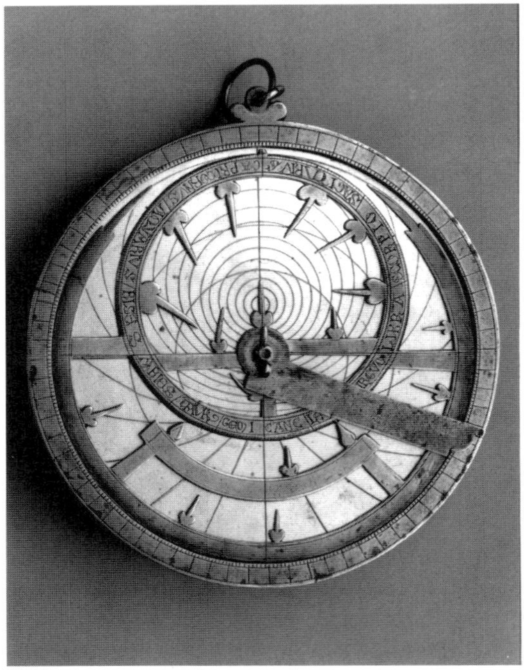

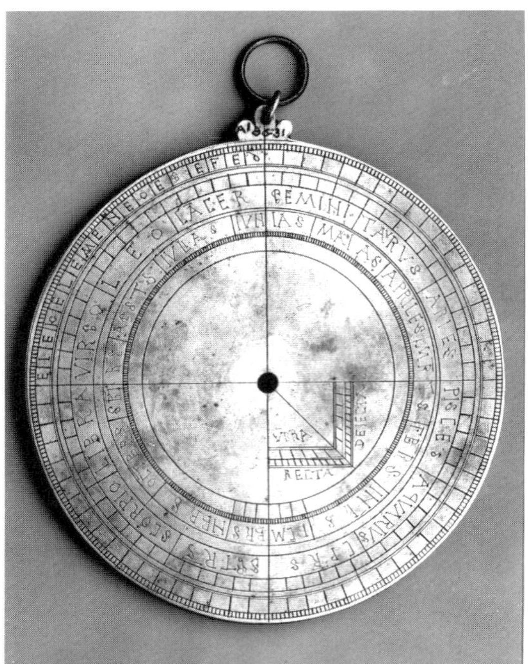

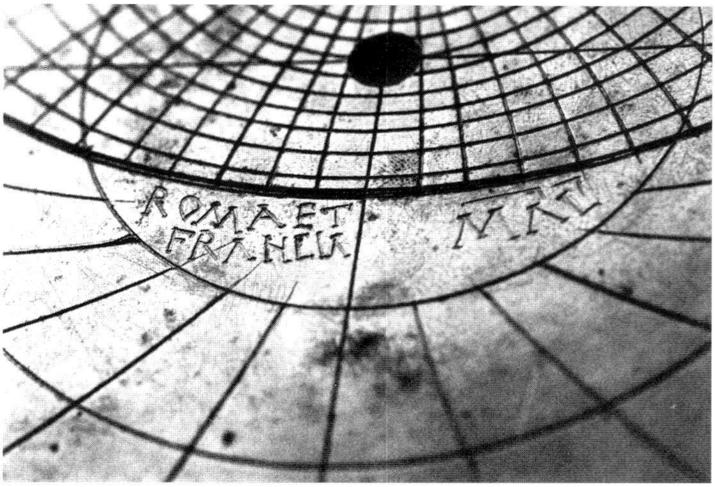

Figs. B3a-c:~ The front and back of the earliest known European astrolabe (#3042), made in Catalonia in the 10th century and a detail of its historically most important plate. The distinctive engraving bears accords with that on contemporaneous Catalan inscriptions. The later engraving on the rete includes the Catalan form VIRGVO, also used by Ramon Llull (d. 1316). No star-names are included, probably because the maker did not know them properly. The alphanumerical notation was used in preference to the clumsy Roman notation, and is derived from the Western Islamic notation. The mater and plates serve latitudes partly related to those of the climates: 36°, 39°, 41°30′, 45° and 47°30′. The plate for latititude 41°30′ bears the additional inscription ROMA ET FRANCIA, which refers to Rome and Catalonia. [Photos courtesy of the Institut du Monde Arabe, Paris.]

5) The astrolabe of Ibn al-Sarrâj of Aleppo dated 1328/29 (#140), the most sophisticated astrolabe ever made. Gunther questioned the date, even though it had been read by a competent Arabist, and he was clearly disturbed by it, not least because it might predate by 250 years the most significant English contribution to the astrolabe of the late 16th century (the "Mathematical Jewel" of John Blagrave).[416] This piece takes the idea of a universal astrolabe, based on a projection that is independent of terrestrial latitude, or fitted with plates for each degree of latitude from the equator to the pole, to unprecedented heights. It can be used

[416] Gunther, *Astrolabes*, I, pp. 284-285 (no. 140). On Blagrave's instrument see *ibid.*, II, pp. 492-501, and, more recently, G. Turner, *Elizabethan Instruments*, pp. 86-89 and 187-190.

universally in five different ways. No European ever made an astrolabe of this complexity or sophistication.[417] See **Fig. B4**.

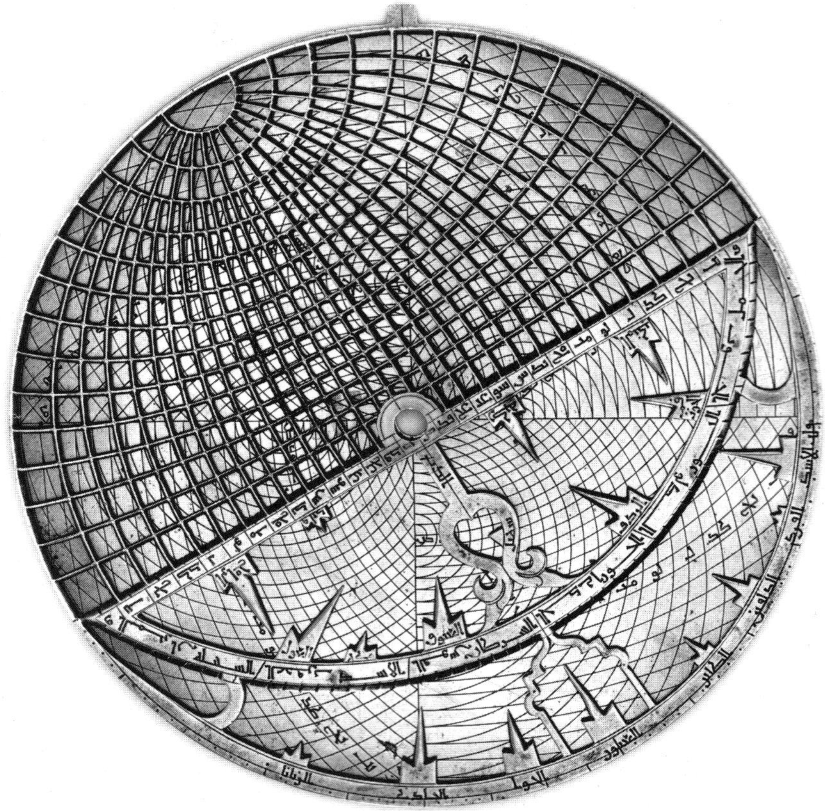

Fig. B4:~ The rete of Ibn al-Sarrâj laid over one of the plates (#140). The rete is based on a mixed north-south projection and the ecliptic arc serves the entire ecliptic. This plate is one of several for each 3° of latitude. Markings for a particular latitude are folded over to fit in a quadrant. The combination boggles the mind but fortunately we have a medieval treatise on the use of all of the components of the piece. [Courtesy of the Benaki Museum, Athens.]

---

417 See the basic description in King, *In Synchrony with the Heavens*, XIVb-5.1. A more detailed description, together with the medieval texts on the different ways to use the instrument, is in Charette & King, *The Universal Astrolabe of Ibn al-Sarrâj*, forthcoming.

6) An astrolabe from 14th-century Spain bearing inscriptions in Hebrew, Latin and Arabic (#4560): see **Figs. B5a-b**. The instrument was apparently made by a Jew, probably in Toledo, engraved by a Christian, and completed by a Muslim, who clearly intended to emigrate or flee with it to Algiers. When this piece first came to light in 1998 it was thought to be of Syrian provenance, completed by a Latin, but the Hebrew inscriptions, or rather, scratchings, were overlooked. This remarkable astrolabe is now published in detail,[418] but its present whereabouts are unknown.

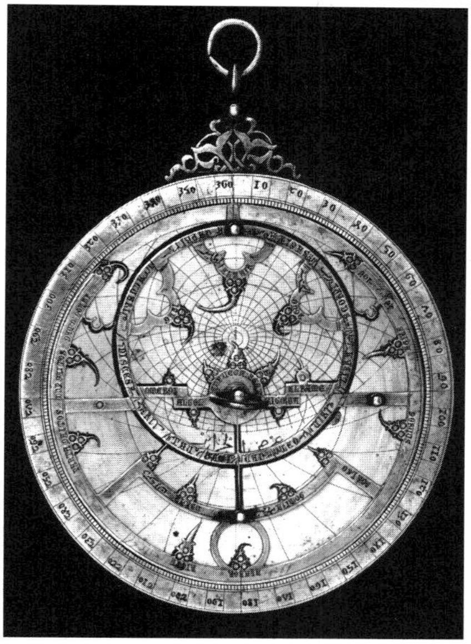

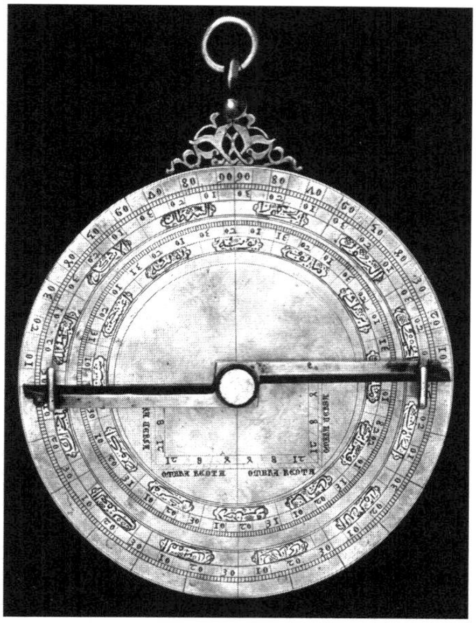

Figs. B5a-b:~ The front and back of a remarkable astrolabe, probably made in Toledo in the 14th century (#4560). The plates bear scratchings in Hebrew, clearly made by a Jew. Most of the engraving was achieved by a scholastic Latin, with some unusual regional influences. The remainder of the engraving, notably on the back and on the shackle, but also on a replacement plate for Mecca and Algiers, was achieved by an Arab named Mas'ûd, who clearly wanted to escape to the safety of *Dâr al-Islâm*, that is, the world of Islam. [Photos courtesy of Christie's of London.]

418 * See King, *In Synchrony with the Heavens*, XV.

7) An astrolabe from al-Andalus *ca.* 1300 with inscriptions in Judaeo-Arabic (#3915), that is in Arabic written in Hebrew script.[425] This piece, which shows resemblances to both Andalusî and Catalan astrolabes, is engraved with a very corrupt version of a Classical Arabic poem. The original compiler was a state secretary to the Buyid court in Baghdad who *ca.* 970 presented the ruler with an astrolabe of 2.5 cm diameter and these three lines of poetry. In the latter, the gift of the astrolabe is likened to giving the ruler "the highest sphere of heaven and all that is within it".[426]

8) The earliest Renaissance astrolabe signed by the leading astronomer in Europe (#640), recognized in 1959 by Derek Price for what is, and later dubbed suspicious or even a fake. This was mainly because no other relevant 15th-century Viennese astrolabes were known in England with which to compare it. See further Appendixes D-V and E.

---

425 See *Christie's Amsterdam 1988 Catalogue*, pp. 88-95, lot 247; and *Khalili Collection Catalogue*, I, pp. 214-217, no. 124; also King, "Cataloguing Islamic Astronomcial Instruments", col. 253, and *idem*, *In Synchrony with the Heavens*, XVII-3.

426 See now Abu Zayed, King & Schmidl, "On astrolabes and poetry from Baghdad to al-Andalus".

There are many other astrolabes I could mention here which are important precisely because they illustrate how little we know about instrument-making in the milieus from which they stem. Regional studies of medieval instruments are lacking: there are, for example, no overviews of medieval English or German or French or Italian astrolabes. A few examples must suffice:

9) The spectacular quatrefoil astrolabe in the British Museum (#290), the oldest astrolabe made in England, dating from *ca.* 1300. This is larger and more imposing than any known later English instruments until the Renaissance.[428]

10) The less spectacular and smaller quatrefoil astrolabe from Einbeck, datable *ca.* 1330 (#2027) and probably made in Erfurt. This is the oldest known complete astrolabe from the German-speaking world, but it displays a highly developed design and considerable technical competence. I shall return to this in Appendix D.

11) Two splendidly-decorated astrolabes from medieval France (#428=-#625 and #2041), the former probably from the 13th century and the latter possibly from the 14th century.[429] Both betray Islamic influence in their rete design. Only the former is properly published.

12) An astrolabe from 14th-century Picardy, with numbers in a notation different from the Roman or the "Hindu-Arabic" (#202). It has now been established that such "monastic ciphers" were introduced in a simplified form from Athens to England in the 13th century, where they were used in at least one Cistercian monastery. A more sophisticated form was developed shortly thereafter by Cistercian monks in the central part of what is now the border region of France and Belgium: these are what we find on the astrolabe. The history of these ciphers

---

428 * On this see Gunther, *Astrolabes of the World*, II, pp. 463-465 (no. 290) and pls. CXXVI-VII; and Ward, *Instruments in the British Museum*, pp. 112 (no. 324) and pl. L.

429 On the former see now van Cleempoel, ed., *Astrolabes in Greenwich*, pp. 115-120. (The two different numbers indicate the provenance and present location.) On the latter see Poulle, *Instruments du Moyen Âge*, B, pp. 12 and 14 (illustrations of back and front); King, *The Ciphers of the Monks*, pp. 394-395; and *idem*, *In Synchrony with the Heavens*, fig. XIIIc-9n.

from Antiquity to the early 20th century has recently been documented for the first time.[430]

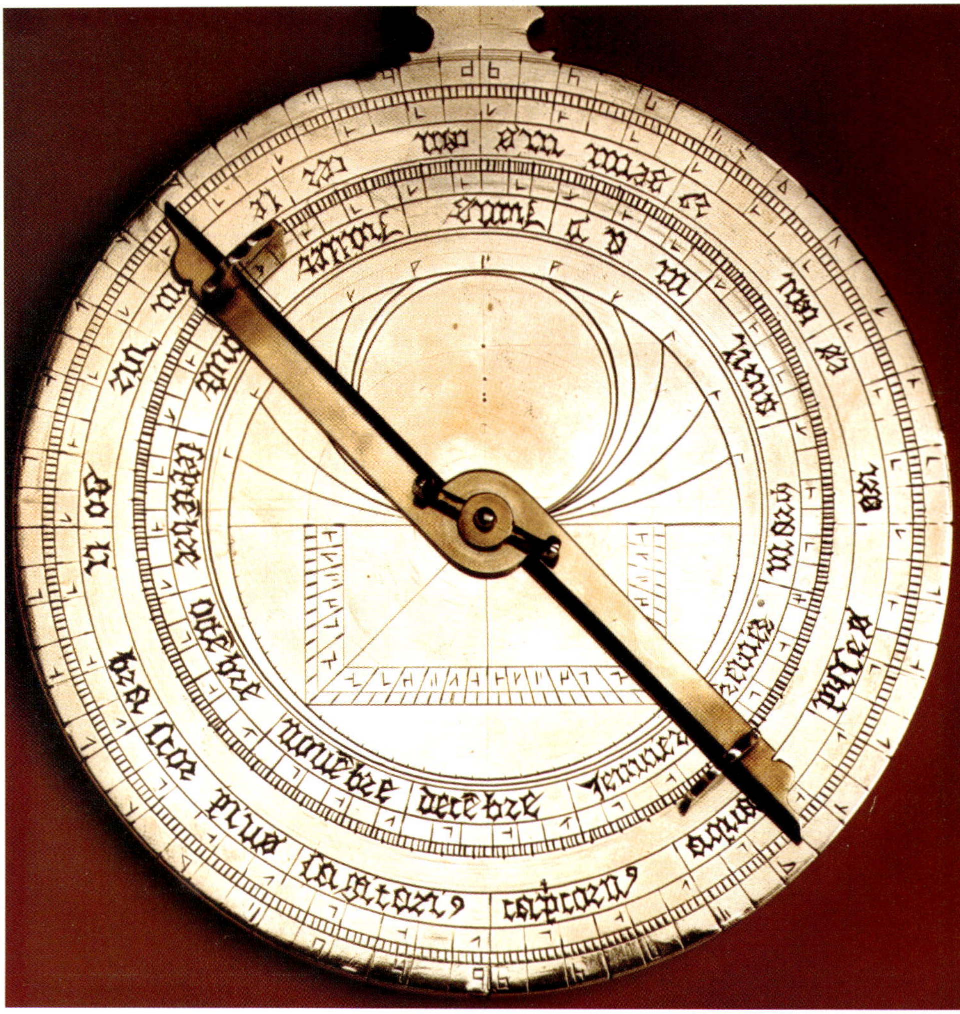

Fig. B6a

[430] See King, *The Ciphers of the Monks*, and on the astrolabe see *ibid.*, pp. 131-151 and 406-419. The first mention of this piece was in Gunther, *Astrolabes of the World*, 349 (no. 202).

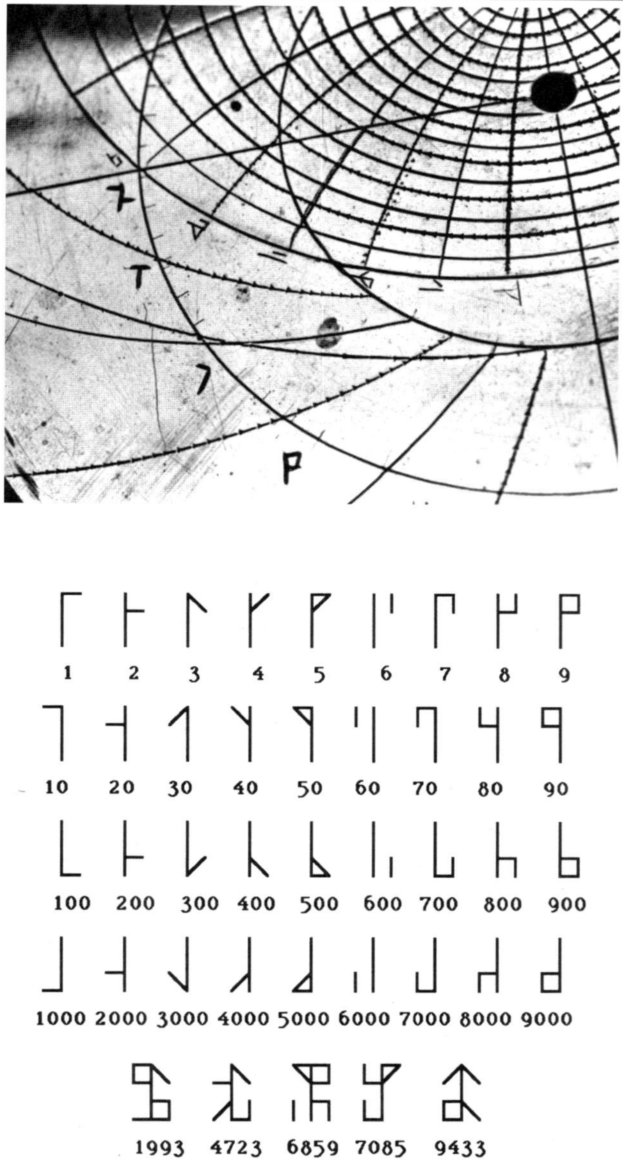

Fig. B6a-c:~ The back of the Picard astrolabe with monastic ciphers (#202), a detail of a plate, and the rationale behind the ciphers. In different forms they were used in limited circles from Spain to Sweden for paginating manuscripts, numbering items in lists, representing year-numbers, and marking the content of wine-barrels. [Photos courtesy of Christie's of London.]

13) A highly sophisticated geared astrolabe (#4601) from "*henau*", the Hainault, probably 14th century, that came to my attention only in 2006.[431] This should be published in detail and compared with three other pieces, also from 14th-century Northern France: firstly, an elegant astrolabe with markings for latitudes 45° and 50° (#461);[432] secondly, the only other geared astrolabe from 14th-century Northern France (#198), with markings for "*engletiere*" and "*hollande*",[433] which is somehow related to a third piece, the Picard astrolabe with all numbers engraved in monastic ciphers (#202 - see no. 11 above). The engraving on the first of these three should be compared with that on the new piece. Both the second and third were made with a lower level of technical competence than the "new" piece.

14) A superbly-made extravagantly-decorated Italian astrolabe from *ca.* 1500 (#171) with an intricate and beautiful rete design. This, now preserved in Hamburg, has wrongly been declared a fake, out of confusion with a fake copy of it now in Chicago.[434]

It is great mistake to think that the only instruments, or the most interesting ones, are astrolabes. Take, for example the works of the early Baghdad astronomer who called himself Nastûlus. He is known by two astrolabes, one dated 927/28, the other undated.[435] A textual source tell us that in the year 893/94 he made an eclipse computer, a instrument far more complicated than a standard astrolabe. In 2006, two important finds were made. Firstly, a treatise by Nastûlus describing a universal sundial on the back of an instrument that is fitted with a set of gear wheels to display the relative motions of the sun and moon. This combined instrument is a Hellenistic invention, but no Greek texts on it are

---

431 See already n. 19 above. The piece belongs to a private collection in France. My thanks to my friend Jacques van Damme for a complete set of images.

432 On this see *Brussels SGB 1984 Exhibition Catalogue*, pp. 38-39 (no. 9), and King, *The Ciphers of the Monks*, pp. 399 and 404.

433 On this see Gunther, *Astrolabes of the World*, II, pp. 347 and pls. LXXX-LXXXI (no. 198) and 349 and further North, "Gears", pp. 370-372 (pp. 167-169 of the reprint), and King, *The Ciphers of the Monks*, pp. 398-399 and 402-403. On its connection to the astrolabe with ciphers (n. 430) see *ibid.*, pp. 410, 415 and 416-417.

434 See Gunther, *Astrolabes*, II, pp. 322-325 (no. 171); and Francis Maddison in *Washington NGA 1992 Exhibition Catalogue*, p. •FFM• (no. •), where the documentation is presented. On the faked copy by the Spitzer workshop see Webster & Webster, *Western Astrolabes in Chicago*, pp. 114-116 (no. 28), also the third item in n. 542 below.

435 On Nastûlus' two astrolabes see King, *In Synchrony with the Heavens*, XIIIb-3.

known; however, Nastûlus' Arabic text is not a translation form the Greek.[436] Secondly, a plate by Nastûlus on which the altitude of the sun at the seasonal hours is shown graphically for all solar longitudes. Whilst tables of such functions were available, the curves on this piece are so accurately drawn that they defy explanation in the light of our present knowledge.[437] Likewise, one could argue that the ring-dial and the equatorium associated with Regiomontanus (see Appendix D-VI-1 and 2) are more technically sophisticated than any of the Vienna astrolabes. A sad reflection on the state of research in the history of instrumentation is that neither of them has been published.

---

436 A description of the new manuscript is available from Sam Fogg, London. On the universal sundial see the references in n. 439 below.

437 See the description in King, "A Remarkable Mathematical and Astronomical Instrument from Baghdad *ca*. 900".

Appendix C – Notes on the 1062 astrolabe in Brescia

Preliminary remarks

We are dealing here with the sole surviving astrolabe with Greek inscriptions from over a millennium of Greek interest in the instrument (4th century to 15th century).[438] On the other hand, several late Hellenistic and Byzantine sundials and geared devices have survived.[439]

The astrolabe (#2) was commissioned by and made for a well-educated high-ranking Greek official, who composed an eloquent poem about the basic function of the astrolabe. It was made with extreme care and competence: the astronomical markings are remarkably accurate. Its very existence to some extent contradicts what we know from textual sources about Byzantine astronomy and astronomical instrumentation in the 11th century:[440] one modern author has even claimed that the poem was engraved because people then no longer knew what an astrolabe was. Muslim craftsmen started making astrolabes in the mid-8th century, at first copying Byzantine examples and then developing Islamic styles and making numerous modifications and improvements.[441] The Byzantine astrolabe shows no Islamic influence whatsoever in its design: claims to the contrary are ill-founded. It was published accurately and in detail in 1926 by the British classicist Ormonde Maddock Dalton,[442] and a new, yet more detailed study in the light of our present knowledge of Byzantine astronomy would be worthwhile. A brief description must suffice here. This is given in italics below, separate from the commentary. The latter is aimed at correcting the several misconceptions about this landmark instrument that are found in the secondary literature. Illustrations are in **Figs. I-7** and **8a-b** and **C1-3**.

---

438 One medieval European astrolabe contains some plates engraved in Greek: see King, *The Ciphers of the Monks*, pp. 290-291.

439 On these see Price, "Portable Sundials in Antiquity", citing various earlier works; Field & Wright, "Gears from the Byzantines", and *eidem*, *Early Gearing*; and Ackermann, "Byzantine Universal Sundial". On the formula underlying this universal sundial see the references cited in n. 390 above.

440 See Tihon, "Traités byzantins sur l'astrolabe", and *eadem*, *Études d'astronomie byzantine*.

441 See the text to n. 386 above.

442 The best description is still Dalton, "Byzantine Astrolabe at Brescia", summarized in Gunther, *Astrolabes of the World*, I, pp. 104-108. Some additional information is in Ferretti Torricelli, "Strumenti scientifici di Brescia", pp. 200-201; and a description by Italo Furlan in Bettini *et al.*, eds., *Venezia e Bisanzio*, no. 35. A derivative account is in *Brussels 1982 Exhibition Catalogue*, p. 176. For new information see Stautz, "Die früheste Formgebung der Astrolabien"; *idem*, *Untersuchungen*, pp. 39-43 and 178-181; and King, *In Synchrony with the Heavens*, XIIIa-4, XIIIb-1.2 and 3.5-6, and XVI-3.

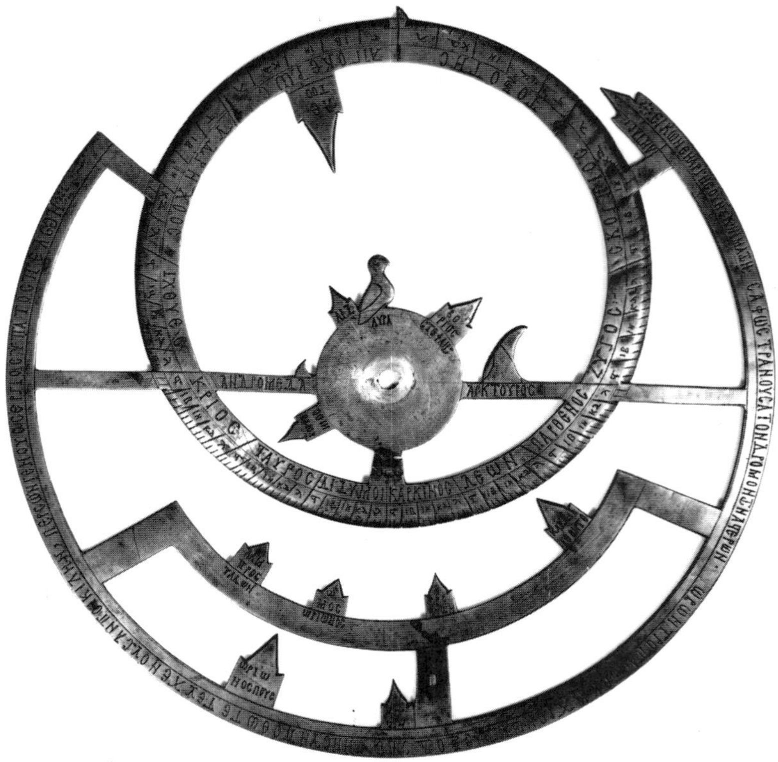

Fig. C1:~ The rete of the 1062 astrolabe (#2), with the distinctive design and the verse around the rim. [Courtesy of the Museo dell'Età Cristiana, Brescia (now the Santa Giulia Museo della Città).]

The instrument has been known since 1844. It was presented in 1844 by Francesco Sailer, Director of the Scuola d'Equitazione in Brescia and of Austrian descent, to the Municipio of Brescia. It later passed to the Museo dell'Età Cristianà, founded in 1882, which was recently incorporated in the Santa Giulia Museo della Città.[443]

The astrolabe is made of brass, and its diameter is 37.5 cm. All medieval astrolabes are made of brass. This is the largest known Eastern astrolabe surviving from before the 13th century.

---

[443] Dalton, "Byzantine Astrolabe", p. 133, n. 1, and Ferretti Torricelli, "Strumenti scientifici di Brescia", pp. 199 and 201, n. 1, taken from a short biography of Sailer published in 1915. The inventory number is 36.

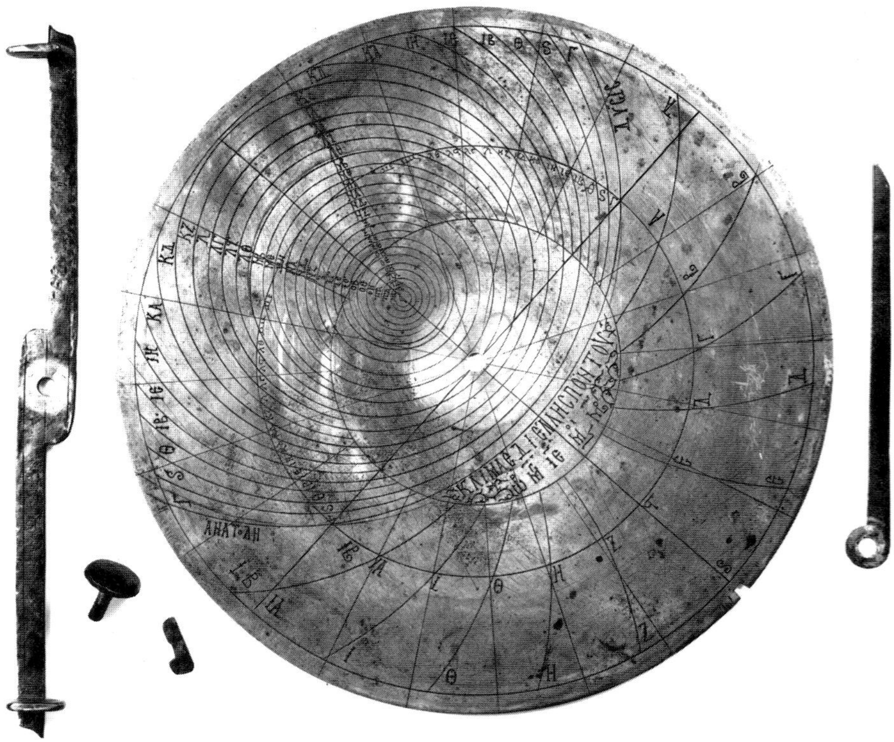

Fig. C2:~ One side of the single plate from the Byzantine astrolabe of 1062 (#2). There are altitude circles for each 6°, and also for each 3° inbetween the solstitial circles, for improved operations with the sun. The latitude served is 40° in the 5th climate, and the Hellespont is specifically mentioned. [Courtesy of the former Museo dell'Età Cristiana, Brescia (now the Santa Giulia Museo della Città).]

All inscriptions are in Greek and all numbers in Greek alphanumerical notation. This is precisely what one would expect on a Byzantine astrolabe. The Greek alphanumerical notation was standard amongst Greek and Byzantine astronomers and instrument-makers.[444]

---

[444] See Gardthausen, *Griechische Paläographie*, II, pp. 353-394; Aaboe, *Episodes in the History of Mathematics*, pp. 103-104; Ifrah, *Histoire universelle des chiffres*, I, pp. 529-540 / *Universal History of Numbers*, pp. 261-274; and King, *The Ciphers of the Monks*, pp. 290-294.

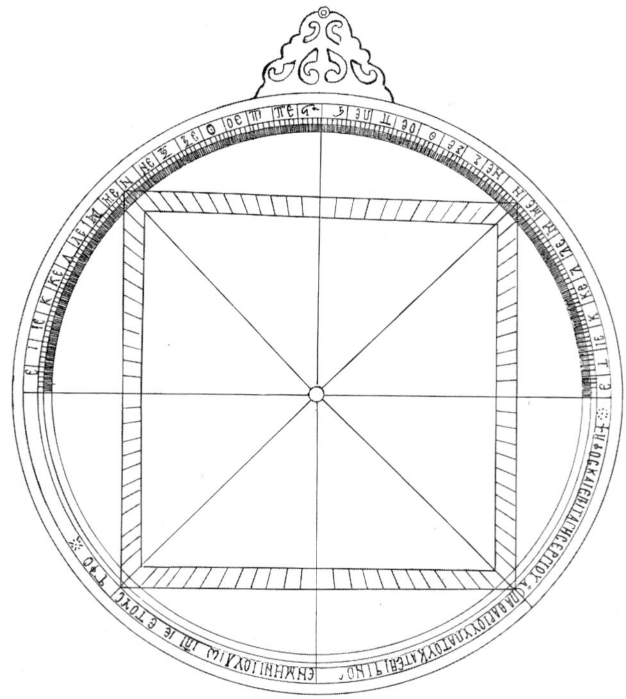

Fig. C3:~ The back of the Byzantine astrolabe, as depicted by O. Dalton in 1926. This lithograph is more revealing than any modern photos. Notice the continuation of the inscription around the rim. The four shadow squares are an absurdity, since one would suffice: they are clearly a later addition by somebody with a *horror vacui*. Notice that in the top right they interfere with the inscription. [From Gunther, *Astrolabes*, I, p. 106.]

There is a poem on the rim of the rete describing the symbolism of the astrolabe. This reads:~

⁎ ΕΙΚΩΝ ΕΝΑΡΓΗϹ ΟΥ(ΡΑ)ΝΟΥ ΚΙΝΗΜΑΤΩΝ
ϹΑΦΩϹ ΤΡΑΝΟΥϹΑ ΤΟΝ ΔΡΟΜΟΝ ΤΩΝ ΑϹΤΕΡΩΝ
ΩΡΩΝ ΤΡΟΠΑϹ ΤΕ ΚΑΙ ΧΡΟΝΩΝ ΔΙΕΞΟΔΟΥϹ

ΗΝ CΥΝ ΠΟΘΩ ΤΕΤΕΥΧΕΝ ΟΥCΑΝ ΠΟΙΚΙΛΗΝ
ΠΕΡCΩΝ ΓΕΝΟΥC CΕΡΓΙΟC ΥΠΑΤΟC ΠΕΛΩΝ ⁜ :~

and was translated by Dalton as:~

"(This is) an image manifest of the heaven's movements,
Making distinct and clear the course of the stars,
The changing of the seasons and the passages of the times,
Which, being an intricate work, with ardent mind fashioned
Sergius the Persian, holding a consul's rank."

It is the astrolabe that is speaking to us, so the insertion "(This is)" would be better rendered "(You are looking at)", or "(You are looking at an astrolabe which is)", or perhaps, more simply and more appropriately, "(I am)". The use of the word εικων, "icon", here rendered "image", is significant, for the astrolabe is precisely a plane representation of something considered wondrous, if not sacred.

It would be difficult to express the basic function of the astrolabe more succinctly.[447] There are asterisks ⁜ at the beginning and end of this inscription,[448] as well as a colon and tittle at the end of it. (Regiomontanus copied the colon and tittle on his 1462 astrolabe: compare **Fig. I-8b** and **I-13**.)

The astrolabe was made for a high-ranking government official named Sergios. The remainder of the inscription on the lower perimeter of the back is not in verse:~

⁜ ΨΗΦΟC ΚΑΙ ΕΠΙΤΑΓΗ CΕΡΓΙΟΥ Α|CΠΑΘΑΡΙΟΥ ΥΠΑΤΟΥ
ΚΑΙ ΕΠΙCΤΙΝΟC ΕΝ ΜΗΝΙ ΙΟΥΛΙΩ ΙΠ|| ΙΕ ΕΤΟΥC \\ΦΟ.

where | represents the abbreviation sign above the A, || the sign above and after ΙΠ and \\ stands for the unprintable letter *digamma*, a survival of hoary Antiquity. This Dalton translated as:~

"Decree and command of Sergios, *protospatharios*, consul and man of science (?), in the month of July, fifteenth indiction, year 6570."

The date, given according to the Byzantine convention,[449] corresponds to July 1062.

---

447 ** An appreciation of this is in North, "The Astrolabe and the Imagination", p. 3.

448 On the asterisk in post-classical Greek see n. 59 above. Those here do not appear to be a Christogram, formed by the letters I (*iota*) and X (*chi*) of the Greek name Iesous Christos, together with the horizontal beam of the Cross. See the article "Christogram" in the *Oxford Dictionary of Byzantium*.

449 On this see Ginzel, *Handbuch der Chronologie*, III, pp. 292-294, and the article "Chronology" in *The Oxford Dictionary of Byzantium*. The date is stated as "month of

Sergios' "decree and command" indicate that somebody else made the astrolabe for him, also engraving the inscription that Sergios himself had conceived. Sergios' family was of Persian origin, but probably many generations back. He was clearly a very well-educated Greek official with imposing honorific titles.[450] Alas, it has not been possible to identify him, although we do have some possible candidates.[451] The second part of the text indicates that the instrument

---

July, 15th indiction, year 6570". The era began on September 1, 5509 B.C. The indiction, a fiscal period of 15 years, is the remainder of the year number when divided by 15, in this case, again 15. The numbers 15 (and various numbers on the plates) and the year number 6570 are given in the Greek alphanumerical notation, on which see n. 446 above. The number 6570 is written *digamma-phi-omikron*, that is, 6000 + 500 + 70.

450 On events in the Byzantine world around that time see de Muralt, *Chronographie byzantine*, II, pp. 1-11. On the ranks of consul and *protospatharios* see Dalton, *op. cit.*, p. 140, n. 1, also the articles "Consul", "Hypatos", "Protospatharios" and "Spatharios" in *The Oxford Encyclopedia of Byzantium*. By the 11th century the title *protospatharios*, "first sword bearer", was more honorary than functional, something like the French "légion d'honneur" (the comparison was made by Dalton). The rank of *hypatos* was slightly higher than that of *protospatharios*. Dalton drew attention to a Persian translator called Sergios known from the 6th century (*op. cit.*, p. 137, n. 1, quoting Krumbacher, *Geschichte der byzantinischen Litteratur*, I, pp. 242-243). By chance, the first attested *protospatharios*, from the early 8th century, was called Sergios.

451 Our Sergios did not find his way into Trapp *et al.*, *Prosopographisches Lexikon der Palaiologenzeit*. Various individuals named Sergios with the title *protospatharios* are listed in the very useful lists of inscriptions on Byzantine seals and statues in the website "Prosopography of the Byzantine World" at

www.kcl.ac.uk/cch/pbe .

Possible candidates for "our" Sergios are the following:~

(1) Sergios Hexamilites, dated during 1066-1082, with various titles on nine seals, including on two the titles of *protospatharios*, imperial *notarios* and *krites* of the hippodrome, and on one, four asterisks.

(2) Sergios, *protospatharios* and *eparchos*, dated between the early 10th century and the year 1204, is named on a bust of St Nikolaos.

(3) Sergios, *protospatharios* and *krites* of Kibyrrhaioton, dated to the first half of the 11th century, is mentioned on another bust of the same saint.

(4) Sergios, *protospatharios epi tou Chrysotriklinou* and imperial *notarios*, dated to the middle of the 11th century, is known from the inscription on a statue of the Virgin and Child:

Μ(ητη)ρ Θ(εου) + Θ(εοτο)κε β(οη)θ(ει) τω σω δ(ουλω) ⁂ Σεργι[ω]
(πρωτο)σπαθα(ριω) επι του Χρ(υσοτρι)κλ(ινου) (και) β(ασιλικω) ν(ο)ταριο,

"Mother of God. Mother of God (Θεοτοκοσ), aid your servant Sergios,
protospatharios *epi tou Chrystoriklinou* and imperial *notarios*."

Note the nice asterisk at the beginning of the second part of the inscription, as on the 1062 astrolabe!

was made *for* him, and was so understood by Dalton. In the modern literature, we sometimes encounter the statement that it was made *by* him, the latter contributing to the myth that the instrument was made by a Persian and hence must display Islamic influence.

The throne is delicately worked with a distinctive perforated design. It should be compared with contemporaneous Byzantine perforated (*ajouré*) metalwork.[452] Its design bears no relationship to the design on any Islamic throne before the 13th century.

The rete has a simple, distinctive design that is strictly Byzantine. The earliest surviving Islamic astrolabe, a much smaller piece (diameter 8.5 cm) from late-8th-century Baghdad (#3702), has an almost identical rete, with the same distinctive single pointer within the upper ecliptic: this is because it was copied from a Byzantine astrolabe.[453] Later, but still early, Islamic astrolabes from 9th and 10th-century Iraq have a related, but more developed rete design.[454] Certain Islamic and medieval European astrolabe retes bear features that seem to attest to more sophisticated Byzantine rete designs.[455]

The star-positions of the pointers for 14 stars are remarkably accurate for the date of construction. The reader is referred to the analysis by Burkhard Stautz.[456] The accuracy of the star-positions proves that the maker had access to a set of star-positions that were reasonably accurate, so that they could not be based on

---

Now the epithet Persios in the astrolabe inscription may have been used for the rhyme. So we mention also:

(5) Sergios Mesopotamites, tentatively dated between the last quarter of the 11th century and the beginning of the 12th, but not identified as a *protospatharios*, is known by a statue of the Virgin Mary inscribed:

Μ(ητη)ρ [Θεου] / Κ(υρ)ιε β(οη)θ(ει) [Σ]εργιο [τ]ο Μεσο[π]οταμ[ιτ(η)],

"Mother of God / Lord aid Sergios Mesopotamites".

It would be a charming coincidence if one or other of the last two mentioned were our man: compare the case of the monk Berselius (n. 157 above), who in addition to his astrolabe marked in monastic ciphers also owned a statue of the Virgin that he himself had commissioned.

452 See, for example, various polycandelons and a suspension lamp in *Brussels 1982 Exhibition Catalogue*, pp. 177-178, Evans *et al.*, *The Arts of Byzantium*, pp. 28-29, and *Paderborn 2001 Exhibition Catalogue*, pp. 216-218.

453 See the text to n. 412.

454 All known examples are described in detail in King, *In Synchrony with the Heavens*, XIIIc. See already n. 410 above.

455 I refer to the quatrefoil on the astrolabe of al-Khujandî (see n. 420) and the rectangular frame with a quatrefoil on an astrolabe from Catalonia *ca.* 1300. These are discussed *ibid.*, XVII.

456 Stautz, *Untersuchungen*, pp. 39-43 and 178-181.

Ptolemy's incorrect value for precession.[457] Here we can suppose Islamic influence. The star labelled AIΞ (Capella, α Aurigae) is incorrectly identified; it should be labelled OPNIΘOY OYPA (Deneb, α Cygni). Ioannis Philoponos (6th century) in his Greek astrolabe treatise suggests that astrolabes be provided with pointers for 17 or more stars: the earliest Islamic astrolabe from the late 8th century has 17 pointers, as do several Islamic astrolabes from the 9th and 10th centuries.[458]

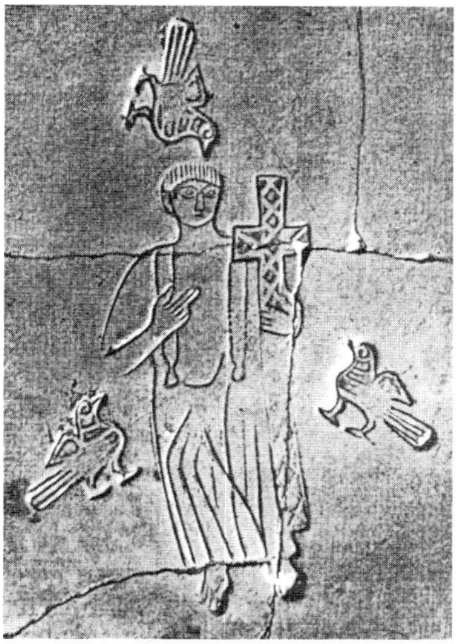

Fig. C3: Three birds surrounding a man carrying a cross on a Byzantine pottery bowl dated to the 5th century, found in Chersonesus (presumably Crimea not Thrace) in 1904. [From Banck, *Byzantine Art in the Collections of the USSR*, no. 24 and p. 336 ("three doves").]

There is a bird on the star-pointer for ΛYPA. This is a Byzantine feature. The bird is deliberately represented either as (1) a sparrow, which would recall a statement attributed to Christ in the *Gospels*[459] that God notices even a sparrow

457 See Pingree, "Greek List of Astrolabe Stars", pp. 476-477, on an incompetent Byzantine table of 19 astrolabe stars for the year 908, which include the 14 stars of the 1062 astrolabe.

458 Segonds, *Philoponos sur l'astrolabe*, pp. 168-169, also Gunther, *Astrolabes of the World*, I, p. 71. See also King, *In Synchrony with the Heavens*, XIIIb-3.5 and XIIIc-1-3.

459 *Matthew*, X:29: "Are not sparrows two a penny? Yet without your Father's leave not one of them can fall to the ground." *Luke*, XII:6: "Are not sparrows five for twopence? And yet not one of them is overlooked by God."

falling from the sky, or (2) a dove, which would signify the Holy Spirit.[460] Why the bird is there is another matter. The earliest Muslim astrolabe-makers in the 8th century happily took over this bird, which is found on the one and only astrolabe from late-8th-century Baghdad (#3702) mentioned above.[461] Contrary to numerous claims in the literature (see below), the bird is not the "falling eagle" (*al-nasr al-wâqi'*) of pre-Islamic Arab star-lore. No surviving Islamic astrolabe from the 9th or 10th centuries has any bird on this star-pointer. The bird on the Byzantine astrolabe bears no resemblance to the eagle or vulture sometimes found on the pointer for Vega on various later Islamic astrolabes.[462] (On a 14th-century Catalan astrolabe (#162), which has a distinctive frame on the rete that is decorated with a quatrefoil, the pointer for CAENT – a Catalan form for Latin CADENS, "falling" – is represented as a cock. Here, I suspect, we are witness to a type of rete that is Byzantine in origin, and the bird, be it a sparrow or a dove, has simply been transmuted.[463]) In the Islamic tradition of pictorial representation of the constellations, associated with the astronomer al-Sûfî (Shiraz, *ca.* 1000), the Lyre is represented as a turtle, as in the classical tradition.[464] In Renaissance constellation images, the lyre was sometimes held in the claws of a bird, always an eagle.[465]

The mater and single plate bear markings for Rhodes, Hellespont and Constantinople. The geographical information explicit and implicit on astrolabes is of considerable historical interest.[466] In the Greek astrolabe text of Philoponos, the seven sides of three plates and the mater are to be engraved for the latitudes of the seven climates of Antiquity, latitudinal bands whose centres and boundaries are defined by the lengths of longest daylight.[467] The earliest Islamic astrolabe, from 8th-century Baghdad, which was modelled after a Byzantine astrolabe, has this arrangement, and so does the earliest known astrolabe from

---

460 Articles "Birds" and "Holy Spirit" in *The Oxford Dictionary of Byzantium*; also, Metford, *Dictionary of Christian Lore and Legend*, p. 232.

461 See King, *In Synchrony with the Heavens*, XIIIb-3.6.

462 For examples see King, *In Synchrony with the Heavens*, XIVd, figs. 1.1, 1.3b, 2.1a, and XVII, figs. 1.4 and 1.7. On the bird see the inconclusive remarks *ibid.*, XVII-1, note 1.

463 King & Maier, "Medieval Catalan Astrolabe", especially pp. 676 and 682-683, also King, *In Synchrony with the Heavens*, XVII-3.

464 al-Sûfî, *Suwaru'l-kawâkib or Uranometry*, fig. 8 facing p. 68.

465 See, for example, Whitfield, *The Mapping of the Heavens*, pp. 68 (Vienna, *ca.* 1440), 71 (Dürer, 1515), 79 (Villa Farnese, 1575), and Stott, *Celestial Charts*, pp. 38-39 (Apianus, Ingolstadt, 1540), 56-57 (Thomas Hood, London, 1590).

466 On this subject see already the text to n. 384 above. The examples given here are all discussed in King, *In Synchrony with the Heavens*, XVI.

467 Segonds, *Philoponos sur l'astrolabe*, pp. 152-155, and also Gunther, *Astrolabes of the World*, I, p. 65. On the climates and their use on astrolabe plates see n. 384 above.

Islamic al-Andalus, datable to the 10th century. Later Islamic astrolabes, including those from the 9th and 10th centuries, have plates for specific latitudes associated with the climates, as does the earliest European astrolabe, from 10th-century Catalonia. Other very early European astrolabes have plates for the seven climates.

The inscriptions on the mater and single plate of the Byzantine astrolabe identify three localities, as well as the climates in which they are found, the associated approximate lengths of longest daylight,[468] and their latitudes:

Climate	Locality	Maximum daylight	Latitude
4	Rhodes	$14^1/_2$ hours	36°
5	Hellespont	15 hours	40°
5	Byzantium	15 hours	41°

The main inscriptions read:

ΚΛΙΜΑ Δ / Ε / Ε ΔΙΑ ΡΟΔΟΥ / ΒΥΖΑΝΤΙΟΥ / ΕΛΛΗΣΠΟΝΤΟΥ

with the additional information given in a set of abbreviations that would challenge any Byzantinist; they cannot be reproduced here. The selection of three specific latitudes reflects a general tendency to break away from the seven climates and restrict the range for the plates.

Byzantine values for the latitude of Constantinople varied between 41° and 45°: Ptolemy put the site at 43° and some Byzantine astronomers put it in the sixth climate (45°) instead of the fifth (41°), where it belongs.[469] There is no need to assume Islamic influence behind the use of the reasonable value 41° for the latitude of Constantinople here,[470] because the same value is given on other late Hellenistic and Byzantine sundials and universal horary devices.[471] The latitude 36° is standard for Rhodes and also serves the middle of the 4th climate, that is, the middle of the inhabited world.[472] The value 40° for Hellespont is not found on any other Byzantine instruments.

The altitude circles are essentially for each 6° of argument. Philoponos states that the altitude circles can be for each 3°, 2° or 1°, according to the wish of the

468 In the Byzantine Church calendar for the month of June (Mercenier, *La prière des églises de rite byzantin*, p. 59) it is stated that there are 30 days with (approximately) 15 hours of daylight and 9 hours of night. This is another way of saying that Byzantium is in the fifth climate.

469 See King, "Byzantine Astronomy", pp. 117-118.

470 Compare Jones, *Byzantine Manual*, pp. 17-18.

471 See Field & Wright, "Gears from the Byzantines", p. 110, and Ackermann, "Byzantine Universal Sundial", p. 18.

472 On the use of 36° for pedagogic purposes in quite different latitudes see King, *In Synchrony with the Heavens*, pp. 115-116, 904, *etc.*

maker, implying that this choice should depend on the size of the instrument.[473] The further implication is that instruments even larger than this one may have been made in Byzantine workshops.[474]

The altitude circles are for each 3° between the solstitial circles. This feature, which provides increased accuracy in operations with the sun, is such a good idea that it is surprising that it was not widely adopted. No Greek or Islamic texts are known to mention this feature, and only one Islamic astrolabe, from 17th-century Isfahan, is known to display it.[475] It remains to be determined whether this is (a) an early Greek innovation; or (b) an early Islamic (9th- or 10th-century) innovation. Certainly, it is not an innovation made in 11th-century Constantinople or in 17th-century Iran.

The plates are marked for the 12 seasonal hours below the horizon and for each of the 24 equinoctial hours. The former markings are standard. The latter markings, radial lines for each 15°, are unusual on astrolabes.

The back bears two altitude scales divided and labelled for each 5° and subdivided for each 1° and then for each $^1/_3$°. The only markings on the backs of astrolabes that are mentioned in Greek astrolabe texts from the 6th century to the 14th are altitude scales. The subdivisions for each $^1/_3$° are not found on surviving contemporaneous Islamic astrolabes. All other markings that one finds on Islamic and medieval European astrolabes are Islamic innovations of the 9th and 10th centuries.

There is a quadruple shadow-square on the back. The quadruple shadow-square (one quarter would suffice), with its scales devoid of the appropriate markings (1-12 on each scale), is a later addition by somebody who did not care that his sloppy engraving interfered with the elegant inscription around the rim mentioning the commission and the date.

This Byzantine astrolabe possibly represents the culmination of Byzantine astrolabe making. The piece preserves for us the distinctive basic design of Greek astrolabes from previous centuries. The throne, the rete, the bird, the form of the inscriptions, the representation of numbers, are fully Byzantine. But there has been a break with the Greek and early Byzantine tradition of Ptolemaic astronomy: if the star-positions were in the Greek tradition, they would be

---

473 Segonds, *Philoponos sur l'astrolabe*, pp. 146-147, also Gunther, *Astrolabes of the World*, I, p. 62.

474 See King, *In Synchrony with the Heavens*, XIVc, for an early-13th-century Syrian astrolabe (#4050) with a diameter of 56 cm and altitude circles for each 1°.

475 In *Geneva MAH 1985 Exhibition Catalogue*, pp. 292-293, this remarkable feature is overlooked. It is stressed and compared with the 1062 astrolabe in *London Sotheby's 16.10.1997 Islamic Art Catalogue*, pp. 64-69 (lot 24).

hopelessly inaccurate. But they are accurate, and the only explanation is to posit Islamic influence. Indeed, the only definite trace of Islamic influence is in the reasonably accurate star-positions. The increased density of the altitude circles for operations with the sun is of uncertain inspiration. The absurd shadow-squares on the back are a later addition. None of the additional features introduced by Muslim astronomers to the standard astrolabe[476] are featured on this one. And none of the known Byzantine treatises on the astrolabe would lead one to expect any more sophisticated astrolabes.

A fine tribute

As already noted, the Byzantine astrolabe was published in 1926 by the British Classicist O. M. Dalton.[477] His description is a fine piece of scholarship; above all, it is respectful of the object itself. Not only did Dalton know what an astrolabe was, he recorded the details on this one with precision, and interpreted the piece, correctly as it happens, within what was then known about the context of Byzantine astronomy and Byzantine interest in astrolabes. In particular, he pointed out that there is no Islamic influence visible on the piece:~

> "The detailed examination of this astrolabe has shown that it is wholly Byzantine. Except for the incidental mention by Sergius of his Persian descent, we find nothing either in the object itself or in its inscriptions at all suggestive of oriental influence."

Derek Price in 1969 wrote of the piece as "a fine large (Byzantine) astrolabe", mentioning the most important details.[478]

Decades of verbal abuse and neglect, or, misled by a bird

Otto Neugebauer, the leading historian of the exact sciences of the 20th century, had little time for instruments, and his pronouncements on this one are, alas, not to be taken seriously. In his influential essay "On the Early History of the Astrolabe" (1947), based almost exclusively on textual sources, he wrote *contra* Dalton and Gunther, who had quoted him:~[479]

> "It is incorrect to say that this (Byzantine) astrolabe shows no Islamic influence. The little bird above λυρα is the "falling eagle" of the Arabic astrothesy [*sic*] ... . The constructor [*sic*] of this astrolabe calls himself

---

476 On these see the references cited in n. 386.

477 See nn. 55 and 445. Dalton's description is in the tradition of excellence already attested in several 19th-century descriptions of Islamic (but not European) astrolabes, on which see King, *In Synchrony with the Heavens*, vol. II, pp. x-xi.

478 Price, "Portable Sundials in Antiquity", p. 262.

479 Neugebauer, "The Early History of the Astrolabe", p. 249, n. 57. His other pronouncements are to be taken very seriously.

> "Sergius the Persian, holding a consul's rank." Also the stylistic argument is not valid. Two early Arabic astrolabes (about 950 and 990 A.D.) show a design almost identical with the Byzantine one [*sic*]."

This enabled Alain Segonds in 1981 to write in the notes to his edition and translation of the Greek astrolabe text of Philoponos (my emphasis):~[480]

> "L'unique astrolabe byzantin que nous possédions est signé par un certain 'Sergios le Persan, de rang consulaire' ; au moins une des étoiles de (l'araignée de l'astrolabe de 1062) montrent clairement une influence arabe [*sic*] ; enfin, la disposition générale de l'instrument rappelle fortement [*sic*] des instruments analogues mais d'origine islamique, du Xe siècle. Autrement dit, **cet astrolabe** est un bon témoin des échanges islamo-byzantins au XIe siècle, et **ne peut passer**, en rigeur de termes, **pour un astrolabe byzantin** [*sic*!!] : c'est plutôt un astrolabe produit à Byzance. Et sans doute cela est-il dû au fait qu'il n'y avait pas d'autre artisan capable de produire des astrolabes à Byzance [!!]."

Again, Anthony Turner in 1994 wrote:~[481]

> "(The 1062 Byzantine astrolabe) has resemblances with the earliest surviving Islamic instruments, shows signs of Arabic influence [*sic*] (use of a bird's head for the Vega star-pointer, incorporation of the shadow square), and carries an inscription stating that it was designed by Sergius the Persian protospatharios, consul and savant. It is therefore [!] an instrument which shows the Greek tradition already incorporating novelties introduced by Islamic scholars. The fact that Sergius felt it necessary to have a four line poem engraved on it to explain what it is [*sic*] may suggest that it had become unfamiliar in 11th century Constantinople [!!]."

In the catalogue of the millennial exhibition at the National Maritime Museum,[482] where the 1062 astrolabe was on display, we read the following fiction:~

> "There is only one known example of an astrolabe that might be called [!!] Byzantine. The text on the instrument is Greek, but the shape of the rete is very similar to Hispano-Moorish astrolabes [!!], with tower-shaped star pointers and the celestial equator appearing as a band covering only one quarter of the rete. This supports the idea that, during the 11th century, 'Byzantine' instruments would have been heavily influenced – if not actually constructed [!!] – by Persian or Arabic craftsmen [!!]."

---

480 Segonds, *Philopon sur l'astrolabe*, pp. 87-88.

481 A. Turner, *Mathematical Instruments*, p. 64.

482 *Greenwich 1999-2000 Exhibition Catalogue*, p. 104 *ad* no. 094.

Sadly, this kind of misinformation, and more besides, has accumulated over the years,[483] and, as a result, the Byzantine astrolabe has not fared well in modern writings on the history of Byzantium, the history of Greek astronomy, or the history of Byzantine astronomy. It is even ignored in works on Byzantine astrolabe texts! I trust that henceforth it will be treated with the respect that it deserves, not least now that I have re-established its Byzantine credentials, and Berthold Holzschuh has established its connection with Bessarion and Regiomontanus!

---

483 For more, see, for example, Field & Wright, "Gears from the Byzantines" (1985), pp. 114-115. Here the piece is said to resemble a certain 12th-century Iranian astrolabe (#1177). In fact, it has nothing whatsoever in common with that piece, which is now illustrated in King, *In Synchrony with the Heavens*, figs. X-4.1.2 and 4.4.2.

Appendix D – The surviving instruments from 15th-century Vienna

Vienna was the leading centre of astronomy and astronomical instrumentation in Europe in the 15th century. This renown is due mainly to four astronomers who worked there in succession: Johannes von Gmunden, Georg Peuerbach, Regiomontanus, and the enigmatic Benedictine monk, Hans Dorn. Over 40 instruments from this milieu and related workshops have survived, including some 20 astrolabes; the author published a list in Vienna in 1994.[484] In addition to the astrolabes, we have horary dials and quadrants, horizontal sundials, and one ring sundial, one albion with a celestial map and a universal horary dial, one torquetum, and one celestial sphere. Of the 20 astrolabes, some 10 pieces have the same or a similar rete design as the 1462 astrolabe. These we label "Regiomontanus-type" astrolabes, although only the 1462 piece is signed by him. This does not mean that Regiomontanus made them with his own hands, or even engraved their astronomical markings. Some of the later ones were possibly made in Nuremberg, where Regiomontanus settled during 1471-75, rather than in Vienna. Others were made in a workshop in Schmalkalden. A significant subgroup of these "Viennese" instruments were made by Hans Dorn; some of these reflect technical innovations proposed by Regiomontanus but not found on the "Regiomontanus-type" astrolabes (not least because they are relatively small). The instruments of Hans Dorn are quite different in style, design and size from those of his predecessor. The prolific instrument-maker Georg Hartmann of Nuremberg, *ca.* 1520-1530, adopted Regiomontanus' distinctive rete design, and even his star positions.[485]

---

484 This preliminary list is in King, "Astronomical Instruments between East and West", pp. 183-188. Many of the astrolabes were identified using the sparse and often unreliable one-line entries in Price *et al.*, *Astrolabe Checklist* (see n. 62 above).

485 That the rete design was copied by Hartmann is stated already in Price, "The First Scientific Instrument of the Renaissance", p. 26, alas without citing any other 15th-century astrolabes from Vienna. (There is an unfortunate and very misleading remark in van Cleempoel, *Astrolabes at Greenwich*, p. 146, that "(Hartmann's) rete pattern can be associated with a group of astrolabes that emerged in Vienna near the end [!] of the fifteenth century, with Hans Dorn as the central figure [!].")

On Hartmann see the useful study Klemm, *Georg Hartmann*. On some astrolabes by him see Gunther, *Astrolabes*, II, pp. 436-440; King, "Die Astrolabiensammlung in Nürnberg", II, pp. 593-597; and Webster & Webster, *Western Astrolabes in Chicago*, pp. 53-57. For the Marian decoration of his astrolabes see n. 498 below. For his treatise on astrolabe construction see now Lamprey, *Hartmann's Sundial and Astrolabe Manual*, pp. 241-289. Far more interesting is his set of drawings of over 80 different varieties of instruments known to him, which are mentioned in Zinner, *Astronomische Instrumente*, pp. 359 and 360-361, and listed on pp. 362-368. These are not as well known as they should be: a facsimile edition with a commentary would be a very useful

Lest the 1462 astrolabe of Regiomontanus be thought to have appeared out of the blue, I present here a list of all of the known instruments from the German-speaking world of the 15th- and early 16th centuries. This is based mainly on my 1994 list of 40-odd instruments from 15th-century Vienna (which included one fake, item VII-2 below), and it includes all of the newly-identified pieces. The objects from the 14th century and before, and from the early 16th century, are included for the sake of completeness, not least since they are so few, but also since some of them are mentioned in the main text and notes to this study. The corpus comprises essentially all known instruments from the German-speaking world before the time of Georg Hartmann. Various medieval astrolabes (from the 13th and 14th centuries) may eventually be added to this list, particularly some with quatrefoil design; for the present, it is often not possible to assign them to any particular region. There are, for example, several medieval astrolabes with a plate for the latitude of Cologne, but these seem to be English or French productions.

In the past fifteen years, in addition to the 1462 astrolabe of Regiomontanus, four out of the ten other "Regiomontanus-type" astrolabes from "the 15th-century Vienna school" and its offshoots have been published in detail:~

- The 1457 astrolabe associated with Georg Peuerbach (#549), preserved in the Germanisches Nationalmuseum, Nuremberg (V-1 below).
- An undated piece from *ca.* 1460 (#256) with an *organum Ptolemaei* on the back, auctioned at Christie's of London in 1998, and now in a private collection (V-5).
- An undated astrolabe (#620) belonging to the Deutsches Museum, Munich (V-7).
- The astrolabe dated 1481 (#530), possibly from Nuremberg (?), belonging to the Observatory at Strasbourg (V-8).

In addition, two of the other astrolabes (#4505 and #4544) identified in 1994 as being related to the Vienna school (but with two different designs, both different from the Regiomontanus-type) have now also been published in detail (see X-3 and X-6 below).

I	***Uncertain date, possibly 11th century***
1	An orthographic projection of the celestial sphere for latitude *ca.* 48°, corresponding to the 7th climate, incised on a circular protrusion atop a marble column, with, on the other side, an

---

to the literature on Renaissance instrumentation. Some extracts are shown in Klemm, *Georg Hartmann*, p. 19, and King, *In Synchrony with the Heavens*, fig. XIIb-10e.

image of a young man kneeling and looking upwards as well as an inscription mentioning Aratus – Regensburg, Stadtmuseum – see Zinner, "Regensburger Lehrgerät", and *idem*, *Astronomische Instrumente*, pp. 167-168 and 589-590, pl. 47, fig. 1, and pl. 48, fig. 1; King, *In Synchrony with the Heavens*, VIII, fig. 3.3 and pp. 868-869, and XIIb-15 and app. B; and the detailed discussion in Wiesenbach, "Astrolab und Astronomie im 11. Jahrhundert", pp. 135-142.

II ***Uncertain date, but preceding 1444***

These three instruments were acquired in Nuremberg by Nicholas of Kues (Cusanus) in 1444.

1 (#589) An astrolabe mater and set of plates serving the seven climates, possibly from the 12th or 13th centuries – Bernkastel-Kues, Cusanus-Stift – published in detail in J. Hartmann, "Die astronomischen Instrumente des Kardinals Nikolaus Cusanus", pp. 40-42, without illustrations; on the latitudes used for the plates see King, *The Ciphers of the Monks*, pp. 412 and 415.

2 (#8502) A torquetum – Bernkastel-Kues – published in detail *ibid.*, pp. 17-28 and pls. 1-3.

3 (#6501) A celestial sphere – Bernkastel-Kues – published in detail *ibid.*, pp. 28-40 and pls. 4-11.

III ***An early-14th-century astrolabe from Einbeck***

1 (#2027) An elaborate quatrefoil astrolabe made for Ludolfus de Scicte (modern Sickte near Brunswick), treasurer of the Stiftskirche in Einbeck, thus datable 1322-1342; diameter 24.9 cm – Cracow, Jagiellonian University Museum, inv. no. 4043(?)-41/V – see Birkenmajer, *Martin Bylica z Olkusza*, pp. 101-106 (in Polish), and the German summary by Karlinski, especially pp. 109-110; Zinner, *Regiomontanus*, pl. 42, figs. 78-79 (front and back); *idem*, *Astronomische Instrumente*, pp. 535-536 (brief description); information on Ludolfus in Härtel, "Ludolfus Borchdorp de Brunswik", pp. 113-116; discussion of quatrefoils in King, *In Synchrony with the Heavens*, XVII-3 and fig. 3.4; detailed description in *idem*, "An Astrolabe from Einbeck *ca.* 1330" (in press).

IV ***15th century, not necessarily Vienna***

1 (#555) An astrolabic plate for latitude *ca.* 46°30′ (locality uncertain) with an earlier horizon for latitude 49°, and a distinctive U-shaped frame on the rete, dated 1468; diameter 28.0 cm –

Nuremberg, Germanisches Nationalmuseum, inv. no. W5 – Gunther, *Astrolabes*, I, pp. 280-281 (no. 137, with an illustration of "The Astrolabe of Regiomontanus"); published in detail in King, "Die Astrolabiensammlung in Nürnberg", II, pp. 589-592 (no. 1.77), also I, p. 105; on the shadow-square and the rete design see King, *In Synchrony with the Heavens*, captions to fig. XIIa-B6 and XIIIa-10.9c; see also Appendix F below.

V ***15th-century astrolabes, mainly from Vienna, although some of the later pieces may stem from other locations, including Schmalkalden (two pieces) and perhaps Nuremberg***

All of these astrolabes have a similar rete design, similar decoration on the thrones, and, with the exception of items 3 and 4, similar Gothic engraving. For some of the thrones see **Fig. D1**, and for an overview of the instruments and their plates see **Tables 10** and **11**.

Astrolabes probably to be associated with Georg Peuerbach

1 (#549) An astrolabe dated 1457, Gothic engraving as on #644, with a lunar volvelle on the back; diameter 12.9 cm – Nuremberg, Germanisches Nationalmuseum, inv. no. WI 129 – see Zinner, *Astronomische Instrumente*, pp. 150 and 465; *Linz 1990 Exhibition Catalogue*, II, p. 32 (no. 29); King & Turner, "Regiomontanus' Astrolabe for Bessarion", pp. 189-190, *etc.*, and figs. 11-13; and the detailed description in King, "Die Astrolabiensammlung in Nürnberg", II, pp. 581-586 (no. 1.75). See also Samhaber, *Die Zeitzither*, pp. 196-197, for some superb images of the front and back, the latter unfortunately backwards. See now Figs. **D2a-b** below.

2 (#644) Undated rete and plates, rete pattern more developed than the others, engraved in the same Gothic hand as #549, fitted to a 14th-century Italian mater; diameter 25.2 cm (mater) – Cracow, Jagiellonian University Museum, inv. no. 4042-40/V – see Zinner, *op. cit.*, p. 296; King & Turner, "Regiomontanus' Astrolabe for Bessarion", pp. 190, *etc.*, and fig. 30; and the detailed description in Burczyk-Marona, "Astrolabium Brozka" (in Polish with an English summary on pp. 113-115).

Table 10

The “Regiomontanus-type” astrolabes: location, date, size, number of stars, date of vernal equinox

(PC = private collection; PLU = present location unknown; size = diameter / thickness in mm; equinox = date in March)

No.	Location	IIC no.	date	size	stars	equinox
1	Nuremberg	#549	1457	129/6	30	10⅓
2	Cracow (rete only)	#644	–	–	34	–
3	Paris PC < Greppin	#452	–	137/7	30	11
4	**England PC < Christie’s**	**#640**	**1462**	**115/4**	**30**	**10½**
5	PLU < Paris PC	#256	–	168/9	27	11
6	Darmstadt	#592	1466	144/7	27	10½
7	Munich	#620	–	132/6	27	10⅓
8	Strasbourg	#530	1481	170/4	27	13
9	Cracow	#4501	1482	152/6	27	12
10	Paris	#516	–	132/5	27	10¼
11	New York	#545	–	133/7	27	10

*****

Table 11

Latitudes represented on the plates of the “Regiomontanus-type” astrolabes

(Note: Georg Hartmann wrote in 1527 that all the astrolabes made by Regiomontanus that he had seen had plates for latitudes 39°, 42°, 45°, 48°, 51° and 54° (see the text to n. 507). These latitudes are printed in bold font.)

No.	Location	IIC no.	latitudes								
1	Nuremberg	#549	32	37	**48**	**51**					one plate missing
2	Cracow	#644	**48**	49	50	**51**	52	67			
3	Paris PC	#452	32	35	37	**42**	**45**	**48**	**51**		
4	PC	#640	**39**	**42 R***	**45**	**48**	**51**	**54**			
5	PLU	#256	30	36	**39**	**42**	**45**	**54**			one plate missing
6	Darmstadt	#592	36	**39**	**42**	**45**	**48**	**51**	**54**	57	
7	Munich	#620	**39**	**42**	**45**	**48**	**51**	**54**			
8	Strasbourg	#530	**42**	**45**	**48**	**51**					
9	Cracow	#4501	**42**	**45**	**48**	**51**					
10	Paris	#516	0	3	6	9	12	**51**	**54**	90	
11	New York	#545	**39**	**42**	**51**	**54**					universal markings**

* ROMAE ** See King, *In Synchrony with the Heavens*, fig. X-5.2.7.

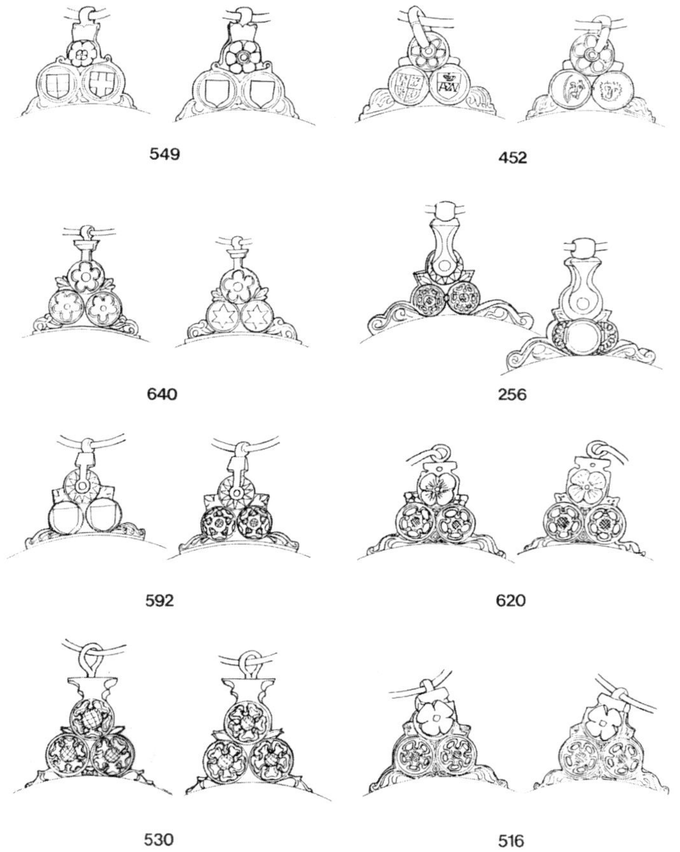

Fig. D1:~ The front and back of the thrones on various astrolabes of the "Regiomontanus type". [Drawn by Helen Waldmann *ca*. 1990.]

Figs. D2a-b:~ The front and back of the 1457 Nuremberg astrolabe. This piece has been known in the literature from the 1930s onwards, and it would have been an obvious choice for an instrument to compare with the 1462 astrolabe. It is usually associated with Peuerbach, mainly because Regiomontanus was only 21 years old at the time. The design of the rete is more complex than that of the 1462 piece. [Courtesy of the Germanisches Nationalmuseum, Nuremberg.]

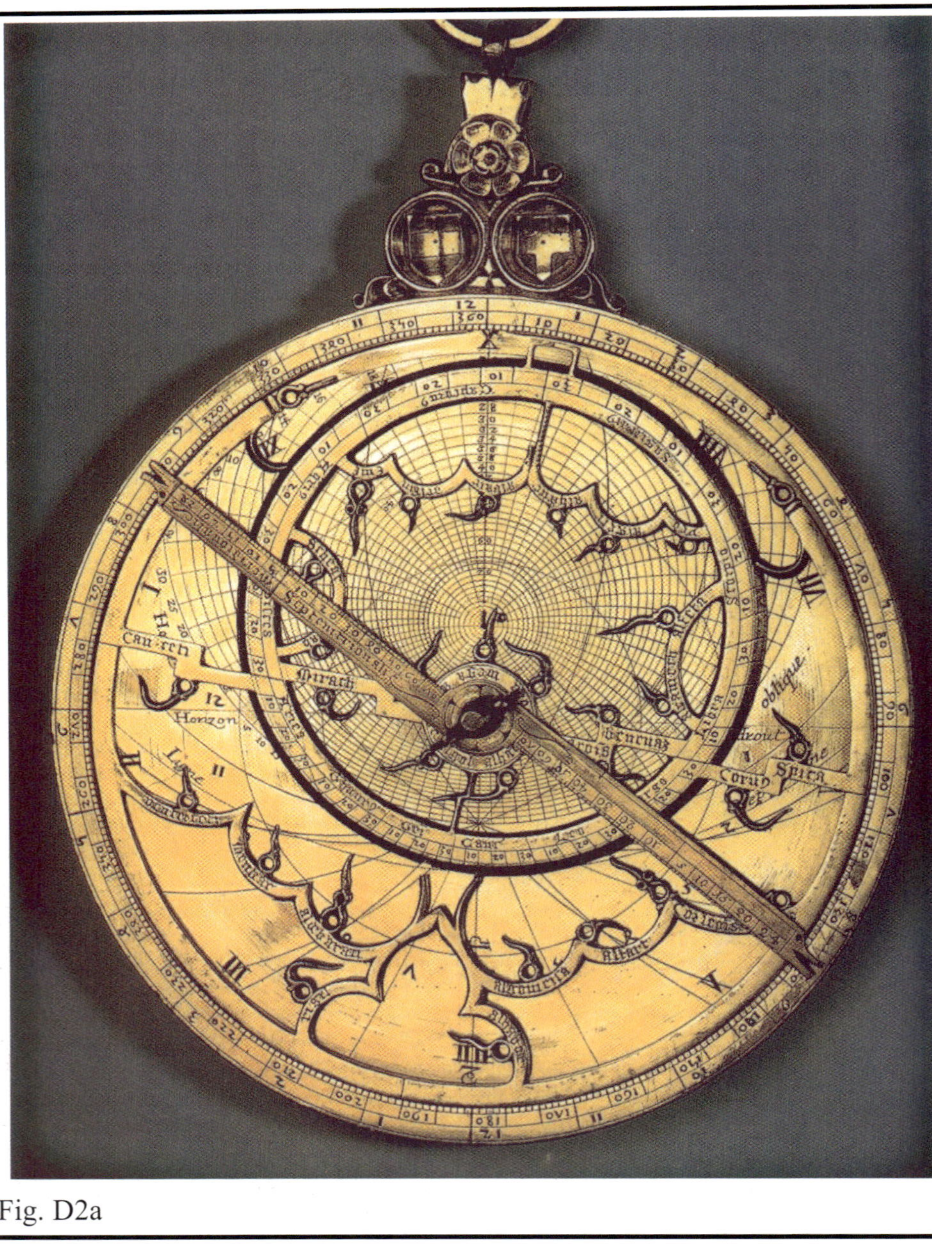

Fig. D2a

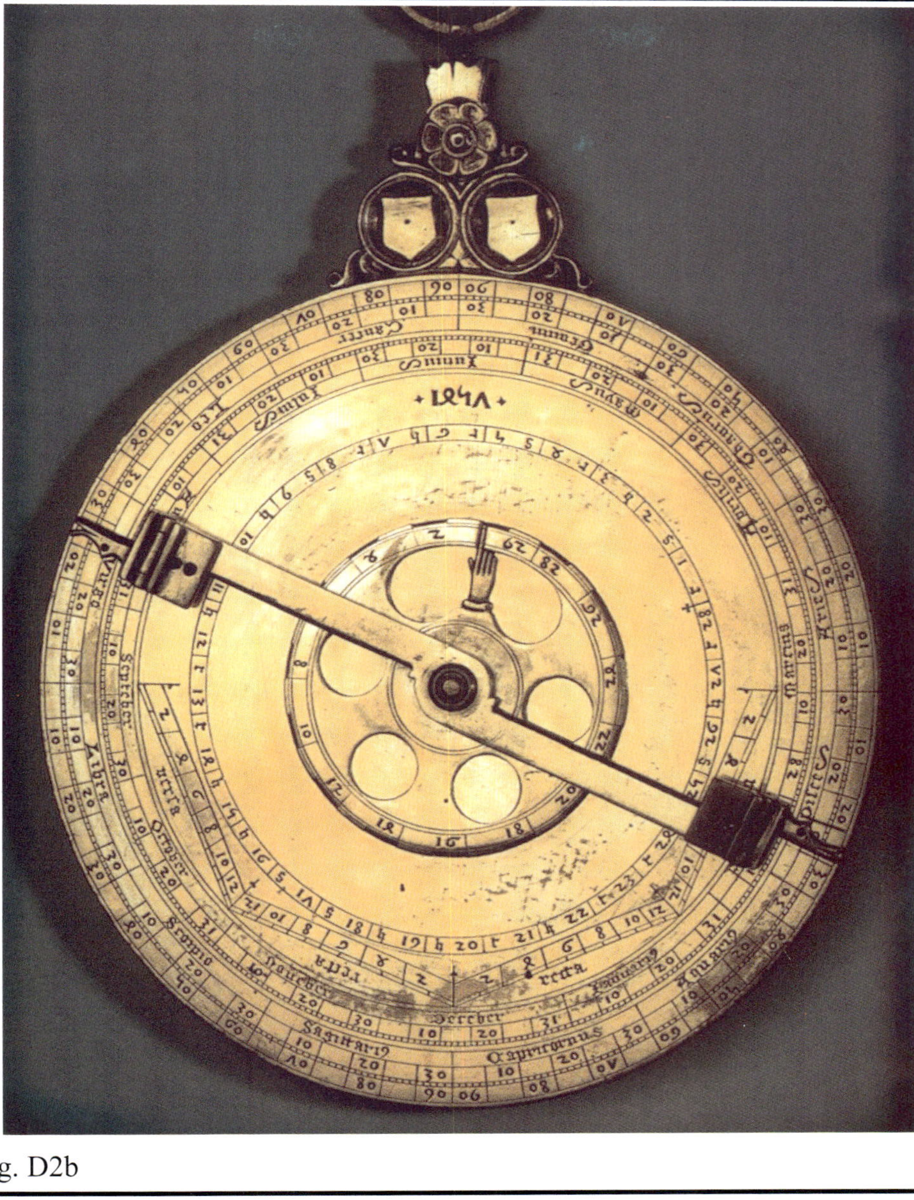

Fig. D2b

An astrolabe attributable to Regiomontanus

3 (#452) Undated, Gothic script as on #549 on the plates and back and Antiqua as on #640 on the rete, datable to *ca*. 1460; diameter 13.7 cm – Paris, private collection – see *Greppin Catalogue*, p. 32 (no. 3122); and King & Turner, "Regiomontanus' Astrolabe for Bessarion", pp. 189-190, *etc*., and figs. 14-17, also **Fig. A2** above.

The astrolabe signed by Regiomontanus

4 (#640) Astrolabe dated 1462, signed by Ioannes (Regiomontanus), dedicated to Cardinal Bessarion, Antiqua script as on the rete of #452, with an *organum Ptolemei* on the back; diameter 11.6 cm – England, private collection – see Zinner, *Regiomontanus*, p. 137; *idem*, *Astronomische Instrumente*, p. 481; and the more detailed descriptions in Price, "The First Scientific Instrument of the Renaissance"; G. Turner, "The Regiomontanus Astrolabe"; King, "Die Astrolabiensammlung in Nürnberg", pp. 586-589 (no. 1.76); *idem*, "Schweinfurter Instrumente", pp. 366-370 (no. 177); and King & Turner, "Regiomontanus' Astrolabe for Bessarion", *passim*.

Related astrolabes with mainly Gothic script

5(#256) Undated astrolabe (with some early-16th-century additions on the back of the movable plate), Gothic script resembling that on #549, with an *organum Ptolemei* on the back, and a compass on the back of the throne; owners' marks dated 1535 and 1537 on the back of the *organum*; diameter 16.8 cm – formerly in the collection of Robert Seligmann, Paris; present location unkown (private collection) – see Gunther, *Astrolabes*, II, p. 434 (no. 256); King & Turner, "Regiomontanus' Astrolabe for Bessarion", pp. 189-190, *etc*., and figs. 18-19; the detailed description in G. Turner, "15th-Century German Astrolabe", and **Figs. D3a-b** below.

Fig. D3a-b:~ This astrolabe (#256) was probably made in Vienna shortly before the 1462 piece. One can see the void above and below the *organum Ptolemaei* on the back which might have inspired Regiomontanus to add an angel and an epigram in the available space on his new astrolabe for Bessarion. [Photos courtesy of Christie's, London.]

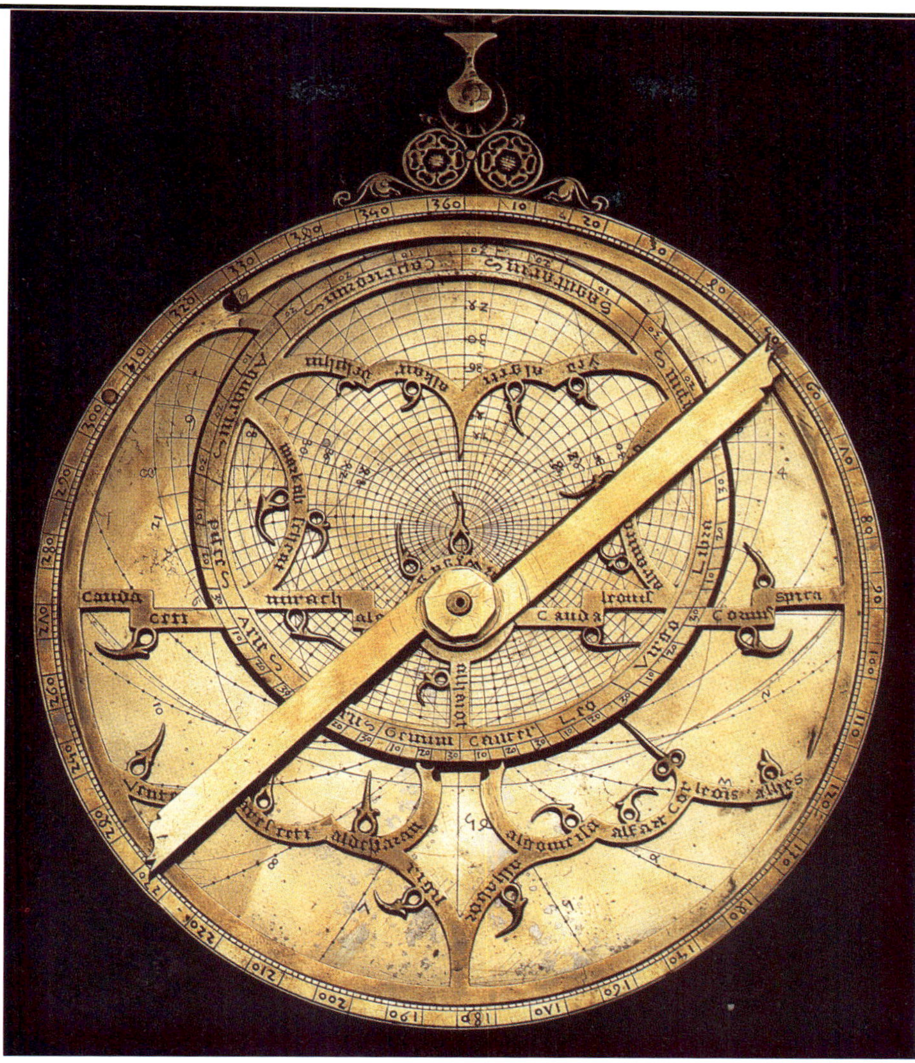

Fig. D3a.

6 (#592) Astrolabe dated 1466, Gothic script; diameter 14.4 cm – Darmstadt, Hessisches Landesmuseum, inv. no. Kg. 25:15 – unpublished; see King & Turner, "Regiomontanus' Astrolabe for Bessarion", pp. 189-190, *etc*., and fig. 20; and Debauvais & Befort, *Astrolabes de Strasbourg*, p. 165.

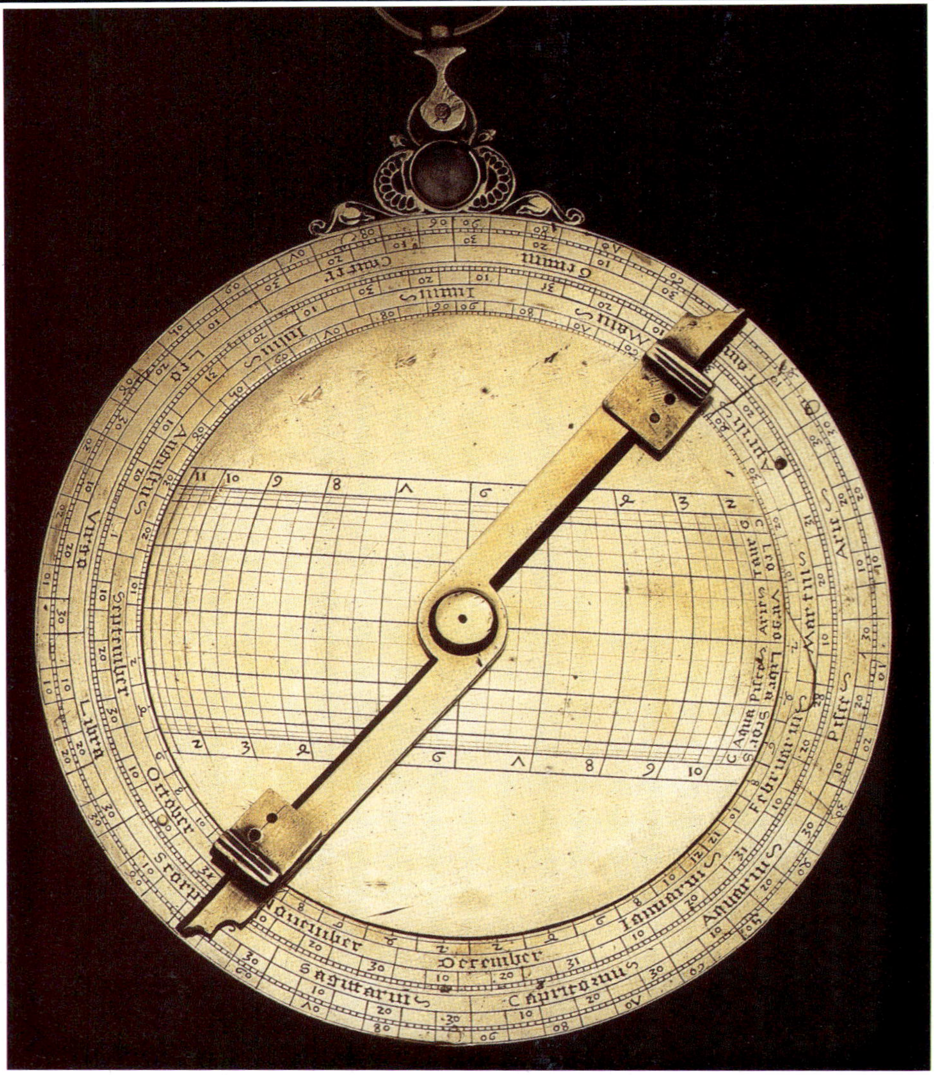

Fig. D3b

7 (#620) Undated astrolabe, Gothic script; diameter 13.2 cm – Munich, Deutsches Museum, inv. no. 70678 – see Zinner, *Astronomische Instrumente*, pl. 41, fig. 1 ("Gotisches Astrolab"); King & Turner, "Regiomontanus' Astrolabe for Bessarion", pp. 189-190, *etc.*, and figs. 21-22; and the detailed description in Stautz, *Die Astrolabiensammlungen in München*, pp. 201-216 (no. 5).

8 (#530) Astrolabe dated 1481, Gothic script as on #4501; diameter 17.0 cm – Strasbourg, Musée Historique, inv. no. IL 74 – see Poulle, "Fabrication des astrolabes", p. 118 (illustrations); Zinner, *Astronomische Instrumente*, pp. 150 and 421 (*sub* Arnold Kunig); *Strasbourg 1987 Exhibition Catalogue*, (illustrations); King & Turner, "Regiomontanus' Astrolabe for Bessarion", pp. 189-190, *etc.*, and figs. 23-24; and the detailed description in Debauvais & Befort, *Astrolabes de Strasbourg*, pp. 130-140, and **Fig. D4** below.

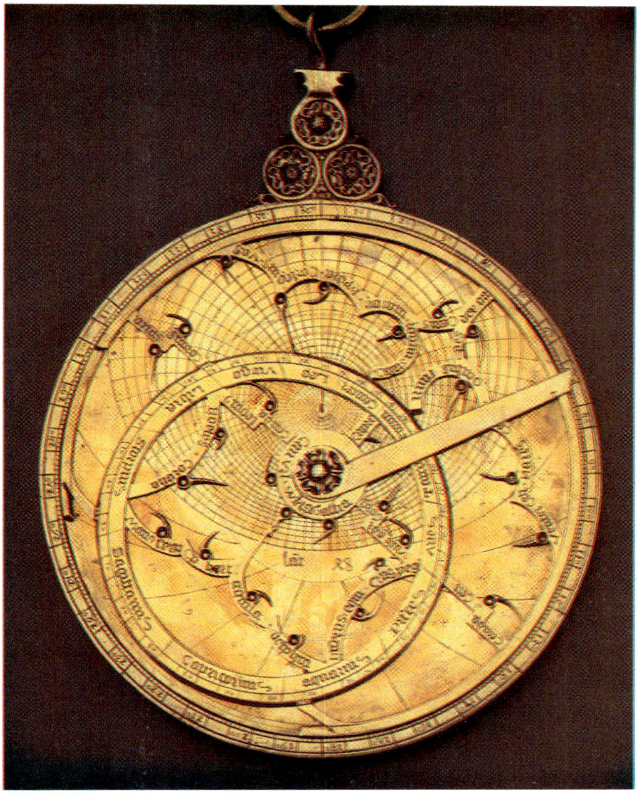

Fig. D4:~ A later example of the "Regiomontanus-type" astrolabes dated 1481 (#530). Such 15th-century astrolabes astrolabes clearly influenced the early-16th-century Nuremberg instrument-specialist Georg Hartmann, who copied their rete design and their star-positions. See **Fig. A5** for the back of this piece. [Image courtesy of Francis Debeauvais, Strasbourg.]

9 (#4501) Astrolabe signed "SWM", dated Schmalkalden, 1482, Gothic script as on #530; diameter 15.2 cm – Cracow, Muzeum Narodove, Czartoryski Collection, inv. no. XIII/753 – unpublished; see King & Turner, "Regiomontanus' Astrolabe for Bessarion", pp. 190, *etc.*, and figs. 25-26; and King, "Schweinfurter Instrumente", pp. 372-373. See also the six astrolabes of Hans Bamann listed below (IX).

Astrolabes with unusual plates and three different contemporary styles of script (both possibly by the same individual)

10 (#516) Undated astrolabe, same three hands as #545; diameter 13.2 cm – Paris, Conservatoire National des Arts et Métiers, inv. no. 3882/1 – unpublished; see King & Turner, "Regiomontanus' Astrolabe for Bessarion", pp. 190, *etc.*, and figs. 27-29.

11 (#545) Undated astrolabe, same three hands as #516; diameter 13.3 cm – New York, Columbia University, inv. no. 27-256 – unpublished; see King & Turner, "Regiomontanus' Astrolabe for Bessarion", pp. 190, *etc.*

VI Miscellaneous instruments (albions, quadrants, ring-dials, cylindrical dials) from 15th-century Vienna

The albion of the English astronomer Richard of Wallingford (d. 1336) was a sophisticated device essentially for calculating planetary longitudes and eclipses (see North, *Richard of Wallingford*, I, pp. 245-401, and II, pp. 127-286). In Vienna before 1442, a commentary was prepared by Johannes von Gmunden, and in 1455 "a much debased version" (North) was prepared by Regiomontanus. Compare Zinner, *Astronomische Instrumente*, pp. 32-33, 37-38, 479, and North, *op. cit.*, II, pp. 271 and 275, as well as North's article "Richard of Wallingford" in the *Dictionary of Scientific Biography*. The sole surviving albion (item 1 below) is from mid-15th-century Vienna but it has never been properly studied. The device should be compared with the equatorium described in the newly-rediscovered, but still unpublished, treatise of the 10th-century Baghdad astronomer Abû Ja'far al-Khâzin (see King, *In Synchrony with the Heavens*, X-5.3). And who brought this piece from Vienna to Rome?

The ring-dial was described by both Peuerbach and Regiomontanus (Zinner, *Astronomische Instrumente*, p. 120). The device was known already in 14th-century Egypt (Charette, *Mathematical Instrumentation in 14th-Century Egypt and Syria*, pp. 115, 142-144, 323-324). The sole surviving example from before 1500 is from mid-15th-century Vienna (item 2) but it has never been published.

Three of these instruments are in ivory, and it was probably from Vienna that the inspiration came for the 16th-century Nuremberg dials described in Gouk,

Ivory Sundials of Nuremberg. (Perhaps the inspiration for the Vienna instruments in ivory came from the Islamic world. Both the horary quadrant for a specific latitude and the cylindrical sundial (items 3 and 4) are Islamic inventions from the 9th century, and, in the 14th century, highly sophisticated instruments were made in ivory in Cairo and Damascus – see King, *In Synchrony with the Heavens*, figs. V-9.2 and X-6.1.3.)

1 (#8507) An albion datable to Vienna, *ca.* 1450-60, fitted with a stereographic celestial map and a universal horary dial, Gothic script as on most astrolabes in V above; diameter 32.7 cm – Rome, Osservatorio astronomico, inv. no. 147/661 – unpublished; see North, *Richard of Wallingford*, III, figs. XXIII-XXIV (illustrations of universal horary dial and planetary scales); G. Turner, ed., *Strumenti*, p. 70 (illustrations of the celestial map and the back, fitted with the removable plate); King, *Mecca-Centred World-Maps*, fig. 8.5.2 on p. 319 (illustration of celestial map); *idem*, *In Synchrony with the Heavens*, fig. XIIb-10h (illustrations of universal horary dial beneath the removable plate); and Calisi, *Strumenti del Museo astronomico di Roma*, pp. 146-147 (no. 294), without illustrations.

2 (#7525) An unsigned, undated ring-dial, from Vienna, *ca.* 1450-60, Gothic script as on most astrolabes in V above, script in black and red; diameter 14.3 cm – Chicago, Adler Planetarium, inv. no. M307 – see Zinner, *Astronomische Instrumente*, pp. 480-481.

3 (#5510) An unsigned ivory horary quadrant dated 1438, marked "aeiou", indicating that it was dedicated to Kaiser Friedrich III (see n. 255 above); radii 9.0/8.0 cm, thickness 5 mm – Vienna, Kunsthistorisches Museum, inv. no. 166 – see Zinner, *Astronomische Instrumente*, pp. 322-323 and pl. 46-1-2; *Linz 1990 Exhibition Catalogue*, II, p. 36 (no. 34) (brief account), with a colour plate of the back only; Samhaber, *Die Zeitzither*, pp. 27 and 81; King, "Astronomical Instruments between East and West", pp. 161 and 187, and *idem*, *In Synchrony with the Heavens*, X-6.3 and fig. 6.3.3, and also **Figs. D5a-b** below.

4 (#7524) An unsigned ivory cylindrical dial dated 1455, Gothic script as on the ivory quadrant of 1438 and most of the astrolabes in V above, script in black and red; height *ca.* 15.0 cm – Munich, Bayerisches Nationalmuseum, inv. no. Phys. 64 – see Zinner, *op. cit.*, p. 123 and fig. 4 in pl. 46; Gouk, *Ivory Sundials of Nuremberg*, p. 28 and fig. 26, for an illustration of one side (in

both associated with Nuremberg rather than Vienna, on account of an "n", alas not visible in the illustrations).

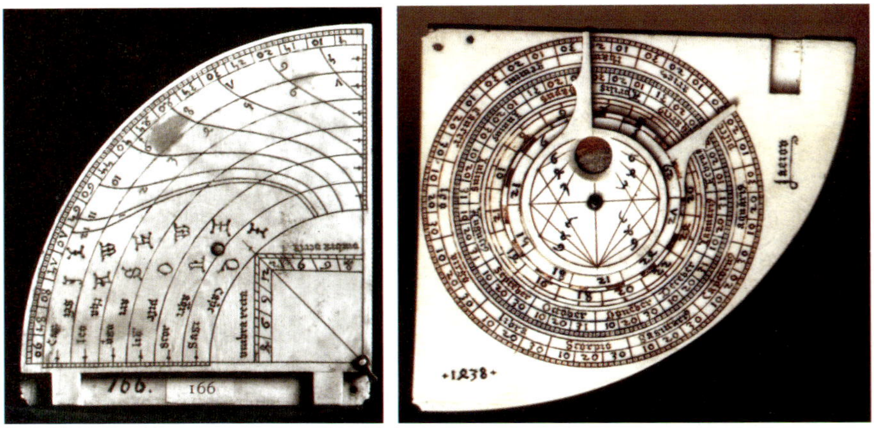

Figs. D5a-b:~ The front and back of the ivory quadrant dated 1438 (#5510). [Courtesy of the Kunsthistorisches Museum, Vienna.]

VII Portable folding sundials from Vienna, 1451-*ca.** ***1465

With the exception of a couple of late-19th-century fakes, the following instruments are probably all from the same workshop as the astrolabes listed above. It is traditional to divide them as productions of Peuerbach (d. 1461) and Regiomontanus; they were, however, probably all made by the same person, that is neither Peuerbach nor Regiomontanus. On most of these see Zinner, *Instrumente*, pp. 464-465 and 481. Three examples are illustrated in *Linz 1990 Exhibition Catalogue*, pp. 35, 37, and 39. Five pieces are illustrated together in Samhaber, *Die Zeitzither*, pp. 188-189.

The "Regiomontanus-type" sundials are all fitted with a magnetic compass and an hour-scale on the base, and all were originally fitted with a string-gnomon which enables the lid to be open so that the string is aligned towards the celestial pole for a specific latitude. They have all sorts of decoration and inscriptions appropriate to Christian and, in particular, Marian devotion, as well as contemporaneous Germanic folklore (notably, the "wilde Männer").[486] The 1453 piece in the British Museum is a fake by the highly dubious workshop of

[486] On these see Bernheimer, *Wild Men of the Middle Ages*, richly documented and illustrated, but inevitably devoid of any references to Viennese sundials, and also *Mainz 2000 Exhibition Catalogue*, pp. 45-51 (tapestries).

Frédéric Spitzer (1815-1890). It is an established fact that some of the instruments amongst the 3369 items from his collection that were auctioned at Paris in 1893 were copies of earlier instruments or had missing parts replaced.[487] The London and Munich pieces (2 and 5 below) stem from that sale;[488] the former is a copy and the latter has an absurd appendage. The second Vienna piece (9 below) is possibly also a fake.

All of these instruments, including the fakes, merit a special study, which should include the similar dials on various instruments of Hans Dorn, on which see already Ward, "The Earliest String-Gnomon Dials" and now items 11-12 below.

1 (#7511) Dated 1451, marked "aeiou", indicating that it was dedicated to Kaiser Friedrich III (see n. 255 above); dimensions (length/width) 79/58 mm – Innsbruck, Tiroler Landesmuseum Ferdinandeum - Zeughausmuseum, inv. no. U5 – see Zinner, *Astronomische Instrumente*, pp. 464-465; *Linz 1990 Exhibition Catalogue*, p. 35 (no. 33, illustrated); and Samhaber, *Peuerbach und Kaiser Friedrich III.*, pp. 72-73 (illustrated).

2 (#7512) Dated 1453 (**FAKE**); dimensions 79/59 mm – London, British Museum, inv. no. 93 6-16 9 – see *Spitzer Collection Catalogue*, A, lot 2787 (not seen); *Spitzer Collection Catalogue*, B, p. 87, no. 28 (misdated to 1473 and mislabelled as "travail italien (?)"), and pl. 1; Zinner, *op. cit.*, p. 465; Ward, "The Eariest String-Gnomon Dials", pp. 800-801; *idem*, *Catalogue of Instruments in the British Museum* (1981), pp. 21-22 (no. 8) and pl. IIa, where there is only a hint of any problem with the piece: "there is some slight doubt as to the authenticity of these dials", referring also to six others. Already in 1956, Derek Price had cast aspersions on it as an "early" fake: see his "Fake Antique Scientific Instruments" (1956), p. 381 (discussion of this piece and one in Munich (item 5 below) with an unfortunate and clearly non-original appendage). So when the piece was presented in the 1981 *Catalogue* without a proper discussion of its authenticity, he reacted accordingly: see his "Instruments in the British Museum" (1982), p. 132, cited below. **This London piece is**

---

487 On the infamous workshop used by Spitzer see A. Turner, "Paris, Amsterdam, London", pp. 33-34. (For another Spitzer copy see the third instrument mentioned in n. 542 below.) Another source of fake instruments in modern collections was the Mensing collection: see Deiman, "Imitations amongst the Mensing Instruments".

488 I have not been able to consult the main *Spitzer Collection Catalogue*, listed as A. Photocopies of extracts from the shorter version, listed as B, are available to me.

the only one that is definitely a fake, and, inevitably, it hails from the highly dubious Spitzer Collection (see n. 487). In 1956, Price had ventured to suggest that another one, preserved in Munich (item 5 below), might also be fake, claiming that both were of "much cruder and apparently more modern construction" than three others in Innsbruck, Vienna and Nuremberg. In 1982, Price announced that: "it is probable that all (or nearly all?) of the similar pieces are dealer-produced fakes that must now be expunged from the prominent place they once occupied as amongst the earliest examples of the first German instrument workshops".

3 (#7513) Dated 1455; dimensions 80/58 mm – Graz, Landesmuseum Johanneum, inv. no. 4525 – see Zinner, *op. cit.*, p. 465; *Graz 1989 Exhibition Catalogue*, p. 9, no. 1 (illustrated); *Linz 1990 Exhibition Catalogue*, p. 37 (no. 35, illustrated); and Samhaber, *Peuerbach und Kaiser Friedrich III.*, pp. 72-73 (illustrated).

4 (#7514) Dated 1455; dimensions 79/59 mm – Paris, Musée de Cluny, inv. no. 18536 – unpublished; see Zinner, *op. cit.*, p. 465 (confused); no information currently available.

5 (#7515) Dated 1456; length 70 mm, with an absurd over-long pointer attached later – Munich, Bayerisches Nationamuseum, inv. no. Phys. 22 – see *Spitzer Collection Catalogue*, A, lot 2788 (not seen); *Spitzer Collection Catalogue*, B, p. 87, no. 29 (misdated 1476, mislabelled as "travail italien (?)"), and pl. 1; Zinner, *op. cit.*, p. 465; von Bassermann-Jordan, *Uhren*, fig. 29 on p. 45 (no comment about authenticity), and *idem*, *Old Clocks and Watches*, fig. 29 on p. 43 ("authenticity unproven", perhaps on account of the hapless additional pointer).

6 (#7516) Dated 1463; dimensions 44/29 mm – Vienna, Kunsthistorisches Museum, inv. no. 122 – see Zinner, *Astronomische Instrumente*, p. 481, pl. 25.2; and *idem*, *Regiomontanus*, figs. 62-63 on pl. 35; also Wattenbach, "Regiomontanus und die astronomischen Instrumente seiner Zeit", fig. 7.

7 (#7517) Dated 1465, dimensions 55/42 mm – Pécs (H), private collection – I have copies of correpondence dated 1987 between Lajos Bartha, Budapest, and Dr. Meinrad Pizzinini, Director of the Zeughaus-Museum in Innsbruck; Bartha was of the opinion that this is a copy of a 15th-century piece; no further information currently available.

8 (#7518) Datable *ca.* 1465, dedicated to Pope Paul II; dimensions 62/44 mm – Nuremberg, Germanisches Nationalmuseum, inv. no. WI 7 – see **Fig. D6** and Zinner, *Astronomische Instrumente*, p. 481; *idem*, *Regiomontanus*, figs. 64-65 on pl. 37; Wattenbach, “Regiomontanus und die astronomischen Instrumente seiner Zeit”, fig. 7; *Linz 1990 Exhibition Catalogue*, p. 39 (no. 37); and Johannes Willers in *Nuremberg 1992-93 Exhibition Catalogue*, II, p. 614-615 (no. 1.98); on the inscription see n. 165 above.

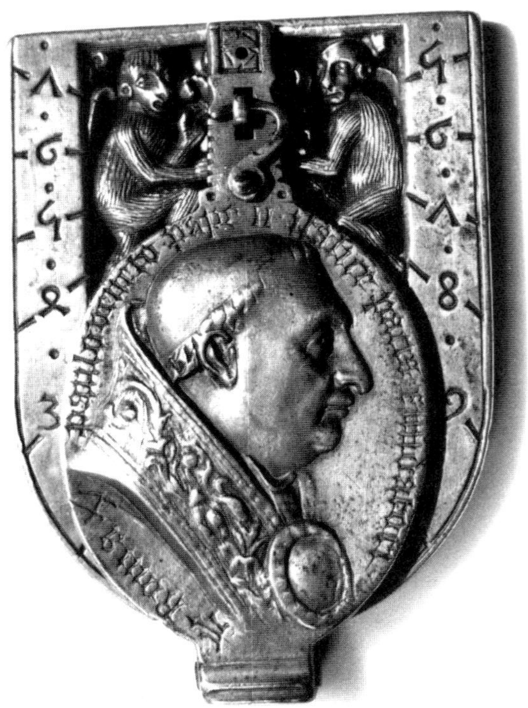

Fig. D6:~ The “Regiomontanus-type” sundial bearing an image of Pope Paul II and a dedication to him (#7518). All of these sundials are decorated with a pair of “wild men”, a popular motif of the time. [Courtesy of the Germanisches Nationalmuseum, Nuremberg.]

9 (#7519) Undated (fake?); dimensions 55/43 mm – Vienna, Uhrenmuseum, inv. no. ? – unpublished.

10 (#7530) A box-shaped sundial with an image of the face of Christ on the veil of St Veronica; 32 × 28 mm – Vienna, Kunsthistorisches Museum, inv. no. F1361 – see Zinner, *Astronomische Instrumente*, p. 481. (This piece was inadvertedly omitted from my 1994 list; I have not inspected it.)

In addition, we should mention the small folding sundial on the horizontal frame of a celestial globe and the much larger horizontal sundial on the base of a torquetum, both attributable to Hans Dorn – see VIII-1-2 below.

11 (#7521) Globe dated 1480 – Cracow, Jagiellonian University Museum, inv. no. 4039-37/V – see Ward, "The Earliest String-Gnomon Dials".

12 (#8503) Undated torquetum; dimensions of base 56.0 × 42.0 cm – Cracow, Jagiellonian University Museum, inv. no. 4038-38/V.

VIII *Instruments by Hans Dorn*

A detailed investigation of all of his instruments would be a useful contribution. A search for more biographical information on him could also be worthwhile.

Two large instruments

1 (#4510) A celestial globe dated 1480, with an astrolabic plate attached to the top and a folding sundial on the horizontal frame (see VII-11 above); total height 124.0 cm – Cracow, Jagiellonian University Museum, inv. no. 4039-37/V – see Zinner, *Regiomontanus*, pl. 45, figs. 89-90; *idem*, Astronomische Instrumente, p. 295; Ameisenowa, *The Globe of Martin Bylika*; Ward, "The Earliest String-Gnomon Sundials"; Gründler *et al.*, eds., *Matthias Corvinus und die Renaissance in Ungarn*, p. 339 (no. 284); and a brief description by Francis Maddison in *Washington NGA 1992 Exhibition Catalogue*, p. 222.

2 (#8503) A torquetum, undated (datable to *ca.* 1485); dimensions of rectangular base (reconstruction): 56.0 × 42.0 cm, overall height 71.0 cm – Cracow, Jagiellonian University Museum, inv. no. 4038-38/V – not yet published; see Zinner, *Astronomische Instrumente*, pp. 296-297; Gründler *et al.*, eds., *Matthias Corvinus und die Renaissance in Ungarn*, pp. 339-340 (no. 285); A. Turner, *Early Scientific Instruments*, fig. 10 on p. 17, good illustration of the ensemble; and a brief description by Francis Maddison in *Washington NGA 1992 Exhibition Catalogue*, pp. 223-224.

Three astrolabes

3 (#492) An unsigned astrolabe dated 1483, fitted with an *organum Ptolemei* and an eclipse calculator (actually part of an albion); diameter 46.0 cm – Florence, Museo di Storia della Scienza, inv. no. 1096 – unpublished, though see *Florence 1954 Catalogue*, pp. 57-59; *Florence 1987-88 Exhibition Catalogue*, p. 14; Goldstein, "Transversal Scale", pp. 103-104; and G. Turner, *Strumenti*, p. 73 (front only). See now **Fig. D7** below.

4 (#642) An unsigned astrolabe dated 1486; diameter 45.1 cm – Cracow, Jagiellonian University Museum, inv. no. 4040-38/V – see Zinner, *Astronomische Instrumente*, p. 296; A. J. Turner, *Instruments*, pls. 25 and 26 on pp. 36-37; Gründler *et al.*, eds., *Matthias Corvinus und die Renaissance in Ungarn*, pp. 338-339 (no. 283); *Washington NGA 1992 Exhibition Catalogue*, pp. 222-223.

5 (#164) An undated, unsigned astrolabe with distinctive rete design including quatrefoils; diameter 32.2 cm – Chicago, Adler Planetarium, inv. no. M28 – misidentified as "Spanish" in both Gunther, *Astrolabes*, II, pp. 311-312 (no. 164) and the small image in pl. LXX, and Webster & Webster, *Western Astrolabes in Chicago*, pp. 49-52 (no. 4), following Gunther; on the attribution and rete design, see King, *In Synchrony with the Heavens*, caption to fig. XVII-5.2.

6 (#4701) An astrological astrolabic plate showing only the houses for a specific latitude and fitted with an ecliptic ring, with a planetary equatorium without movable parts on the back; diameter 42.0 cm – Milan, Pinacoteca Ambrosiana, inv. no. ? – published in Tomba, "Gli astrolabi della Pinacoteca Ambrosiana", pp. 11-14, with illustrations of the front and back, instrument recognized as being of late 15th-century German provenance (Bamann or Dorn); see also North, *Horoscopes and History*, frontispiece, illustration of front only, "16th or late 15th century, no provenance suggested, although markings are said to be for a latitude in S. Italy (Naples or Salerno).

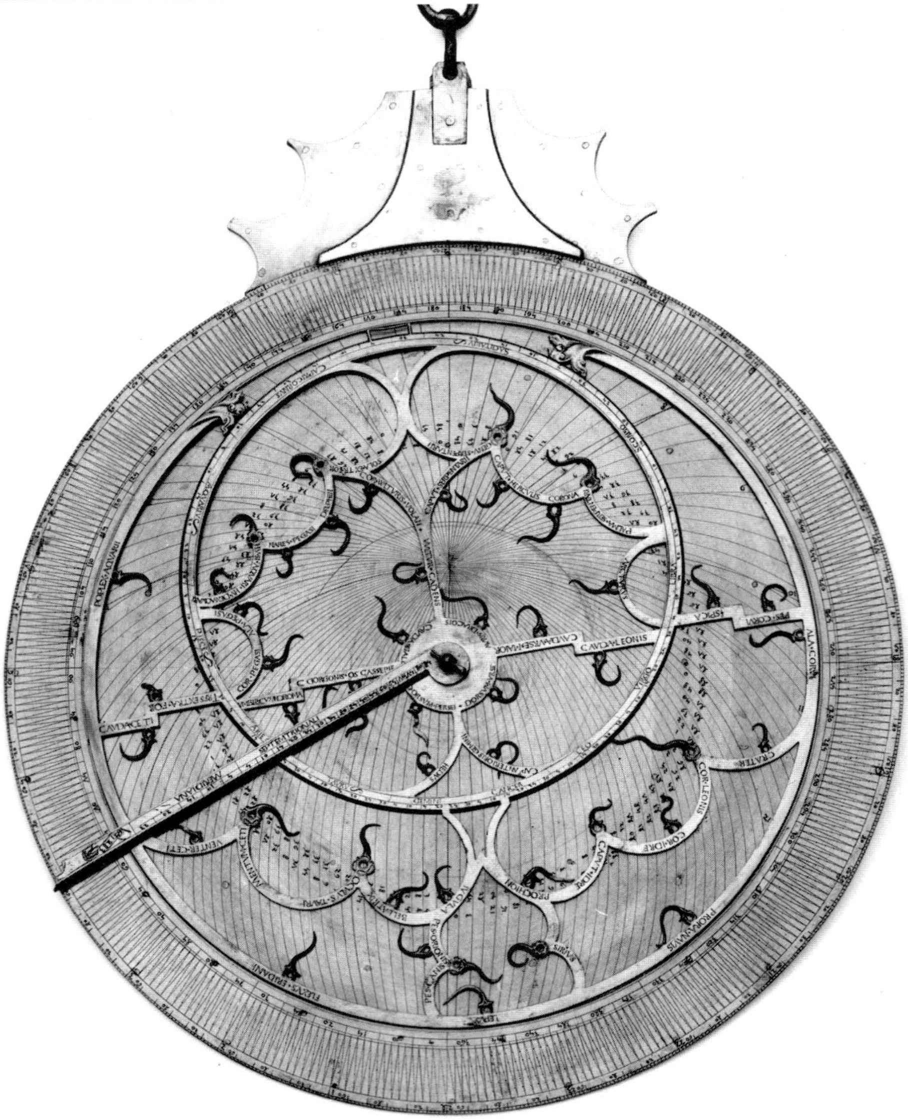

Fig. D7:~ The splendid Florence astrolabe of Hans Dorn (#492) dated 1483. His instruments are quite different in style and size from those of Regiomontanus, but they are equally clearly inspired by them. [Courtesy of the Museo di Storia della Scienza, Florence.]

7 (#-) An astrolabe possibly attributable to Hans Dorn in medieval style with a large trefoil on the rete, with astrolabic markings only for latitude 45°, diameter 23.8 cm – Milan, Museo d'Arti applicate, inv. no. 261 – published in brief in Tomba & Brusa, *Strumenti scientifici del MAA*, pp. 20-21, no. 16, and figs. 35-36 on p. 40.

See also the astrolabic plate on the celestial globe dated 1480 and attributable to Hans Dorn, listed above (item 1). Two astrolabic plates for a single latitude in Florence and Salzburg are closely related to Dorn's work, if not actually by him (see King, "Astronomical Instruments between East and West" p. 190).

Compendia

6 (#250/8611) A compendium (multi-functional instrument) with an astrological astrolabic plate signed by Hans Dorn, dated 1491; overall length 22.6 cm – London, British Museum, inv. no. 94 6-15 1 – see Zinner, *Astronomische Instrumente*, p. 293; Ward, *Instruments in the British Museum*, p. 74 (no. 207) and pl. XXVIII (a); G. Turner, ed., *Strumenti*, p. 120 (illustrations).

7 (#8615) *Caveat*: there may be two such instruments in Oxford, and these details may be confused. We mention here an undated, unsigned universal equinoctial dial fitted in a square frame that is to be suspended vertically; dimensions 9.5 × 9.5 cm – Oxford, Museum of the History of Science, inv. no. G425 – see Maddison, "Medieval Scientific Instruments", fig. 41 and n. 137 on p. 43 (dated *ca.* 1480); A. Turner in *Amsterdam 1990 Exhibition Catalogue*, p. 105, *ad* no. 197 on p. 104 (dated *ca.* 1480), with a different illustration (neither illustration shows all parts of the instrument). (The universal equinoctial ring dial was known to Regiomontanus: see Zinner, *Astronomische Instrumente*, pp. 117-118.)

8 (#5524/8513) A universal horary dial (of the kind described by Regiomontanus) and an horary dial for latitude [4]9°30′, attributable to Hans Dorn; dimensions 16.1 × 10.3 cm – Bielefeld, Kunstgewerbesammlung der Stadt, Stiftung Huelsmann, inv. no. HW 97 – see the detailed description by Dirk Syndram in *Bielefeld Catalogue*, pp. 69-70 (no. 9); an illustration of the universal dial in *Greenwich 1999-2000 Exhibition Catalogue*, p. 119; and, on the attribution, see King, *In Synchrony with the Heavens*, captions to figs. XIIb-10c-d. Syndram dated it to the last quarter of the 15th century but was

deceived by the latitude 39°30′ engraved on the piece, which he thought might be for Lisbon or Valencia. Perhaps Dorn was very old when he engraved this; certainly, the Latin and German inscriptions show signs of a failing memory.

Three other unsigned compendia associated with Dorn (these need to be compared with the London and Oxford pieces)

9 (#8612) Dated 1476 – Stuttgart, Württembergisches Landesmuseum, inv. no. 12364 – unpublished.

10 (#8613) Dated 1479 – Chicago, Adler Planetarium, inv. no. M288 – unpublished.

11 (#8614) Undated – Hamburg, Museum für Kunstgewerbe, inv. no. 1893.2 – unpublished.

IX Six astrolabes by Hans Bamann of Schmalkalden, 1484-1493

The astrolabes are small and rather crudely fashioned (see already #4501 above, which was also made in Schmalkalden).

1 (#633) Unsigned astrolabe, dated 1484, simplified rete; diameter 9.4 cm – Schweinfurt, Stadtarchiv, AvS Praetorius-Saxonius, inv. no. 3 – Zinner, *Astronomische Instrumente*, p. 244; published in detail in King, "Schweinfurter Instrumente", pp. 370-372 (no. 180).

2 (#604) Astrolabe signed "Johannes bamann Smalkaldensis 1487"; diameter ? – present location unknown; formerly in Schloß Kreuzenstein near Vienna (the castle administration claims that it was destroyed by a fire in 1915, but Zinner saw it in the 1930s) – unpublished; see Zinner, *Astronomische Instrumente*, p. 244; and King, "Astronomical Instruments between East and West". p. 190, and *idem*, "Schweinfurter Instrumente", p. 373.

3 (#248) Astrolabe mater signed "hb" and dated 1490, simplified rete, owner's mark on mater: "Andrea Leonardi, Catania 1899 Luglio"; diameter ~14.0 cm – Oxford, Museum of the History of Science, inv. no. IC248 – see Gunther, *Astrolabes*, II, pp. 423-424 (no. 248) and pl. CIV; Zinner, *op. cit.*, p. 244; and King, *op. cit.*, p. 373.

4 (#249) Astrolabe signed "hb" and dated 1491, simplified rete; diameter 10.1 cm – Munich, Bayerisches Nationalmuseum, inv. no. Phys. 21 – see Rohde, *Instrumente*, fig. 117 on p. 92 and p. 94: "plump und barbarisch"; Gunther, *Astrolabes*, II, p. 424 (no. 249), after Rohde; Zinner, *op. cit.*, p. 244; King, *op.*

cit., p. 373; and the detailed description in Stautz, *Die Astrolabiensammlungen in München*, pp. 217-227 (no. 6).

5 (#540) Astrolabe signed "hb" and dated 1491, simplified rete; diameter ~10 cm – Oldenzaal (NL), Historisch Museum "het Palthe-Huis" – unpublished; see Zinner, *op. cit.*, p. 244; and King, *op. cit.*, p. 373.

6 (#4502) Astrolabe signed "hb", dated 1493, unusual rete; diameter 13.5 cm – present location unknown, formerly in the collection of Tullio Tomba, Milan – unpublished; see *ibid.*, p. 373.

X Early-16th-century German astrolabes

These four pieces are all by the same maker in the tradition of the 15th-century Vienna astrolabes with distinctive Y-shaped frames inside the ecliptic and below it.

1 (#173) Diameter 19.7 cm – Oxford, Museum of the History of Science, inv. no. IC173 – see Gunther, *Astrolabes*, II, pp. 327-328 (no. 173), misidentified as Italian and dated *ca.* 1500; a slightly later dating may be more appropriate).

2 (#453) Diameter 19.7 cm – Leonard Linton Collection – see *Greppin Catalogue*, p. 33 (no. 3124: "français vers 1500"); and a detailed description with fine illustrations in *Linton Catalogue*, p. 133 (no. 186).

3 (#4505) Diameter 19.7 cm, with a leather case, possibly original, dated 1542 – Munich, Deutsches Museum, inv. no. 1991-229,1 (astrolabe) + 2 (case) – published in detail in Stautz, *Astrolabiensammlungen in München*, pp. 229-241 (no. 7).

4 (#4543) Diameter 19.5 cm (according to the catalogue) – Rome, Osservatorio Astronomico, inv. no. 148/662 – unpublished; see Calisi, *Strumenti del Museo astronomico di Roma*, pp. 118 and p. 147 (no. 299).

These two both have a vaguely rectangular frame on the rete, developed from #644 (V-2 above).

5 (#491) Diameter 13.4 cm – Florence, Museo di Storia della Scienza, inv. no. 1110 – see Poulle, "Instruments astronomiques", A, pp. 20-21 (illustration of front and back); King, "Astronomical Instruments between East and West", p. 186; also Samhaber, *Die Zeitzither*, p. 199.

6 (#4544) Diameter 18.7 cm – Stuttgart, Württembergisches Landesmuseum, inv. no. L1985-200 – see King, *op. cit.*, p.

186; and the detailed description in Oestmann, *Stoeffler's Himmelsglobus*, pp. 59-60 (no. 28).

XI ***Miscellaneous***

1 (#403) An astrolabe on paper and vellum by Hanns Herghamer dated 1492 – Oxford, Museum of the History of Science, inv. no. ? – unpublished; not mentioned in Zinner, *op. cit.*

XII ***Fixed sundials***

Various sundials in Germany and Austria predate 1550. The interested reader may consult Zinner, *Sonnenuhren* (Europe) and Schwarzinger, *Sonnenuhren* (Austria). One particularly interesting (double) example must suffice here:

1 (#8550/51) two vertical sundials dated respectively 1487 and 1509, the earlier one with Gothic numerals more sophisticated than the later one, which has Roman numerals – Regensburg, south wall of the Cathedral – see Zinner, *Sonnenuhren*, p. 164 and fig. 33 on pl. XI; and King, *The Ciphers of the Monks*, pp. 286-287 and fig. B3 on p. 288.

Appendix E – Notes on the 1462 astrolabe

Description

Since the instrument has been published in detail several times,[489] a very brief description is presented in italics below, deliberately separated from the commentary. The illustrations are in **Figs. I-9-10** and **I-13**.

The instrument has been known since 1848. The 1462 astrolabe first came to light in "modern" times in the year 1848, when it was presented to an Englishman, Dr. William Somerville, in Rome.[490] The source of this acquisition, and the occasion, are unknown. Somerville (1771-1860) was an inspector of hospitals in Scotland and later (from 1816) Inspector of the Army Medical Board in London. In 1812, he became the second husband of his cousin, the well-known scientist Mary Somerville (b. Jedburgh, Scotland, d. 1872, Naples),[491] In London they moved in the leading scientific circles of the time, and were, for example, friends of the astronomer John Herschel (b. Slough, England, 1792, d. Hawkhurst, Kent, 1871).[492] On account of William's ill-health, in 1836 the couple move to Italy, first to Florence, then to Naples. Mary was "a valued guest and brilliant part of the Italian scene of the Risorgimento", "not only ... offered access to Italian libraries and scientific facilities, but she was given membership in six of the leading Italian scientific societies (1840-1845)".[493] Mary dedicated her 1848 book *Physical Geography* to John Herschel. At some time the astrolabe passed to Herschel, then eventually to one of the executors of his will, Edward Hardcastle, a leading Manchester businessman. Herschel's will mentions only "my scientific books and instruments". The astrolabe remained in his family until 1958, when Commander M. H. Hardcastle

---

489 Price, "The First Scientific Instrument of the Renaissance"; G. Turner, "The Regiomontanus Astrolabe" (auction catalogue for Christie's); King, "Die Astrolabiensammlung in Nürnberg", II, pp. 586-589; Turner & King, "Regiomontanus' Astrolabe for Bessarion"; also King, *In Synchrony with the Heavens*, XIIIa, figs. 10.6a-e, with captions adjusted in proof during the week the acrostic was recognized.

490 For more details see King & Turner, "Regiomontanus' Astrolabe for Bessarion", pp. 174-176.

Jeremy Black, in his *Italy and the Grand Tour*, vividly describes the experiences of British residents and travellers in Italy *before* the disastrous invasion of the French at the end of 18th century.

491 See the article "Somerville, Mary Fairfax Grieg" by Elizabeth C. Patterson in the *Dictionary of Scientific Biography*.

492 See the article "Herschel, John ... ", in the *Dictionary of Scientific Biography*. Herschel was on the committee choosing scientific instruments for the Great Exhibition of 1851 in London, and it is a pity that he did not already have access to this splendid piece.

493 Cited from Patterson (see n. 491).

placed it on extended loan to the National Maritime Museum at Greenwich until 1985. In 1956, Hardcastle had loaned the piece to Derek de Solla Price, who took it to the United States for "study and exhibition in Europe and America" and published it in 1959. It was auctioned at Christie's of London in 1989. See below on its subsequent fate.

The astrolabe is made of brass, and has a diameter of 11.6 cm. All medieval and Renaissance astrolabes are made of brass, although a tendency to use astrolabes from printed kits began in the 16th century. This piece is smaller than the nine other astrolabes of the "Regiomontanus type" with original maters (one has a larger mater that is not original); their diameters range from 12.9 to 17.0 cm.

All engraving is in Antiqua script. The Antiqua script here replaces the Gothic script that had previously been used on Vienna instruments, such as on the 1457 Nuremberg astrolabe associated with Georg Peuerbach (#549 – see Appendix D-V-1). One other astrolabe of the Regiomontanus type, in a private collection in Paris, has an identical Antiqua script on the rete but Gothic elsewhere (#256 - see Appendix D-V-5). The 1462 Urbino astrolabe (#4506 – see Appendix G) also has Antiqua script. The script was featured in the *Alphabetum Romanum* by Felice Feliciano in Verona, 1460.[495] This was a time when German scholars were discovering the Antiqua script in Italy: for example, it was used in a heavily decorated form in Cicero's *De oratore* printed by Conrad Sweynheym and Arnold Pannartz at Subiaco in 1465.[496]

Some numbers still have their Gothic forms. Habits are hard to break. All of the Regiomontanus-type astrolabes have the old forms for the "4", the "5" and the "7".[497] The 1462 Urbino astrolabe (#4506 – see Appendix G) has the new forms, but still uses "Z" for "2".

The astronomical markings on the astrolabe are very carefully and accurately engraved. This is typical of all the instruments of the Vienna school.

There is religious decoration on the throne and on the rete. The rosettes with five petals on the throne, the lily and lily fronds on the rete are Marian

---

495 * See the bibliography for a facsimile edition of the author's manuscript. For some extracts see Feld, "Constructed Letters and Illuminated Texts", p. 365.

496 *Ibid.*, pp. 357-359.

497 On Hindu-Arabic numeral forms in the Middle Ages see Hill, *Arabic Numerals*. For various other examples see King, *The Ciphers of the Monks*, pp. 286-288 and 309-317. On Regiomontanus' tenacity in using the Gothic forms see Zinner, *Regiomontanus*, pp. 48-49.

symbols.[498] The six-pointed stars on the back of the throne may also be Marian symbols.[499] Or they may be related to Regiomontanus' mark of ownership, which he inscribed on 17 surviving manuscripts: this is a common symbol of the Cross, with the cross on a symbolic hill for Golgotha and a six-pointed star on either side of it.[500] Any potential religious significance of the crown on the pointer for the star Cor Leonis (Regulus, "little king") on the rete is still obscure. It could also be a trademark of *Regio*montanus, involving a word-play on his name. The same feature is found on other related astrolabes, notably, the one with engraving in an identical hand (#452): see **Fig. A2**.

The rete has a distinctive and elegant design. The rete with an ecliptic ring and pointers for 30 stars is standard for 10 out of about 20 astrolabes from 15th-century Vienna. The design seems to have been influenced from Italy: see the three astrolabes discussed in Appendix G. A small circular frame on the pointer for Vega indicates the pole of the ecliptic.[501]

The star positions are accurate for the epoch of the astrolabe. A preliminary analysis of the star-positions on all of the 11 Regiomontanus-type astrolabes has been conducted by Burkhard Stautz.[502] The implication of their accuracy is that

---

498 See the articles "Lilie" and "Rose" in both Bäumer & Scheffczyk, eds., *Marienlexikon*, and *Lexikon der christlichen Ikonographie*; also Metford, *Dictionary of Christian Lore and Legend*, pp. 161 and 214.

We have come a long way since Robert Gunther in 1932 used the expression "Tudor roses" to describe the Marian five-petal rosettes on the astrolabes of Georg Hartmann, who was vicar of St Sebald's in Nuremberg *ca.* 1525: see his *Astrolabes of the World*, II, p. 437. Even Derek Price ("The First Scientific Instrument of the Renaissance", p. 29), refered to them as "tudor". Anthony Turner in 1985 (in *The Time Museum Catalogue*, 1:1, p. 128), calls Hartmann's rosettes "formal flowers" with the comment "the resemblance to 'Tudor' roses is probably fortuitous". In Webster & Webster, *Western Astrolabes in Chicago*, p. 56, they become "a Hartmann trademark", and this is correct if one keeps in mind that Hartmann was also a vicar. He put them on his astrolabes because Regiomontanus had used Marian rosettes to decorate his astrolabes. In van Cleempoel, *Astrolabes at Greenwich*, p. 143, Hartmann's Marian rosettes become "sculpted circles, or roundels, which give the appearance of flower petals".

499 Article "Stern" in *Lexikon der christlichen Ikonographie*; also Metford, *Dictionary of Christian Lore and Legend*, p. 232.

500 Zinner, *Regiomontanus*, pp. 293-295, also Folkerts, "Die mathematischen Studien Regiomontans in seiner Wiener Zeit", p. 190. On the use of a kind of six-pointed star (formed by I and X, the initials of Iesous Christos in Greek) see the article "Christogram" in *The Oxford Dictionary of Byzantium*.

501 This was not noted in previous descriptions. On the same feature on 10th-century Islamic astrolabes see King, *In Synchrony with the Heavens*, XIIIc-3.1.

502 Stautz, *Untersuchungen*, pp. 111-115 and 274-278.

the Viennese astronomers had access to Ptolemy's coordinates adjusted to *ca.* 1460 by a good value for the precession. This merits further investigation.[503]

There is a diametral rule for measuring declinations. This is graduated for each 5° and labelled for each 10° (up to 30° N and 20° S). North and south are labelled MERIDIES and SEPTEMTRIO.[504]

The three plates are marked for each 3° of latitude from 39° to 54°, with Rome being specifically mentioned on the plate for 42°.[505] All three plates are marked with altitude circles and azimuth circles for each 5° of argument. The range of latitudes results from a deliberate attempt to make the instrument "universal".[506] In Nuremberg in 1527, Georg Hartmann wrote that all the astrolabes made by

---

503 Burkhard Stautz found that the positions correspond to the Ptolemaic ones corrected with a precessional value of 1° in 68 years. Regiomontanus in his correspondence with Bianchini in 1464 mentioned the parameter of 1° in 66 years of al-Battânî (*ca.* 910) alongside Ptolemy's inaccurate 1° in 100 years (Curtze, "Regiomontan's Briefwechsel", p. 263, translated in Swerdlow, "Regiomontanus on the Critical Problems of Astronomy", p. 171). The tradition of the Toledan *Alphonsine Tables*, in which the Vienna school was steeped, knew of another, better value, 1° in 70 years, which they associated with al-Sûfî (984), although in fact he advocated 1° in 66 years (Chabás & Goldstein, *The Alphonsine Tables of Toledo*, p. 235). The value may have been taken from Ibn Yûnus (*ca.* 990), who derived the excellent value 1° in $70^1/_4$ years. It is unlikely that the value that Stautz found yielded the best fit with the star-pointers on the 1462 astrolabe was simply the average of 66 and 70. And it is very curious that Plethon, a correspondent of Bessarion *ca.* 1446, used 1° in *ca.* 68 years (Tihon & Mercier, *Pléthon – Manuel d'astronomie*, p. 248). A detailed analysis of the star names and star positions on all of the Regiomontanus-type astrolabes, as well as a search for the underlying star-table, is a task for the future. The new star-table prepared by John of Gmunden in Vienna is apparently still unpublished (see Kunitzsch, *Typen von Sternverzeichnissen*, n. 8 on p. 5, *idem*, "Star Catalogue Appended to the Alphonsine Tables", p. 92, and *idem*, "The Stars on the Astrolabe", p. 45). Note, however, that Johannes Stoeffler reproduced a star-catalogue of Regiomontanus, which was also used by Hartmann, and this explains why the star-positions on his astrolabes correspond to a date *ca.* 50 years before his time (see Zinner, *Regiomontanus*, pp. 197-198, also Kunitzsch, "The Stars on the Astrolabe, p. 45). For Hartmann's star-table see Lamprey, *Hartmann's Practika*, p. 274.

504 The word *septemtrionalis* (with "m" for "n") was used by Regiomontanus in his letter of 1465 to Jakob von Speyer in Urbino: see Curtze, "Regiomontan's Briefwechsel", pp. 319 and 320. This orthography was probably favoured because seven stars (six of the Plough and the Pole Star) define north.

505 Regiomontanus mentions that the latitude of Rome is 42° in his 1471 letter to Christian Roder in Erfurt: *ibid.*, p. 330.

506 On universal solutions in the history of astronomy see King, *In Synchrony with the Heavens*, VIa-b, and on astrolabes, *ibid.*, XVI-2-4.

Regiomontanus that he had seen had plates for each 3° of latitude from 39° to 54°.[507]

On the back of the mater, inside the altitude scales, there is an ingeniously-conceived solar / calendrical scale. The organization of the scale is unusual indeed. The mode of labelling the months and zodiacal signs is not found on any other known astrolabe, for very good reasons. In 1989, Gerard Turner unfortunately stated that it was incorrectly labelled.[508] We now know, thanks to Robert Dietz, that the reason behind the idiosyncratic labelling of the scales was to link certain of the letters to those in the epigram, as part of Regiomontanus' very personal message to Bessarion.[509] The equinox is at March 10.5, and 8 out of the 10 "Regiomontanus-type" astrolabes have March 10-11 (see **Table 10** in Appendix D-5). The date March 10.5 corresponds to an epoch of *ca.* 1525 using the astronomical calendar (noon to noon) and *ca.* 1590 in the civil reckoning (midnight to midnight), and, as Gerald Turner has shown, such discrepancies reveal how little we know about the way in which such calendar scales were constructed.[510] Inside the upper half of these scales are two secondary altitude scales, both serving the markings of the *organum Ptolemaei.* The alidade is a poor replacement, as can be seen from other instruments of this type.

Inside these scales below the horizontal axis are two others for the shadows to base 12 digits.[511] Both scales show tangents up to argument 45° and cotangents from 45° up to 90. The scales are divided and labelled for each 2 digits up to 12, subdivided for each single digit. The Viennese instrument-makers experimented with different kinds of shadow-scales.[512]

Inside the back there is fitted a rotatable plate bearing an orthographic projection of the heavens between the two solstitial limits. The markings serve as a universal horary device, for finding the time of day from the altitude of the sun at any time of the year and at any latitude. This is an observational device, and to be used, it must be rotated so that the vertical axis points to the latitude of locality on the innermost altitude scale. In the modern literature the projection is

---

507 Cited in Zinner, *Astronomische Instrumente*, p. 480, and King & Turner, "Regiomontanus' Astrolabe for Bessarion", p.186, n. 40. For the text within the context of Hartmann's treatises on instruments see now Lamprey, *Hartmann's Sundial and Astrolabe Manual*, p. 245.

508 See the text to n. 109 above.

509 See the text to nn. 111 above and 523 below.

510 See n. 389 above.

511 On shadow-scales and base 12 see nn. 391 and 392 above.

512 See King, "Die Astrolabiensammlung in Nürnberg", II, pp. 583-584, on the shadow scale on the 1457 Vienna astrolabe, cunningly shaped in the form of half of a regular octagon.

unfortunately, and incorrectly, called the "de Rojas" projection after an author who wrote on it in 1550.[513] The underlying theory – an orthographic projection of the celestial sphere – was known in Antiquity and in the Islamic world, and is depicted on an 11th-century pillar from the monastery of St Emmeram in Regensburg, which Regiomontanus might have visited.[514] The device seems to be mentioned for the first time in Vienna in 1441, and Regiomontanus apparently copied a treatise on it in 1457.[515] Regiomontanus appropriately called it *organum Ptolemaei*. He clearly preferred this kind of horary markings to the standard horary markings on the backs of earlier medieval European astrolabes, not least because the latter – consisting of a pair of universal horary quadrants – provide only an approximate solution to the same problem which is not at all satisfactory in European latitudes.[516]

Above the astronomical markings is the image of an angel, represented as a beautiful young man with a fine pair of wings. Below them is an inscription that is associated with the angel. The ensemble forms the main subject of Part I of the present study. The fact that Regiomontanus signed the astrolabe, if only as "RGO IO", does not mean that he made the instrument with his own hands. In this case, however, it seems highly probable that he alone was responsible for engraving at least the angel and the inscription and the idiosyncratic markings on the solar / calendrical scales. But if he engraved the angel, he also engraved the markings of the *organum Ptolemaei*. And if he engraved the inscription, he also was personally responsible for all of the engraving on the entire instrument.

The hours are engraved 1-12 twice in divisions around the outer rim in a later hand. The forms are more modern but could still be late-15th-century.[517]

Two fine tributes

In 1959, a brief description of the astrolabe was published by Derek J. de Solla Price as "The first scientific instrument of the Renaissance". Price made the conjecture that the "angel" might be a self-portrait of Regiomontanus.[518] He was

---

513 Zinner, *Astronomische Instrumente*, pp. 131-134, and King, *In Synchrony with the Heavens*, XIIb-15 and 16. See also Maddison, "Hugo Helt and the Rojas Astrolabe Projection". On aspects of the history of this projection see Dupré, "Galileo, Mathematical Instruments and Orthographic Projection".

514 On this remarkable object see item I-1 in Appendix D.

515 Zinner, *op. cit.*, p. 132.

516 See n. 394 above.

517 See King & Turner, "Regiomontanus' Astrolabe for Bessarion", pp. 177-178 and figs. 5-6.

518 Price, "The First Scientific Instrument of the Renaissance", pp. 28-29. Ernst Zinner (*Regiomontanus*, 2nd edn. (1968), p. 137) independently (?) made the same suggestion; it is not clear how he knew about the astrolabe.

criticized for this at the time, perhaps unduly harshly,[519] but he was correct in realizing that the image was important, and that is was not the Cardinal. Scholars in Europe saw no problem with the instrument or its inscription.[520] Another fine description was prepared in 1989 by Gerard L'E. Turner for Christie's.[521] Turner's description contained one unfortunate error, namely, that the calendrical scale was incorrectly divided.[522] This supposed "defect" was later cited as one of several reasons why the astrolabe could not be genuine.

The unfortunate events of 1989-1990 and the rescue operation

> "Faking, as distinct from the copying of a master by his apprentices and pupils, was virtually unknown amongst antique scientific instruments until late in the nineteenth century when such objects began to attract the attention of scholars and collectors." Derek J. de Solla Price, "Fake Antique Scientific Instruments" (1956), p. 381. [This statement is still valid today.]

The astrolabe was auctioned at Christie's of London in 1989, with a special catalogue containing the more detailed description by Gerard Turner. In the auction catalogue, no information of consequence was provided on related instruments, such as the obvious one, the 1457 astrolabe in Nuremberg associated with Peuerbach: the two pieces are clearly from the same workshop.[523] In any case, the piece was sold for £229,000 to a New York collector, Daniel Wolf. Nobody had ever countered Price's description of the piece, in which he attributed it to Regiomontanus, but now it seemed that the field was in danger because this dubious piece had achieved such an astronomical price. Maybe "Ioannes" was not Regiomontanus but Hans Dorn. Rumours were rife – "So-and-so doesn't like this or that" – and the press

---

519 This was countered by in Scholderer, "Regiomontanus' Astrolabe" (1960), p. 209, who pointed out that it would have been sacrilegious for Regiomontanus to depict himself in this way: for more, see n. 108 above.

520 For example, Zinner, *Astronomische Instrumente* 2nd edn. (1967), p. 481, and *idem*, *Regiomontanus*, 2nd edn. (1968), p. 137, neither with illustrations; and Zorzi, *La Libreria di San Marco*, (1987), p. 54, and 56-57, with an illustration of the back of the astrolabe.

521 G. Turner, "The Regiomontanus Astrolabe".

522 *Ibid.*, p. 4. See also the text to nn. 109 and 508-509.

523 The 1457 astrolabe in Nuremberg (#549) was illustrated by Zinner, *Regiomontanus* (1938, repr. 1968), pl. 40, figs. 70-71; it is described in some detail in *idem*, *Astronomische Instrumente* (1967), p. 465. The undated astrolabe with an *organum Ptolemaei* on the back (#256) was briefly mentioned in Gunther, *Astrolabes of the World* (1932), II, p. 434. Both have recently been described in detail: see items V-1 and V-5 in Appendix D.

coverage of the auction was either sceptical[524] or derisive.[525] Daniel Wolf consulted a group of British scholars who were convinced that Price and Turner

---

524 The serious report by Anne Crane in the *Antiques Trade Gazette* (London), 28 October, 1989, is here abridged:

> "Under the Protection of the Divine Bessarion Called Cardinal I Arise in Rome ... 1462".
>
> The sale of scientific instruments held by Christie's South Kensington at the end of September raised a total of more than half a million pounds, the bulk of which was contributed by astrolabes.
>
> The most expensive of these, at £190,000, was afforded its own separate illustrated 12-page catalogue which described it as "The Regiomontanus Astrolabe, dated 1462. A unique 15th century European brass astrolabe."
>
> Any instrument made by Regiomontanus ... ... is bound to arouse considerable interest as a Renaissance object of considerable importance.
>
> Christie's attribution was made on the basis of a dedicatory inscription ... ... , On this basis Christie's conclude that the astrolabe was made to the order of* Cardinal Bessarion by Johannes Müller (Regiomontanus) in that year.
>
> There is an established connection between scholar and cardinal ... . The date on the astrolabe therefore places its manufacture within the period when Regiomontanus was staying in Rome.
>
> The subsequent history of the instrument can only be traced back as far as the mid 19th century, although it does begin in Rome, ...
>
> This impressive looking catalogue makes out a strong case for the astrolabe's dating and attribution but not everyone in the scientific fraternity, including some museum specialists, appear to be as cast-iron in their confidence as Christie's.
>
> Dr. J. V. Field, a curator in the Resource and Information Services Division of the Science Museum, says that she is unhappy about the association with Regiomontanus because there are mistakes in the Latin on the rete and on the label and one does not expect mistakes from Regiomontanus. "As a humanist he was a leading proponent of the use of correct classical Latin", she added.
>
> Also it is interesting to note that, between leaving the Maritime Museum and appearing for sale at Christie's, the astrolabe was taken to Sotheby's who decided not to offer it for sale.
>
> When the astrolabe was offered at Christie's on September 28, there did not appear to be a marked, if indeed any, interest from the major British institutions who might be candidates for such a purchase and the astrolabe was knocked down to Daniel Wolf, an American, who had left his successful £190,000 bid on commission."
>
> * "to the order of": *sic*, but precisely as stated in the catalogue (G. Turner, "Regiomontanus' Astrolabe", p. 9)

525 Joseph Roy (pseudonym) in *L'Exprès* (Paris), 13-19 October, 1989, wrote (exclamation marks in square brackets are in the original):

had exaggerated the importance of the instrument. The piece was considered more or less in isolation: the absurd angel and the kitchen Latin of the epigram, the fact that Regiomontanus did not even sign his own name, and the fact that there was nothing with which to compare the piece except the (irrelevant) instruments of Hans Dorn, were all very disturbing.

In March, 1990, an imposing 7-page report on the astrolabe was sent from the Museum of History of Science in Oxford to Daniel Wolf. I refrain from citing

---

Sous le signe des étoiles – Du télescope de grand-papa à l'astrolabe du XVe siècle, des prix à décrocher la lune.

... . Entre un curieux astrolabe. Christie's South Kensington a fait, pour lui, les choses en grand. Catalogue spécial avec l'objet en couverture et en couleur sur fond noir, titré << The Regiomontanus Astrolabe dated 1462 >>. À l'intérieur, toutes les pièces sont photographiées en détail. Et, au dos même, on peut admirer, justement, le dos de l'astrolabe. L'instrument n'en finit pas d'intriguer. Étonnant, cet ange qui tient la projection des méridiens et des parallèles. Surprennante, cette dédicace en latin de cuisine au Cardinal Bessarion, qui, en VF, [a] donne à peu près ceci : << Sous l'égide du divin [!!!] Bessarion appelé "de Cardine" [?!] j'apparais[b] comme l'oeuvre de Ioanes.[c] 1462. >> Incompréhensibles, les trois cercles ajourés du << trône >>, dont deux enferment une étoile à six branches, que l'on ne dénomme pas encore étoile de David : à cette époque, c'est un symbole ... musulman ![d] Bref, il est vraiment stupéfiant, cet astrolabe de << 1462 >> dont les chiffres ressemblent bien à ceux du XVe siècle, alors que l'allure de l'ensemble ferait juger qu'il date de 1520 ou de 1530,[e] n'était la dédicace curieuse, dans le fond comme dans la forme << divin Bessarion >> force vraiment la note.

Un affreux soupçon envahit l'esprit. L'histoire de l'objet remonte à 1848, quand une touriste anglaise de passage à Rome[f] le reçoit en cadeau. Ladite histoire commencerait-elle, d'aventure, dans la décennie qui a précédé ce passage ? Heureusement, le marchand Daniel Wolf, qui a vendu en bloc sa collection de photographies au Getty, est un homme qui va droit au but. << C'est un chef-d'oeuvre, n'est-ce pas ? >> a-t-il demandé à l'experte, comme celui-ci me l'a raconté. Il y eut un pudique instant de communion artistique. Pour 229 900 livres, 2,5 millions de francs payés par Wolf, les deux hommes pourront continuer ce long silence à deux.

[a] "en VF" : c'est-à-dire, en << vrai français >>. [b] notice that "à Rome" has been omitted. [c] *sic*, with one "n". [d] see the text to n. 500 of the present study for the two stars on the owner's mark of Regiomontanus! [e] the writer is referring to astrolabes of Georg Hartmann, dated around this time, since the design of the piece resembles these. [f] Mary Somerville was hardly a "touriste de passage" at Rome. Besides, the astrolabe was presented to her husband.

527 * Various scholars of Renaissance Latin in Oxford, Cambridge, London and Belfast had been consulted. More recently, others in Germany found nothing extraordinary about the epigram. It is quite remarkable that the epigram, for over 50 years available to such scholars in publications in English, for over 35 years in German publications, and for close to 20 years in an Italian publication, has attracted no serious attention from Neo-Latinists. See n. 4 above.

the report; Daniel Wolf passed it to Christie's. The most serious problems raised in Oxford and London (not all mentioned in the report) were as follows (italicized), with my comments:

(1) *This is a Hartmann-type astrolabe supposedly dated 1462.* Hartmann copied the design of his astrolabes from those of Regiomontanus.

(2) *There are no known instruments to compare it with.* There are at least ten astrolabes with which one can compare it. They all have the same or similar rete design. One (#549), dated 1457 and with a more elaborate rete design, was well-known at least in the German literature on astronomical instruments. Another (#452) has identical engraving to the 1462 piece. In all, some 45 instruments are known from 15th-century Vienna.

(3) *The star-positions correspond too well to the supposed date of the piece.* See the comments in n. 503 above.

(4) *The astrolabe bears a "de Rojas-type" projection on the back.* The projection was described by Regiomontanus about a century *before* de Rojas. It is perhaps time to abandon the appellation "de Rojas" for this projection.

(5) *The above-mentioned grid is on a removable, rotatable plate.* And so it must be, if one wants to use it.

(6) *The calendrical scale is incorrectly marked.* No, it is not. It is correctly marked but in a very idiosyncratic way so that some of the labels correspond to the angel and certain letters of the epigram.

(7) *The script is Antique and not Gothic. One would expect Gothic on a 15th-century German instrument.* All the Vienna instruments before Hans Dorn have Gothic engraving except this one and another datable to *ca.* 1460 (see **Fig. A2**), which is also by Regiomontanus. All of the later instruments of Hans Dorn have Antiqua script.

(8) *The Latin dedication is "suspicious" and composed in "kitchen" Latin.* We are dealing with what is perhaps the most ingenious epigram of the Renaissance.

(9) *The Ioannes in the inscription refers to a Ioannes other than Regiomontanus, perhaps Hans Dorn of Vienna.* This reveals lamentable ignorance of the instruments of Hans Dorn, which are different in design and size. Even the Hans Dorn instrument in Oxford had not been identified (see n. 529 below).

(10) *The ribbon on which the inscription is engraved is not "happy".* Then neither are those featured on 15th-century Viennese representations of the Annunciation.

(11) *The angel depicted on the back is absurd and not in keeping with what we know about the decoration of German instruments of the period.* The angel is very similar to that on Regiomontanus' ephemerides (see **Fig. I-4**), and it is intended to represent the angelic St Bessarion.

(12) *The metal composition does not correspond to the 15th century.* In one analysis, it "did", although it should be emphasized that no other 15th-century German instruments have ever been analyzed.

(13) *Regiomontanus never made any astrolabes.* Zinner, *Astronomische Instrumente*, pp. 479-483, had already published all the textual references to the various large-scale instruments of Regiomontanus.

(14) *Regiomontanus would never have made an astrolabe for Bessarion.* He did, and we have it.

In brief, the astrolabe was branded "suspicious", and possibly even an early-19th-century Italian fake. Scholars were unhappy with the angel, with the supposedly inaccurate calendar scale, and with the "kitchen Latin" of its inscription.[527] It was suggested that the "Ioannes" might be Hans Dorn, of whom one instrument has long been available in Florence for wicked fakers to imitate (see **Fig. D7**), although his instruments are distinctly different in style and in decoration and in size, and he had no special relationship to Bessarion and he was not in Rome in 1462.[528] (However, an unusual instrument attributable to Hans Dorn that is preserved in Oxford had not been recognized for what it is,[529] and illustrations of an albion possibly by Peuerbach or Regiomontanus and now in Rome had been published in Oxford in 1976 without its Viennese connection being recognized.[530]) Again, no comparison was made with any other relevant 15th-century Vienna instruments.[531] However, chance also played a role, for a would-be 15th-century Viennese pocket sundial preserved in London happens to

---

528 On Dorn and his instruments see now Appendix D, items VIII. A definitive study of all of his known instruments would be a valuable contribution.

529 On this Oxford instrument see now Appendix D, item VIII-7.

530 North, *Richard of Wallingford*, III, pls. XXIII-XXIV. On this remarkable object see now Appendix D, item VI-1.

531 In the Oxford report, an astrolabe in the Conservatoire national des arts et métiers, Paris (#516), was introduced as a "16th-century" German piece, the dating on the authority of the Conservatoire. In fact, this piece is from 15th-century Vienna, and is one of three "Regiomontanus-type" astrolabes that were in Paris in 1990, the two others (#452 and #256) being at that time in very accessible private collections. All of these are considered in the 1994 King & Turner study: see Appendix D, item V-10, also 3 and 5.

be a fake.[532] In the Oxford report, no comparison was made with any other "real" fake astrolabes from early-19th-century Italy, not least because none is known. It is curious that one problem not alluded to in the report was the fact that the 1462 piece was totally unlike "The Astrolabe of Regiomontanus" illustrated by R. T. Gunther in 1932 on the basis of authoritative German publications.[533] Yet another problem was that the 1462 piece looked, if only at first sight, like one of the many surviving *Volksastrolabien* of the well-known Georg Hartmann of Nuremberg, but was dated some 60 years before his time. It was not realized in 1989 that Hartmann simply copied the elegant rete design of the Vienna school (which was a good idea) and even the star-positions (which was not a good idea). Also, no mention was made of the excellent workmanship of the piece. In brief, the report was sincere, but completely misguided; also, there could be no going back, because, once the report was circulated, the damage was done. In any case, when the purchaser, Daniel Wolf, received the report, he promptly returned the astrolabe to Christie's and got his money back.

In the early 1990s things started to improve for Regiomontanus' 1462 astrolabe. In 1992-93 it was exhibited at the Germanisches Nationalmuseum in Nuremberg, together with the equally imposing 1457 astrolabe associated with Peuerbach.[534] In 1992, at the Institut für Realienkunde des Mittelalters und der Frühen Neuzeit at Krems in Austria, the author gave a lecture to an audience of Byzantinists, Islamicists and Medievalists on the utility of scientific instruments as historical sources. In the ensuing discussion, one expert on *Realienkunde* – the study of material objects – in the audience asked "Sehr schön, und ... ?" "Very nice, but so what?" Prompted by this question, I included in the conference proceedings, published by the Austrian Academy of Sciences, a list of some 40 instruments from 15th-century Vienna,[535] and another of all medieval instruments housed in Austrian museums, or stolen from them in the early 1940s

---

532 See item VII-2 in Appendix D. The London piece (#7512) is one of at least ten of such dials that survive in museums in Europe, and the only one that is clearly a fake. Price had ventured to suggest that another one, preserved in Munich, might also be fake, claiming that both were of "much cruder and apparently more modern construction" than three others in Innsbruck, Vienna and Nuremberg. In 1982, Price announced that all (or nearly all?) of the similar pieces are dealer-produced fakes. Then Gerard Turner in the Christie's 1989 auction catalogue (p. 11) made the unfortunate sweeping comment that "these string-gnomon horizontal dials with magnetic compasses have anachronistic features typical of nineteenth-century reconstructions". All fuel for the fire.

533 On this see now Appendix D, item IV-1, and Appendix F.

534 See King, "Die Astrolabiensammlung in Nürnberg", II, pp. 581-588, for detailed descriptions of both pieces.

535 King, "Astronomical Instruments between East and West", pp. 183-188, already cited in nn. 43 and 484 above.

and now housed elsewhere.[536] See now Appendix D for an updated list of all known instruments from the German-speaking world before *ca.* 1500. This list provides a context for the 1462 astrolabe of Regiomontanus that was altogether lacking in 1989-90.

In 1993-94 Regiomontanus' astrolabe was included in the exhibition "450 Jahre Copernicus *De revolutionibus*" held at the Bibliothek Otto Schäfer in Schweinfurt.[537] Later in 1994 it was featured at the exhibition "Bessarione e l'Umanesimo" held in the Biblioteca Nazionale Marciana in Venice.[538] That same year, the 1462 astrolabe was reassessed – albeit without any new insights into the inscription – in the light of ten other related astrolabes from the Vienna school[539] that had mostly never been seriously studied previously.[540] It then passed quietly to a private collector in England and is available for exhibitions. A few scholars in England are apparently still not convinced of its authenticity.[541] In the meantime, not one single early-19th-century Italian fake astrolabe has come to light.[542]

---

536 *Ibid.*, pp. 189-191. As far as I know, there has been no reaction to my assertion that some of the instruments in Austrian collections (seen by Ernst Zinner in the 1930s) are now housed in museums elsewhere.

537 See my chapter on scientific instruments in *Schweinfurt 1993-1994 Exhibition Catalogue*, pp. 361-381, especially pp. 366-370 (no. 179). The catalogue includes an instrument dated 1484 by Hans Bamann of Schmalkalden (pp. 370-373), with information on other instruments by the same maker. See now items IX-1-6 in Appendix D.

538 The catalogue is listed under *Venice 1994 Exhibition Catalogue*. See, in that volume, Rigo, "Gli interessi astronomici del cardinal Bessarione", pp. 109-110, and the same author's description of the piece on p. 472 (no. 86). A similar study to that mentioned in the next note is published on pp. 340-367, where unfortunately some of the captions are muddled.

539 King & Turner, "Regiomontanus' Astrolabe for Bessarion".

540 See n. 523 on the exceptions.

541 John North, in "Scrutinizing Astrolabes" (2004), pp. 485-486, states that each of the claims made in King & Turner (1994) associating the 1462 astrolabe with Regiomontanus could be countered by a sceptic, but does not mention the ten related astrolabes from the same Vienna workshop, none of which was known to the authors of the Oxford report of 1990. His comments merit citing in full (my emphasis):

> "Here the two authors are ... concerned with a conjectural [!!] reconstruction of the history of one instrument, which famously carries a dedication to the great Cardinal Bessarion, and a date of 1462 ... . Bessarion's is a name to be conjured with [??], but the problems to be faced here are numerous. In the context of the Sotheby's [*sic*, *leg.* Christie's] sale of 1989 in which the instrument appeared, they are not without economic relevance, albeit of a non-historical sort. They include the questions of whether the instrument belongs to the date it bears; whether the inscription belongs to that date, and was not added

> later; whether "opus Ioannis" means a work made by John, as opposed to one owned by John, or one following a design by John; whether this John is to be identified with Regiomontanus; and whether the inscription is a clever piece of punning Renaissance Latin. The two authors opt for affirmative answers in all these cases. (They avoid the less serious questions of whether the late Derek Price was overenthusiastic in calling this the "first scientific instrument of the Renaissance", or in considering the winged human figure on it as a Regiomontanus self-portrait.) Their discussion is thorough, and the circumstantial detail they introduce is invariably interesting, especially on metallurgical questions. **It will never be possible to convince a sceptic of the truth of their conclusion, since every argument they offer can be met with a counter-hypothesis.** They are able to quote, for instance, V. Scholderer's opinion that the majuscules are of much the same style as those Regiomontanus cut for his smaller type, when he came to print at Nuremberg in 1474. The *advocatus diaboli*, however, might ask whether this is strictly true – of the 'N', and 'R', for example. Having been given an answer to that, the sceptic will ask whether it is not obvious that anyone faking a Regiomontanus instrument would naturally have copied his typeface. This is a form of study where probabilities are as subjective as is the punning use of language. Perhaps "de cardine" is a pun, not on cardinal points, but on teasels."

For non-native speakers of English who may have problems with this last remark, we note that "card" is an implement originally consisting of teasel heads set in a frame, for raising the nap on cloth, or a toothed instrument for combing out fibre. A teasel is a plant of the genus Dipsacus, the prickly heads of which are used for teasing cloth. To tease is to separate the fibres of wool, *etc*., but also, at least since the 17th century, to irritate by persistent action. On the etymologies, see Onions, *English Etymological Dictionary*, p. 145b and 906b. On carding, see Osbourne, *Decorative Arts*, p. 820.

542 A miniature medieval Italian astrolabe preserved in Oxford (#162) that was acquired in Florence *ca.* 1900 (see Gunther, *Astrolabes*, II, pp. 319-320) was deemed a bit suspicious by this author because it is so unusual and also because it is non-functional. In fact, it is genuine, and probably dates from *ca.* 1300, but the maker was copying, not necessarily at first hand, a sophisticated Islamic astrolabe that he did not understand. The latitude-dependent markings are for 24°, the second climate, which implies that the original had a set of plates for each of the climates. The first detailed description of this piece in King, "A Remarkable Italian Astrolabe from *ca.* 1300", overlooks the fact that it stems from the same workshop as the even smaller Italian astrolabe depicted in the painting "St Jerome in his Study" by Jan van Eyck (d. 1441). See now *idem*, *In Synchrony with the Heavens*, XIIId-2.7-2.8, and also nn. 60 and 401 above.

On a Renaissance Italian astrolabe (#4507) in the Museo Nazionale del Bargello, Florence, that has been deemed suspicious see Miniati, *Florence Bargello Catalogue*, pp. 48-54, no. 22: "sec. XIX (?)", and King, *op. cit*., X, figs. 4.1.4 and 4.1.3.

For another late-19th-century copy by the Spitzer workshop (on which see n. 487), see Webster & Webster, *Western Astrolabes in Chicago*, pp. 114-116 (no. 28). This is a copy of a spectacular (and genuine) Italian astrolabe from *ca.* 1500 (#171) that is fortunately still extant and preserved in Hamburg: see n. 434 above.

I pay tribute to two specialists on scientific instruments, Derek de Solla Price and Gerard Turner, who recognized the 1462 astrolabe for its merits as a scientific work of art and – because it was so expertly made – saw no reason to doubt its authenticity.

The fact that one does not understand various aspects of a complex historical artefact is no reason to brand it a fake. It happens that Derek Price was the first scholar to write seriously on fake instruments anyway.[545] He was no longer alive when this instrument was declared dubious or even a fake; I knew him well enough to know that he would have been furious, not least because he knew of some possibly related pieces. Price knew the language of instruments,[546] as is evident from his 1959 study as well as his ground-breaking analysis of the Antikythera mechanism in 1974.[547] The source of his expertise is easy to identify: as a graduate student, he had prepared a catalogue of the rich collection of scientific instruments in the British Museum.[548]

Gerard Turner, another scholar who knows the language of instruments, was certainly very happy in 1990 to learn of the existence of ten other astrolabes

---

545 ** See Price, "Fake Antique Scientific Instruments" (1956). All of the examples he cited are from the end of the 19th century.

On the other hand, Price occasionally declared genuine instruments as fake. See n. 532 for his unfortunate remarks about the Vienna sundials: probably only the London one is a fake, but he dismissed the whole lot.

To cite another example, Price did not like the astrolabe in the Metropolitan Museum of Art in New York whose inscription indicates that it was made by a Yemeni sultan (#109). Whoever heard of an astrolabe made by a sultan in, of all places, the Yemen? It took the publication of a notice about a book on astrolabe contruction by the same sultan, and remarks by the sultan's teachers about five astrolabes made by him, one the New York piece, to allay all fears that this was a fake. See King, *In Synchrony with the Heavens*, XIVa.

Also, Price did not like a medieval English horary quadrant in the British Museum because it had symbols for the zodiacal signs engraved on it. Gerald Turner later showed that these were added by the 16th-century English astronomer Charles Whitwell. Compare Price, "Instruments in the British Museum", p. 131, and Turner, "Whitwell's Addition to a 14th-Century Quadrant".

546 For an obituary notice see G. Turner, "Derek John de Solla Price – 1922-1983", and for a tribute to Price's writings on astronomical instrumentation, see King, *In Synchrony with the Heavens*, introduction to XIIb.

547 See n. 19 above.

548 See his "Instruments in the British Museum".

from the same Vienna school, one with identical engraving on the rete and another with an *organum Ptolemaei* on the back. For this younger colleague of his, the documentation of these pieces provided a very exciting excursion from the Islamic Renaissance into the European Renaissance.

It took considerable effort to reinstate the 1462 astrolabe to its rightful place in the history of 15th-century instrumentation. Nevertheless, it (like the Byzantine astrolabe) has been deliberately omitted from several recent overviews of the history of the astrolabe.[549] Over the years the piece has been discussed numerous times in seminars at Frankfurt, and then, in 2005, ...[550]

549 See, for example, the brief histories of the astrolabe in J. Evans, *Ancient Astronomy*, pp. 155-158, and Schechner Genuth, "Astrolabes: A Cross-Cultural and Social Perspective".

550 The reason why the 1462 astrolabe was brought up yet again in the Seminar was to discuss the comments of John North (see n. 541) in the light of the 1990 Oxford report and the 1994 King & Turner study of the piece within the context of ten other "Regiomontanus-type" astrolabes.

Appendix F – The so-called "astrolabe of Regiomontanus" dated 1468

In 1932 Robert Gunther illustrated as "The Astrolabe of Regiomontanus" an astrolabic plate dated 1468 (#555) that is preserved in Nuremberg. (See Gunther, *Astrolabes of the World*, I, pp. 280-281 (no. 137), and item IV-1 in Appendix D.) The same attribution occurs more recently in Wattenberg, "Regiomontanus und die astronomischen Instrumente seiner Zeit", fig. 4, and p. 350, where it is stated that the piece stems from the astronomer's time in Hungary, with latitude-dependent markings for latitude 50°. In fact, the markings are for latitude *ca.* 46.5°, but there is an earlier horizon engraved for *ca.* 49°. See also King & Turner, "Regiomontanus' Astrolabe for Bessarion", p. 187. The piece may have belonged to Regiomontanus but it was not made by him. Gunther's description confuses it with a spectacularly beautiful 12th-century Islamic astrolabe (#137) engraved with circus figures that is also preserved in Nuremberg. Both instruments are now catalogued in detail.[562] They are mentioned here only to

---

562 ** King, "Die Astrolabiensammlung in Nürnberg", II, pp. 589-592 and 570-574. On the Islamic piece see also *idem*, *In Synchrony with the Heavens*, XIVb-2.

save possible confusion in the future,[563] some of which may be generated by a Florida-based outfit named Clipart ETC, now circulating images of "Regiomontanus' astrolabe".[564]

Fig. F1a-b:~ The front and back of the German astrolabe dated 1468 (#555) that probably belonged to Regiomontanus, although he surely did not make it. [Courtesy of the Germanisches Nationalmuseum, Nuremberg.]

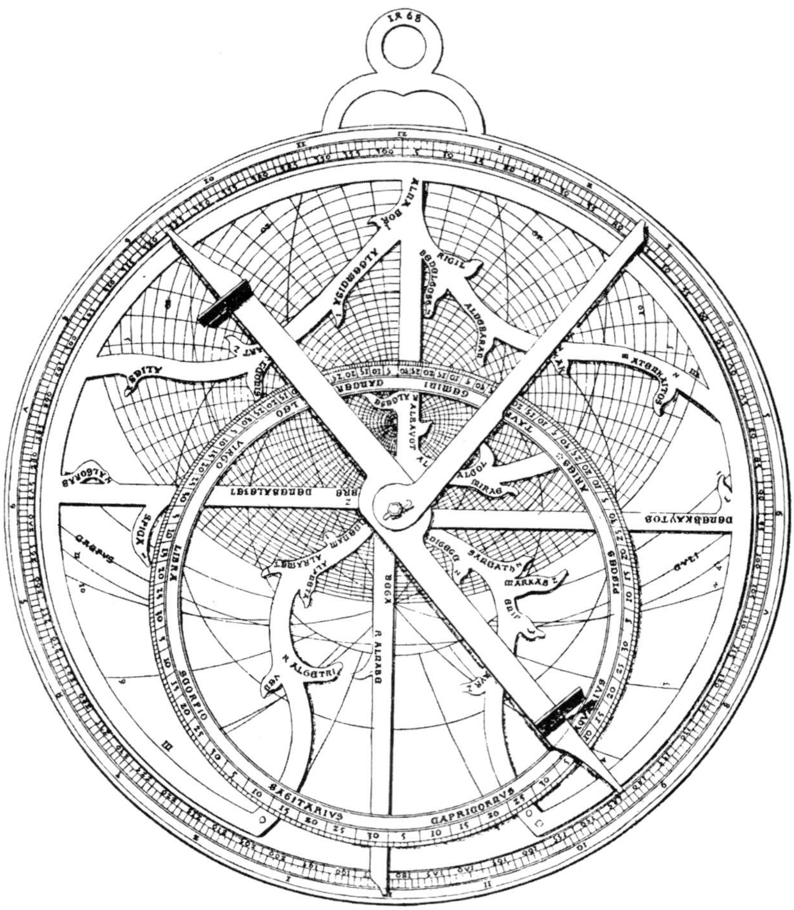

FIG. 129. ASTROLABE OF REGIOMONTANUS. 'A.D. 1468.'

---

563 In Samhaber, *Die Zeitzither*, p. 199, a different Islamic astrolabe, this one from 11th-century al-Andalus, is illustrated as the "Astrolab des Regiomontanus, 1468".

564 See http://etc.usf.edu/clipart/25100/25160/astrolabe_25160.htm .

Fig. F1a

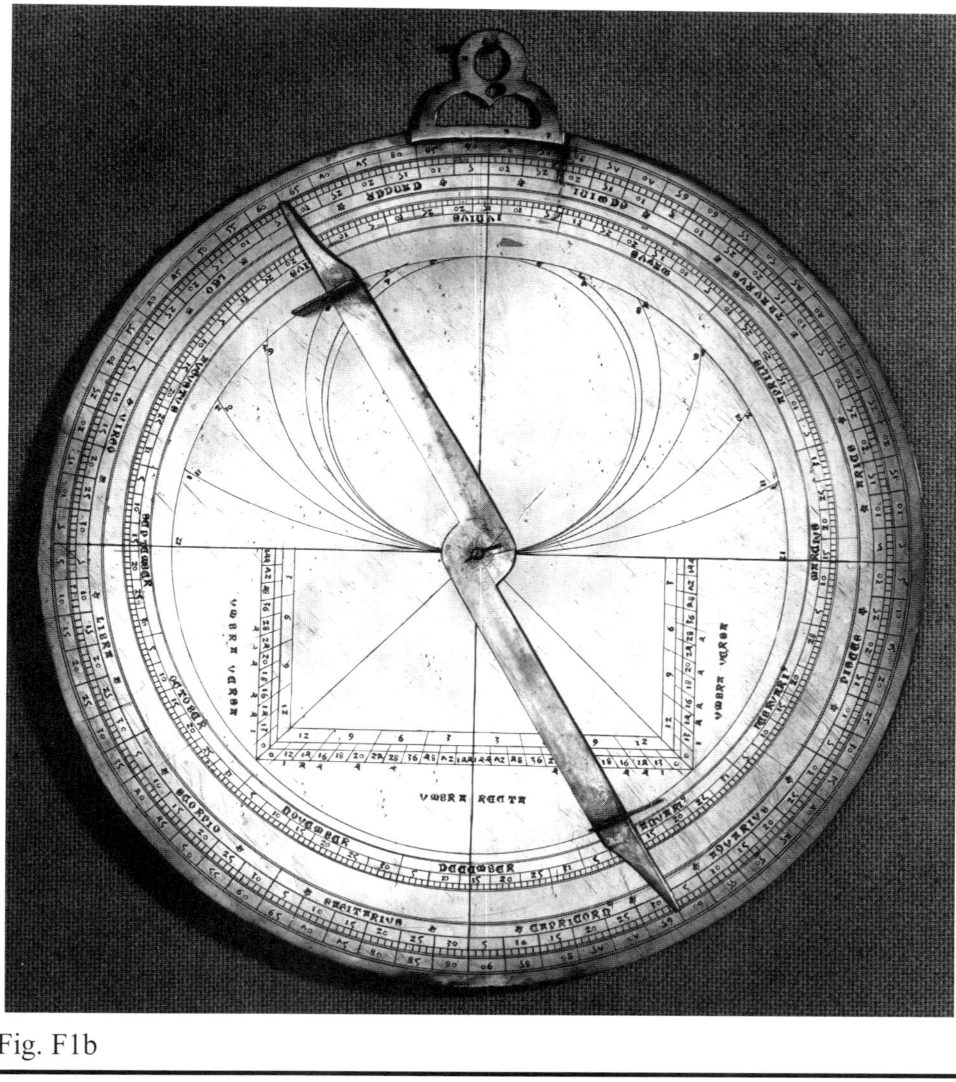

Fig. F1b

Appendix G – Notes on the 1462 Urbino astrolabe

When a historian of science looks at the history of Urbino, the emphasis can be refreshingly different from the usual.[565] Yet there are surely more materials available for the study of the history of astronomy and astrology at the Court of Federigo than have been exploited to date, and the purpose of this note is simply to draw attention to what has been written already, and, in particular, to some astrolabes from 15th-century Urbino.

A single report – and there are surely many others – indicates that important events in the life of Federigo, such as the propitious moment to depart for a battle with a neighbour, were based on astrological computations and the appropriate astrological prognostications. This report, from March 11, 1460, relates that he left Urbino "last night at 7 hours and three minutes", "at a time determined by astrology" (*per puncto de astrologia*).[566]

It is surely significant that no astronomers or astrologers are mentioned in the book of staff rules at the Ducal Palace.[567] In this, the duties of even the least minions are outlined, down to the pages and the cooks. The librarian is mentioned, but no others linked with scholarship of any kind. Perhaps the astrologers were so close to the Duke that they need not have been mentioned amongst his staff. Chief among them was Federigo's second in command, Ottaviano Ubaldini dalla Carda.[568]

A very mysterious figure is the German Jakob von Speyer, Iacobus Spierensis, of whom we know very little other that he was the court astrologer in Urbino in 1465, a position that he apparently held until 1481. Susech of Castel Durante, a former page in the Duke's service, drew up a list of his courtiers and servants, and under "Maestri del Duca in Gramatica, Logica e Filosofia", he listed "Maestro Iacomo Astrologo"; our Jakob was a priest with the Church of Sant'Antonio outside the Porta Valbona of Urbino (the present-day Convento

---

565 See Vetrano, "Renaissance grandeur and scientific culture in Urbino", pp. 17-22 of *idem*, *Il Gabinetto di Fisica dell'Università di Urbino*, and the various articles in *idem*, ed., *La scienza del Ducato di Urbino*, all with English translations.

566 Cited in Bonvini Mazzanti, *Battista Sforza Montefeltro*, p. 74.

567 See the various studies and an edition in Eiche, ed., *Ordine et officij*.

568 On astronomy and astrology at the Court of Urbino see further Michelini Tocci, "Federico da Montefeltro e Ottaviano Ubaldini dalla Carda"; Apa, "Cultura figurativa e dibatto sull'astronomia ad Urbino nel secolo XV"; Guerrini, "Federico da Montefeltro e Sisto IV"; and Garbini, "Corti e comete: lo scritto di Nicolò Conti sulla cometa del 1456". On the more general context see also Pepe, "Astronomia e matematica nelle università italiane del Quattrocento".

dei Cappuccini).[569] Regiomontanus corresponded with him from Rome and then from Viterbo in 1465.[570] It is clear that by then they had not yet met, if they ever did. Susech also listed "Maestro Paolo Astrologo", who was the Dutch astronomer-astrologer Paulus de Middelburg (1445-1533), formerly lector in astrology at the University of Padua, then (1486) abbot of San Cristoforo, Castel Durante, later (1504) bishop of Fossombrone; he apparently became Jacob's successor in Urbino in 1481 and in 1512 he was a member of the commission established to reform the calendar.[571]

Amongst the portraits of scholars in the *studiolo* of Federigo in the Ducal Palace,[572] Ptolemy of Alexandria and Euclid of Megara are the representatives of serious mathematical sciences.[573] Also symbolic of Federigo's interest in astronomy and astrology are the images of an armillary sphere and an astrolabe in one of the intarsias in the same room: see **Figs. G1a-b**. The former accurately displays the principal circles of the celestial sphere, but little historical information can be derived from it. However, the astrolabe, depicted in minute detail, is more revealing.[574]

569 Clough, "Federigo da Montefeltro's Patronage of the Arts", pp. 131-132.

570 On Jakob von Speyer see already nn. 31, 96, and 119 above, also nn. 580-581 below.

571 See Struik, "Paulus van Middelburg"; also Zinner, *Regiomontanus*, p. 242; North, "The Western Calendar", pp. 95-96; and Clough, "Federigo da Montefeltro's Patronage of the Arts", pp. 132-133.

572 On the *studiolo* in Urbino see Cheles, *The Studiolo of Urbino*, and Kühnast, *Studien zum Studiolo zu Urbino*, and on the studiolo in Gubbio see Winternitz, "Quattrocento Science in the Gubbio Study". On the intarsia and the paintings see Dal Poggetto, *La Galleria Nazionale delle Marche*, pp. 132-136 and 127-131.

573 I have not found the image of Ptolemy in the secondary literature. For the image of Euclid see Dal Poggetto, *op. cit.*, fig. 127 on p. 129, and also Brilli, *Alla ricerca di Piero della Francesca*, p. 203.

574 See King, "The Astrolabe in the Urbino Intarsia", for a detailed account of the three astrolabes – Urbino intarsia, Florence and ex-Moulins. I find it revealing that this study, published in Urbino in 2001, does not make it into a very recent 13-page bibliography on Federigo's Palace with 14 other entries for 2001 alone (see *Urbino 2005 Exhibition Catalogue*).

For more information on the stars represented on the medieval astrolabe (and also the other two) see *idem*, "Star-Names on Three 14th-Century Astrolabes", pp. 320-325 (the other two here are Spanish and French). For an overview see also *idem*, *In Synchrony with the Heavens*, figs. XIIIc-10.4a-c, also van Cleempoel, "Representations of Astrolabes in Western Art", p. 107.

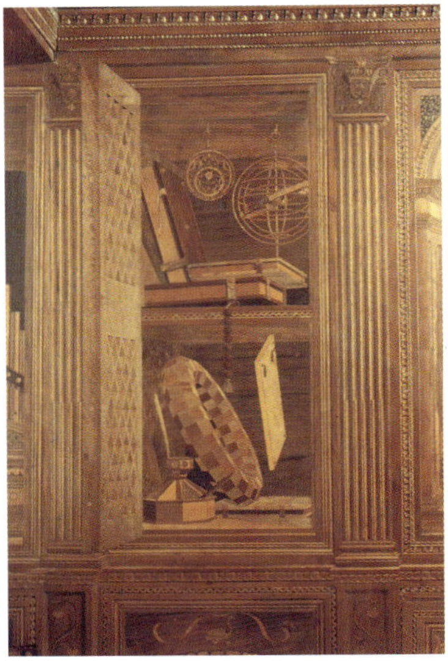

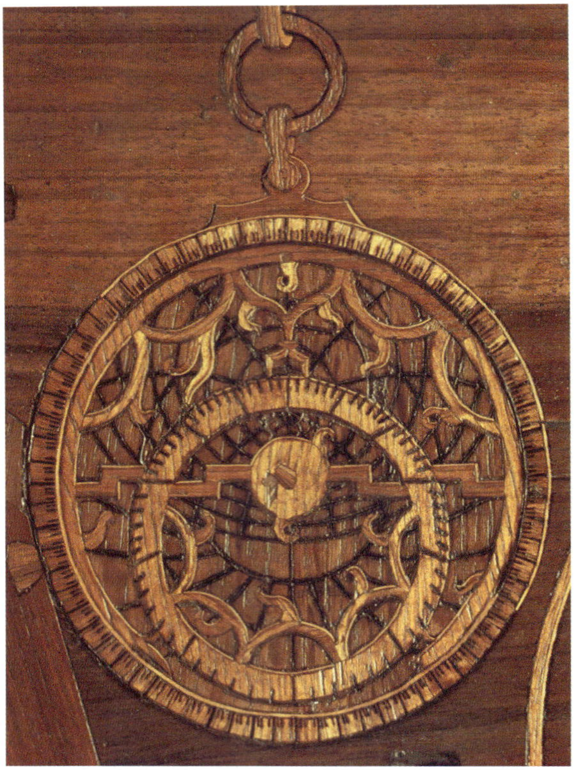

Figs. G1a-b:~ The astrolabe in the intarsia of the studiolo of Federigo da Montefeltro. [Courtesy of Professor Roberto Mantovani, Urbino.]

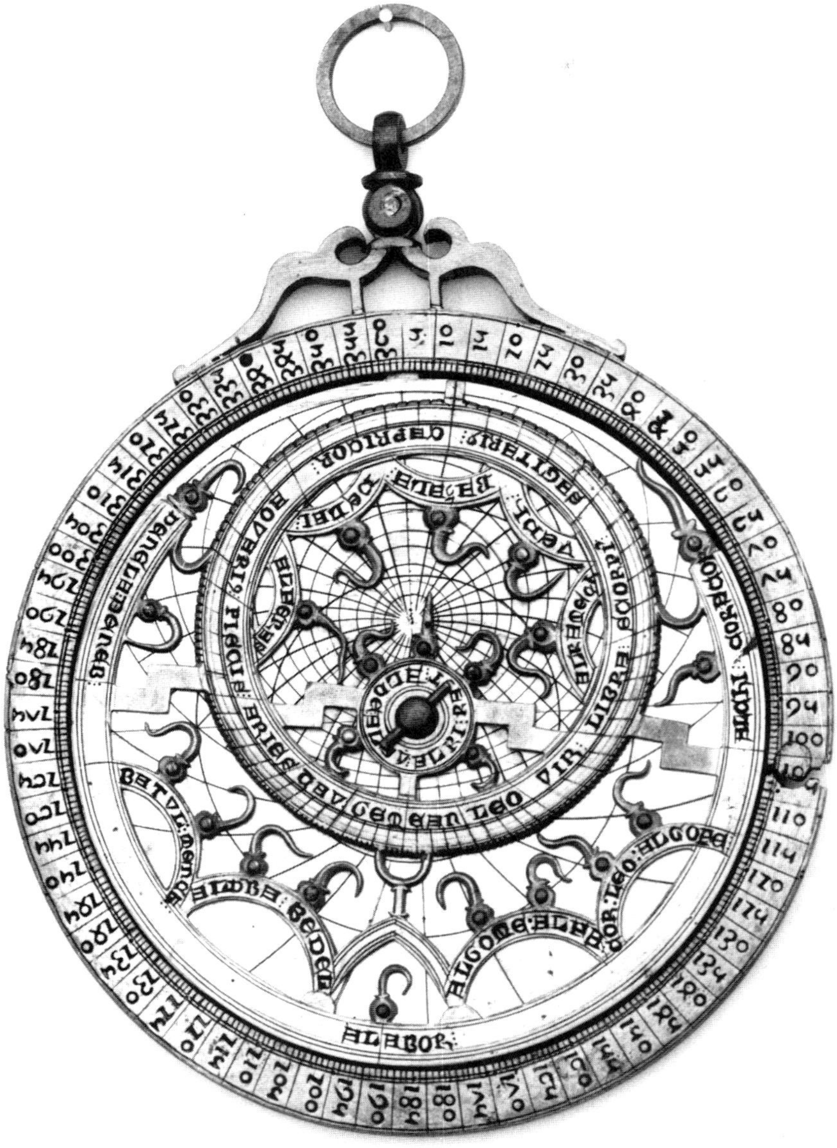

Fig. G2:~ The front of a "medieval" Italian astrolabe (#493). [Photos courtesy of the Museo di Storia della Scienza, Florence.]

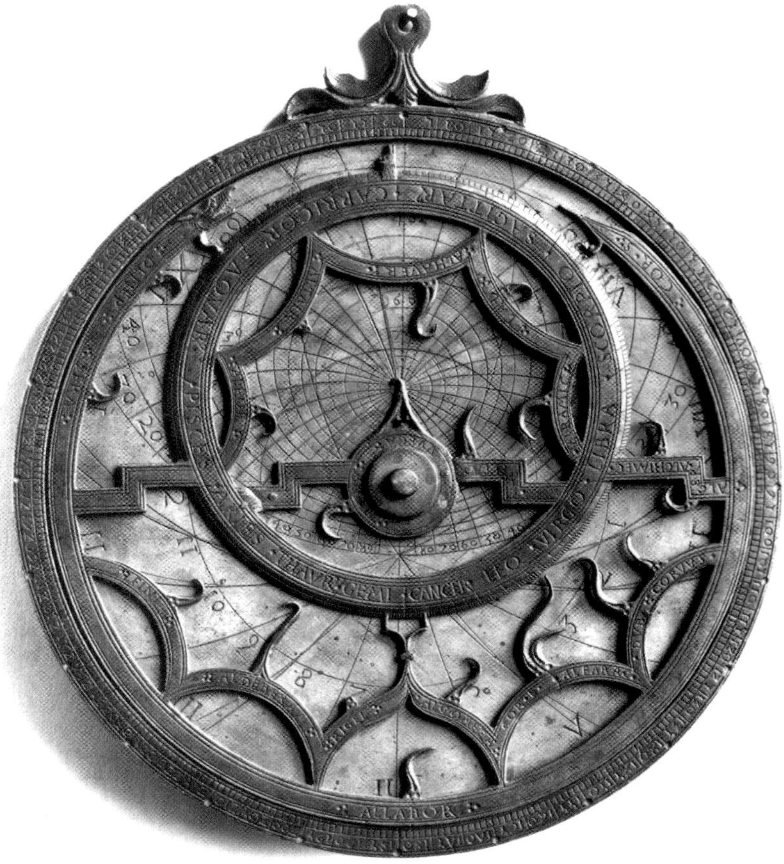

Fig. G3:~ The front of the 1462 Urbino astrolabe (#4506), with two details. [Photos courtesy of the Museum of the History of Science, Oxford.]

The distinctive design of the rete is found on a medieval Italian astrolabe (#493) preserved in the Museo di Storia della Scienza in Florence: see **Fig. G2**. An identical rete design is found on a Renaissance astrolabe (#4506) that was stolen in 1977 from a French provincial museum, now the Musée Anne de Beaujeu in Moulins, Allier, and has not resurfaced: see **Fig. G3**.[575] The Antiqua engraving

[575] It is mentioned in a catalogue of the Moulins museum in 1896, and in the card index of M. Bertrand, the *conservateur* of the museum from 1893 to 1912. It was possibly acquired at an auction at Gannat (Allier) where the estate of the distinguished French admiral and naval historian Jean-Pierre-Edmond Jurien de la Gravière (1812-1892) was dispersed. (His father Pierre-Roch (1772-1849) had been a vice-admiral.) There may be

on the last of these and the incorrect forms THAVRVS and VNBRA have led to the hypothesis that the maker was a German. Two additional inscriptions on the back read "1462 VRBINI K" and "KP", and the meanings of these two abbreviations are not clear. A further complication is that the only astronomical markings that are visible (and the only ones recorded in a 1973 catalogue) are for latitude 37°, serving Sicily, suggesting that it was destined for use there, and only there.[576]

The only relevant name that one could abbreviate with one letter as "K" and with two letters as "KP" would be surely "Koenigsperg". Although one might have expected a Latin name from Ioannes Germanus de Regio Monte, who may have passed near Urbino in late 1461, he gave his name as Ioannes de KvnigsPerg in the dedication of the *Epitome* to Bessarion,[577] and signed his name as Ioannes Germanus de Regiomonte in 1463.[578] Besides, the engraving of these letters appears to be different from that on the rest of the instrument, and it is also different from that on the 1462 Regiomontanus astrolabe.

However, there is another possibility, namely that – although it does not look like it – the "K" might be a ligature for "IC", so that "KP" could stand for "ICP".[579] Then we may be a step closer, for Federigo's court astronomer was the German Jakob von Speyer.[580] He signed his name and was addressed as IACOBVS DE SPIRA or SPIRENSIS in his correspondance with

an article mentioning this purchase in *Bulletin de la Société d'Émulation du Bourbonnais* 40 (1937), which we have not seen. We owe this information to the kindness of Anne Adrian, *Conservateur du patrimoine* at the Musée Anne de Beaujeu (letter of 11.03.2005).

On the father and son and their world-wide exploits see the articles in *Dictionnaire de Biographie Française*, XVIII (1994), cols. 1039-1040 and 1038-1039, by J.-G. Kirchheimer. From these notices, it is not clear that either visited Sicily. The son was also the author of *Les Chevaliers de Malte et la marine de Philippe II*, and *Les Corsaires barbaresques et la marine de Soliman le Grand*, both Paris, 1887, neither mentioned in *op. cit.*

576 See *Paris BN 1973 Exhibition Catalogue*, p. 75 (no. 176), with illustrations on p. 57, and King, "The Astrolabe in the Urbino Intarsia", p. 132.

577 Zinner, *Regiomontanus*, p. 81.

578 *Ibid.*, p. 320.

579 This was suggested by Kurt Maier of Frankfurt in 1991-92.

580 See n. 570 and the references there cited.

Regiomontanus.[581] More convincing of an association with Jakob would be an abbreviation such as "IAC SP".[582]

We cannot claim to have resolved the problem of these inscriptions, and, alas, it is not yet possible to examine the stolen 1462 Urbino astrolabe. It is not too much to hope that it will eventually resurface.

---

581 Curtze, "Regiomontan's Briefwechsel", pp. 292 (*Ioannes Germanus ad Iacobum de Spira, astronomum domini Federici Comitis Urbinatum*, from Rome, 1465), 302 and 309 (from *Iacobus Spirensis* in Urbino to *Ioannes Germanus* in Rome, 1465).

582 Cappelli, *Lexicon abbreviaturarum*, pp. 170-171, gives "Ia" and "Iac" as attested abbreviations of Iacobus, whereas "IC" is essentially used for Iesus Christus. On IACO for Iacobus see n. 277 above.

Appendix H – On astronomy in the Islamic Renaissance

It is customary to use the term Renaissance only in relation to "our" reappropriation of the Greek and Latin heritage. When "others" appropriate things like Indian trigonometry and Iranian astrological doctrines – even when these are subsequently adopted by "us" and we do not know that they were originally Indian or Iranian – we are less interested.

At least for this author, there was an earlier Renaissance in astronomy and mathematics, and also other sciences, as well as humanism in general, in the world of Islam. It will never be possible to rewrite history with the Renaissance beginning in Baghdad in the 8th-10th centuries. In this study we should keep in mind that *astronomy and astronomical instrumentation in the Islamic world were more advanced than in Europe until the 15th century*. There were people with interests and abilities similar to those of Regiomontanus and Bessarion (and Alberti) in Baghdad and other centres in the 9th and 10th centuries.[583] And we should judge them by (a) the astronomical theories they developed; (b) the sophistication of the mathematics they developed; (c) the astronomical tables they compiled; and (d) the astronomical instruments they conceived and used. We should not judge them by the fact that most of their works were not known in medieval or Renaissance Europe and hence had no great influence in European astronomy. If they, rather than the Europeans, had invented the art of printing, the course of history would have been very different.

The most staggering renaissance in astronomy in human history took place under the early Abbasids (late 8th-10th centuries), and in this the Muslims adapted to their own needs all that they wanted from pre-Islamic Arabian, Greek, Persian and Indian sources.[584] Considerable attention was paid to what I

---

583 For their names and works see Sezgin, *Geschichte des arabischen Schrifttums*, vols. III-VII and X-XII. For what happened in the field of mathematical astronomy see the overview in King, "Islamic Astronomy", also *idem* & Samsó, "Islamic Astronomical Handbooks and Tables".

I mention Alberti to draw attention to the recent publication Mrayati, Mir ʻAlam & Tayyan, *New Contributions to the History of Cryptology and Origins of Arab Cryptography and Cryptanalysis*,

584 For more references see "The Renaissance of Astronomy in Baghdad in the Ninth and Tenth Centuries – A List of Publications" at

www.uni-frankfurt.de/fb13/ign/astronomy_in_baghdad/bibliography.html

and also n. 19 (Mecca-centred world maps), the text to n. 42 and n. 154 (earlier Islamic instruments more sophisticated than that of Regiomontanus). See, in particular, Pingree, "The Greek Influence on Early Islamic Mathematical Astronomy", and "Indian Influence on Sasanian and Early Islamic Astronomy and Astrology"; Sabra, "Appropriation and Naturalization of Greek Science in Islam"; and Gutas, *Greek*

call "astronomy in the service of Islam". Astronomy from the 11th century onwards flourished mainly in regional schools.[585] And why Islamic science did not progress after the 16th century is another matter, the most satisfying explanations being that all of the problems of medieval astronomy had been solved, some many times over, lack of communication between the different schools, and lack of access to printing (in other words, the manuscript tradition essentially lasted into the 20th century).[586]

The most complicated tables of Regiomontanus, the *Tabula primi mobilis*, were of functions already tabulated by Habash in the 9th century.[587] The astrolabe of the astronomer al-Khujandî, made in Baghdad in 984/85 (see **Fig. B2**), was technically more sophisticated and arguably more beautiful than that of Regiomontanus.[588] The universal astrolabe of Ibn al-Sarrâj (see **Fig. B5**), made in Aleppo in 1328/29, would have blown Regiomontanus' mind, had he seen it.[589] I have hypothesized that Regiomontanus' most sophisticated instrument, the *Uhrtäfelchen*, or universal horary dial for timekeeping by the sun for all latitudes, which he admits he adopted from earlier sources, was a 9th- or early-10th-century Baghdad invention.[590] The strongest evidence for this is that from that milieu we have a related, but more complicated universal device by Habash al-Hâsib for timekeeping for all latitudes by the fixed stars.[591] Also, an 18th-century Arabic illustration of a universal dial for the sun appears to be copied from an earlier Islamic one.[592]

Thought, Arabic Culture. On the Buyid renaissance (late-10th-mid-11th centuries) see Mez, *The Renaissance of Islam*, and Kraemer, *Humanism in the Renaissance of Islam*. See also Darling, "The Renaissance and the Middle East", in which the Buyid and Timurid (15th century) renaissances are mentioned.

585 An extensive bibliography by region is given in the article "Zîdj" [astronomical handbooks and tables] in *Encyclopaedia of Islam*, 2nd edn., XI, pp. 507-508.

586 Much nonsense has been written on this subject, inevitably mainly by persons who have no conception of the richness of Islamic astronomy. To understand this decline it is useful to consider the astronomical tradition in one particular region – see my *Astronomy in Medieval Yemen* (although there, alas, the very earliest, highly sophisitcated works from the 10th century are not extant) – or the various contributions to one particular aspect of astronomy – see my *In Synchrony with the Heavens*, I (dealing with astronomical timekeeping. In both cases, astronomy progessed from the 9th century to the 13th and 14th centuries and then declined until the 19th century.

587 King, *In Synchrony with the Heavens*, I-9.1 and 10.2.

588 *Ibid.*, XIIIc-9, also the text to n. 420 above.

589 *Ibid.*, XIVb-5.1, and the text to n. 423 above.

590 *Ibid.*, XIIb.

591 Charette & Schmidl, "Universal Plate for Timekeeping by the Stars by Habash", also King, *op. cit.*, XIIb-12.

592 *Ibid.*, fig. XIIb-11a.

Competent Arabic translations of the *Almagest* were available in the 9th century, and from the outset, and right through to the 15th century, there were serious attempts by Muslim scholars to correct and to improve upon Ptolemy's work. The latter were mainly unknown in the West until the second half of the 20th century, and most of them would have been unknown in the new Ottoman capital.[593] Not only were Copernicus' models copied from Ibn al-Shâtir (Damascus, *ca.* 1360), but Copernicus misunderstood his predecessor's most complicated model, that of Mercury.[594] For decades we have been searching for "the missing link" to Regiomontanus' statement that eventually led to the so-called Copernican revolution.[595]

Other aspects of astronomy were cultivated in the new Ottoman capital. Some of the achievements of Mamluk Syrian and Egyptian scholars were known there, and Turkish scholars already in the 15th century made new contributions to astronomical timekeeping and instrumentation.[596] No astrolabes survive from that milieu, although we do have two astrolabes made for the Sultan Bâyazît II *ca.* 1505.[597]

---

593 See various studies on Islamic solar, lunar and planetary theory collected in Kennedy *et al.*, *Studies in the Islamic Exact Sciences*; Hartner, "Ptolemäische Astronomie im Islam und zur Zeit des Regiomontanus"; Saliba, *A History of Arabic Planetary Theories*; and King & Samsó, "Islamic Astronomical Handbooks and Tables", pp. 69-78.

594 See Swerdlow, "Copernicus's *Commentariolus*", p. 504

595 See now Ragep, "Freeing Astronomy from Philosophy" (2001) and "'Alî Qushjî and Regiomontanus" (2004), and Saliba, *Islamic Science and the Making of the European Renaissance*,

596 King, *In Synchrony with the Heavens*, II-14.

597 See further King, *In Synchrony with the Heavens*, XIVe, and *idem*, "Two Astrolabes for Sultan Bayazit II".

Bibliographical notes

Note: No claim is made that these notes are in any way exhaustive. For all of the subjects treated, much more information could be given. What is needed will depend on the reader. For some of the subjects treated here, the best available literature may be in any of the languages English, German, Italian or French.

Dramatis personae

Alberti. The literature is vast. The article by Bertrand Gille in *Dictionary of Scientific Biography* deals only with his scientific works. On his cryptographical writings see Meister, *Geschichte der italienischen Kryptographie des XV. Jahrhunderts*, and *Die Geheimschrift im Dienste der päpstlichen Kurie*, also Alberti, *De cifra*. For his treatise on painting see Alberti, *Della pittura*. For an insight into current interest in him see:

ourworld.compuserve.com/homepages/mpaoli/ .

Archimedes. See Heath, *The Works of Archimedes*, the article "Archimedes" by Marshall Clagget in the *Dictionary of Scientific Biography*, and, on the medieval and Renaissance traditions, the same author's *Archimedes in the Middle Ages*. See also Rose, *The Renaissance of Mathematics in Italy*, p. 94, Zinner, *Regiomontanus*, p. 91, and Banker, "A Manuscript of the Works of Archimedes in the Hand of Piero della Francesca", on the manuscripts of Bessarion, Regiomontanus and Piero.

Bernardino Ubaldini, son of Ottaviano Ubaldini, who died in his late teens of the Plague in 1458. He and Buonconte (see below) became infected whilst travelling together.

Bessarion. The standard works are Mohler, *Bessarion*; Labowsky, *Bessarion's Library*; Monfasani, *Byzantine Scholars in Renaissance Italy*; and numerous studies in the volume *Venice 1994 Exhibition Catalogue*: Fiaccadori, ed., *Bessarione e l'Umanesimo*. See also Loenertz, "Bessarione"; Labowsky, "Bessarione"; and Zorzi, "Cenni sulla vita e sulla figura di Bessarione". A detailed bibliography is in Bianca, *Da Bisanzio a Roma*, pp. 175-216. On his year of birth see Monfasani, *op. cit.*, VI. On Bessarion and Regiomontanus in Italy see Rose, *The Italian Renaissance of Mathematics*, pp. 90-117, and Rigo, "Bessarione e Regiomontano". On Bessarion and astronomy see also Rigo, "Gli interessi astronomici del cardinal Bessarione". On Bessarion and the Turkish threat see Carretto, "Bessarione e il Turco", and Bisaha, *Creating East and West*, pp. 108-114. On images of Bessarion see Lollino, "L'iconografia di Bessarione". See also Regiomontanus below.

St Bessarion. The articles "Bessarione" by Joseph-Marie Sauget in *Bibliotheca sanctorum*, III (1963), cols. 140-141, and "Bessarion, Saint" by Lucien Regnault in *The Coptic Encyclopedia*, II (1991), p. 379. For his veneration as an angel see the translation of passages from the Byzantine liturgy in Anonymous, "Die Engelstellen im Triodion der byzantinischen Liturgie", p. 4, and also, for example, Bohác *et al.*, eds. *Apostolos*, p. 650. For the monk Basileios Bessarion's 1423 treatise on his patron saint see Ioannou, "Opuscule du cardinal Bessarion".

Buonconte da Montefeltro, natural son of Federigo da Montefeltro, died of the Plague in 1458. See Franceschini, "La morte di Buonconte"; Clough, "Cardinal Bessarion and Greek at the Court of Urbino", pp. 161-163; Monfasani, *Byzantine Scholars in Renaissance Italy*, IX, pp. 95-98; and Ginzburg, *The Enigma of Piero*, pp. 82-85 and 117-118.

Euclid. See the article by Ivor Bulmer-Thomas in the *Dictionary of Scientific Biography*. His *Elements* are accessible in English in Heath, *Euclid's Elements*. See also Folkerts, "Regiomontans Euklidhandschriften", "Piero della Francesca and Euclid", and "Luca Pacioli and Euclid".

Federigo da Montefeltro. The literature is abundant. Essential reading is Denis Mack Smith's splendid essay, *Federigo da Montefeltro*. A respected study is Tommasoli, *La vita di Federico da Montefeltro*. An early work in French is by de La Sizeranne (1927), with an undated bibliography in the 1979 Italian translation, *Federico di Montefeltro*. See also the various studies in the three volumes of Cerboni Baiardi *et al.*, eds., *Federico di Montefeltro*, dealing with politics, art and culture. A very useful survey on the Internet is Anonymous, "Urbino in the Age of Federico da Montefeltro". A new popular study is Lauts & Herzner, *Federico da Montefeltro*. For Federigo's relations with Bessarion see Monfasani, *Byzantine Scholars in Renaissance Italy*, IX.

Luca Pacioli. Article by S. A. Jayawardene in the *Dictionary of Scientific Biography*, also Crosby, *The Measure of Reality*, pp. 210-223. See more recently *Sansepolcro 1994 Exhibition Catalogue*; and *Sansepolcro 1994 Conference Proceedings*, especially Folkerts, "Luca Pacioli and Euclid", and Jayawardene, "Towards a Biography of Luca Pacioli".

Georg Peuerbach. Article "Peurbach" by C. Doris Hellman and Noel M. Swerdlow in *Dictionary of Scientific Biography*, Supplement. Also Samhaber, *Die Zeitzither – Georg von Peuerbach*, and *idem*, *Der Kaiser und sein Astronom*.

George of Trebizond. See Monfasani, *George of Trebizond*, and *idem*, *Collectanea Trapezuntiana*, and also Bádenas, "Byzantine Intellectuals at the Ottoman Court", pp. 4-5, and Bisaha, *Creating East and West*, pp. 153-156. On his scientific work see Rose, *The Italian Renaissance of Mathematics*, pp. 41-42 and 101-102; and Shank, "Regiomontanus against George of Trebizond".

Giovanni Bacci. I have relied mainly on Ginzburg, *The Enigma of Piero*: see his index on p. 153.

Ludovigo Gonzaga, Marquis of Mantua, life-long friend of Ottaviano Ubaldini dalla Carda. There is no article "Gonzaga, Ludovico III" in *Dizionario biografico degli Italiani*, vol. 57; see Lavin, *Piero's Flagellation*, pp. 53-62.

Matthias Corvinus, King of Hungary (*reg.* 1458-1490). See, for example, Gründler *et al.*, eds., *Matthias Corvinus und die Renaissance in Ungarn.*

Ottaviano Ubaldini dalla Carda, nephew of Federigo da Montefeltro, his chief counsellor and treasurer of state, passionately interested in astrology. See Michelini Tocci, "Federico da Montefeltro e Ottaviano Ubaldini dalla Carda", and *idem*, "Ottaviano Ubaldini della Carda ... "; Lavin, *Piero's Flagellation*, pp. 62-67, also Curtze, "Regiomontan's Briefwechsel", pp. 292, 301, 309.

Piero della Francesca. The literature is abundant: recent bibliographies are in Calvesi, *Piero della Francesca*, pp. 227-236, and Wood, ed., *The Cambridge Companion to Piero della Francesca*, pp. 257-262. A recent overview is in the article "Franceschi, Piero dei" by R. Lightbown in the *Dizionario biografico degli Italiani*. On his early life see Banker, *The Culture of San Sepolcro during the Youth of Piero della Francesca*. For detailed discussions of all of his paintings and frescoes and superb illustrations see, in addition to the early works of Roberto Longhi and Kenneth Clark, for example, Battisti, *Piero della Francesca*, Calvesi, *Piero della Francesca*, and also Maetzke, *Piero della Francesca*. A useful guide to the "Piero trail", richly illustrated with historical maps and early lithographs of important localities, as well as the usual paintings, is Brilli, *Alla ricerca di Piero della Francesca*. On Piero and Urbino see the various studies in *Urbino 1992 Exhibition Catalogue*. On his mathematical treatises see various studies in *Arezzo-Sansepolcro 1992 Conference Proceedings*; also Daly Davis, *Piero della Francesca's Mathematical Treatises*, and Folkerts, "Piero and Euclid". For various editions see Calvesi, *op. cit.*, p. 223 *ad* pp. 68-80. For an overview see Field, *The Invention of Infinity*, pp. 62-113. It is appropriate that there is an article "Francesca, Piero della" by William P. D. Whiteman in the *Dictionary of Scientific Biography*. Field, *Piero della Francesca: A Mathematician's Art*, offers a new approach by a historian of science but does not present any mathematical investigations of his most mathematical paintings. Surveys of previous opinions on **"The Flagellation"** are in Gouma-Peterson, "Piero's Flagellation", pp. 217-218; Lavin, "*Monarca della Pittura*: Piero and his Legacy", especially the appendixes; and Calvesi,

Piero, pp. 105-116 and 207-208. A bibliography published in 2005 is in Marchi, "Piero della Francesca : *La Flagellazione*", p. 185. Particularly important studies are Gouma-Peterson, "Piero's Flagellation"; Lavin, *Piero's Flagellation* and *eadem*, *Piero*; Ginzburg, *The Enigma of Piero*; and Calvesi, *Piero*. Relevant and provocative information is in Ronchey, "Malatesta/Paleologhi", pp. 543-557, and *eadem*, *L'Enigma di Piero*, with additional material available at the associated website. An important study of the perspective is Wittkower & Carter, "The Perspective of Piero della Francesca's 'Flagellation'". See now the discussion in Andersen, *The Geometry of an Art*, pp. 34-80. On animated three-dimensional reconstructions see Criminisi, *Accurate Visual Metrology*, pp. 140-142 and colour pls. 4-5, and his various websites, also Field, *op. cit.*, pp. 176-177. For the basic geometry of the painting see:

www.artemis.jussieu.fr/hermes/hermes/actes/ac0102/flagellation.gif ,

alas without attribution or commentary. (Not all of the construction lines given there are meaningful.)

Ptolemy. Article by Gerald Toomer in *Dictionary of Scientific Biography*. For the text of the *Almagest* and a commentary see the publications of Toomer and Olaf Pedersen listed under astronomy below.

Regiomontanus. Article by Edward Rosen in *Dictionary of Scientific Biography*. The biography of Pierre Gassendi (1651) is not fully reliable: for an English version see Hughes, *Regiomontanus on Triangles*, pp. 11-18. The standard modern works are Zinner, *Regiomontanus*, available in German and English, and various articles in Hamann, ed., *Regiomontan-Studien*; a new work of a more popular nature is Mett, *Regiomontanus*. On his prowess as a poet, see Zinner, *Regiomontanus*, pp. 40, 73, 78, 130, 276 and 309. On his correspondence see Curtze, "Regiomontan's Briefwechsel", with a German translation of the non-mathematical sections in Zinsmeister, "Der Briefwechsel Regiomontans", some, but not all of which are included in Mett, *Regiomontanus*, pp. 69-84. On Regiomontanus and Bessarion in Italy, see Rose, *The Italian Renaissance of Mathematics*, pp. 90-117. See also Shank, "The Classical Tradition in Fifteenth-Century Vienna"; Swerdlow, "Astronomy in the Renaissance", *idem*, "Recovery of the Exact Sciences", and especially *idem*, *The Renaissance of Astronomy in the Age of Humanism* (forthcoming), for a detailed survey of the contents of the *Epitome*. See also Shank, "Regiomontanus against George of Trebizond", on his writings against Bessarion's and his own favorite adversary. On the instruments made by Regiomontanus, some known only from textual sources, see Zinner, *Astronomische Instrumente*, pp. 479-483, and Wattenberg, "Regiomontanus und die astronomischen Instrumente seiner Zeit". On his 1462 astrolabe in the context of ten related astrolabes with the same distinctive design from the same workshop see King & Turner, "Regiomontanus' Astrolabe for Bessarion", A-B.

Vangelista, son of Ludovigo Gonzaga, died as a teenager (of meningitis?) between 1456 and 1460. See Lavin, *Piero's Flagellation*, pp. 70-71.

Topics

Byzantine topics. See *The Oxford Dictionary of Byzantium*, especially articles "Abbreviations", "Angel", "Argyropoulos, John", "Astrolabe", "Astronomy", "Basil" (personal name), "Basileus", "Bessarion", "Bessarion Reliquary", "Birds", "Calendar, Church", "Christogram", "Chronology", "Constantine I the Great", "Constantinople", "Constantinople, Patriarchate of", "Constantinople, Siege and Fall of", "Consul", "David I Komemnos" (last emperor of Trebizond), "Ferrara", "Ferrara-Florence, Council of", "Filelfo, Francesco", "Filioque" (on one of the main doctrinal bones of contention at Ferrara-Florence), "George Trapezountios", "Helios", "Hellespont", "Hypatos", "John (on the personal name)", "John VIII Palaiologos", "Kosovo Polje", "Mehmed II", "Monograms", "Protospatharios", "Rhodes", "Spatharios", "Thomas Palaiologos", "Trebizond", "Triodion", "Union of the Churches", "Varna, Crusade of". The extensive article "Himmelsmächte, Erzengel und Engel", in Wessel, ed., *Reallexikon zur byzantinischen Kunst*, contains nothing on angelic saints.

Christian topics. See *Lexikon der christlichen Ikonographie*, especially articles "Lilie", "Rose", and "Maria – Marienbild", and *Dictionnaire d'archéologie chrétienne et de liturgie*, especially articles "A Ω", "Abréviations", "Acrostiche", "Alphabet numérique", "Chiffre", "Monogramme"). On aspects of Mariology see Bäumer & Scheffczyk, eds., *Marienlexikon*. Metford, *Dictionary of Christian Lore and Legend*, is always useful. On Eastern monasticism see Laboa, ed., *Atlante storico del monachesimo*. On the lives of saints the most reliable information is usually in *Bibliotheca sanctorum* and *Bibliotheca sanctorum orientalium*.

Classical topics. See Pauly-Wissowa, *Realencyclopädie*, or *Der kleine Pauly*, especially articles "Abecedarii", "Akrostichis" or "Akrostichon", "Satorquadrat", "Sol", and "Τεχνοπαιγνια", and the articles "Acrostic", "*deus/divus*", "Helios", "Hermes", "Jupiter", "Sol", in *The Oxford Classical Dictionary*. On Ptolemy and Theon see under "Astronomy" below.

Islamic topics. See *The Encyclopaedia of Islam*, especially articles "Asturlâb" (astrolabe), "Hirmis" (Hermes), "Iblîs" (Satan), "Idrîs", "Istanbul", "Mehemmed II" (the Ottoman Sultan Fâtih Mehmet), "'Othmân I" (after whom the Ottoman dynasty was named); "Tarabzun" (Trebizond); also Pitcher, *Ottoman Empire*. On the scope of the Renaissance in science in Islamic society to *ca.* 1050 see Sezgin, *Geschichte des arabischen Schrifttums*, III (medicine), IV (alchemy), V (mathematics), VI (astronomy), VII (astrology), and X-XII (geography and cartography). On Muslim contributions to cryptography see Mrayati, Mir 'Alam and Tayyan, *Arabic Origins of Cryptology*.

Medieval topics. See *Lexikon des Mittelalters*, especially the articles "Achrostichon", "AEIOU", "Astrolab", "Christusmonogramm", "Figurengedichte", "Geheimschriften", "Hrabanus Maurus", and "Monogramm", and *Dictionary of the Middle Ages*, especially the articles "Acrostics-Wordplay", "Astrolabe", "Astrology", "Astronomy", "Astronomy, Islamic", "Hrabanus Maurus".

Renaissance topics. In the *Oxford Dictionary of the Renaissance* articles "Accolti family", "Alberti, Leon Battista", "Algebra", "Arithmetic", "Argyropoulos", Astrology" and "Astronomy", "Bessarion", "Geometry", "George of Trebizond", "Golden Section", "IHS", "Perotti, Niccolò", "Perspective", "Piero della Francesca", "Regiomontanus",

Abbreviations in medieval Latin and Greek. See the article "Abréviations" by H. Leclercq in *Dictionnaire d'archéologie chrétienne et de liturgie*. On abbreviations in ancient and Byzantine Greek see, for example, Gardthausen, *Griechische Paläographie*, II, pp. 319-352; and Guarducci, *Epigrafia greca*, I, pp. 398-406; and also the website listed under Halsall, "Greek Abbreviations". The standard work for medieval Latin is Cappelli, *Lexicon abbreviaturarum*. For an overview see, for example, Bischoff, *Latin Palaeography*, pp. 150-168.

Acrostics and configured letters. See, for example, the articles "Acrostiche" in *Dictionnaire d'archéologie chrétienne et de liturgie*, "Acrostics-Wordplay" in *Dictionary of the Middle Ages*, "Akrostichon" in *Lexikon des Mittelalters*, and "Akrostichon" in *Historisches Wörterbuch der Rhetorik*. Older works such as Dornseiff, *Das Alphabet in Mystik und Magie*, pp. 146-151, are still useful. For a substantial bibliography on acrostics in Greek and Latin see Damschen, "The Ancient Acrostic – A Bibliography". For the classical tradition see Courtney, "Greek and Latin Acrostichs", and Vogt, "Das Akrostichon in der griechischen Literatur". For Byzantine liturgical poetry see Krumbacher, "Die Akrostichis in der griechischen Kirchenpoesie", and Weyh, "Die Akrostichis in der byzantinischen Kanonesdichtung". A universal history of configured letters, with many acrostics, is Ernst, *Carmen figuratum*, with earlier introductions to the subject in *idem*, "Zahl und Mass in den Figurengedichten der Antike und des Frühmittelalters", and "The Figured Poem". Yet more examples are in *Wolfenbüttel HAB 1987-1988 Exhibition Catalogue*. For acrostics with five vertical axes, albeit not concealed, see Ernst, *Carmen figuratum*, pp. 202-209 and 549-550.

The astrolabe. For an introduction to the theory and use of the instrument see North, "The Astrolabe", and *Greenwich Astrolabe Booklet*. The standard work on astrolabes is Gunther, *Astrolabes of the World*, long out of date but still very useful. On the use of Islamic and European astrolabes as historical sources see King, "Astronomical Instruments between East and West", and *idem*, *In Synchrony with the Heavens*, XIIIa: "The neglected astrolabe – A supplement to the standard literature on the favourite astronomical instrument of the Middle Ages", which features both the 1062 and the two 1462 astrolabes. See also Appendixes A-G of this study.

For introductions to the history of **astronomy** from Greek Antiquity to the Renaissance see various articles in C. Walker, ed., *Astronomy before the Telescope* (1996), namely, Toomer, "Ptolemy and his Greek Predecessors"; Jones, "Later Greek and Byzantine Astronomy"; King, "Islamic Astronomy"; Pedersen, "European Astronomy in the Middle Ages"; and Swerdlow, "Astronomy in the Renaissance". For a reliable translation of and commentary on Ptolemy's *Almagest* see Toomer, *Ptolemy's Almagest*, and Pedersen, *A Survey of the Almagest*. For more on early astronomy see also J. Evans, *The History and Practice of Ancient Astronomy*.

Astronomy in 15th-century Europe, particularly Vienna and Italy. In addition to the literature on Peuerbach and Regiomontanus see Rose, *The Italian Renaissance of Mathematics*, pp. 5-117; Shank, "The Classical Tradition in Fifteenth-Century Vienna"; Swerdlow, "The Recovery of the Exact Sciences of Antiquity", "Astronomy in the Renaissance", and *The Renaissance of Astronomy in the Age of Humanism* (forthcoming). On German instruments see Zinner, *Astronomische Instrumente*. On the Vienna workshop see *ibid.*, pp. 322-323 (Johannes von Gmunden), 463-465 (Georg Peuerbach), 479-483 (Regiomontanus), and 292-297 (Hans Dorn), as well as Wattenberg, "Johannes Regiomontanus und die astronomischen Instrumente seiner Zeit". For a list of some 40 instruments from the Vienna workshop that survive to this day, see King, "Astronomical Instruments between East and West", pp. 183-188. On the Nuremberg connection see also *Nuremberg 1992-93 Exhibition Catalogue*.

Cryptography in 15th-century Italy. The standard works are Meister. *Geschichte der italienischen Kryptographie des XV. Jahrhunderts* and *Die Geheimschrift im Dienste der päpstlichen Kurie*. On the medieval precedents from all over Europe see Bischoff, "Übersicht über die nichtdiplomatischen Geheimschriften des Mittelalters".

The divine proportion. For a study within the history of mathematics, see Herz-Fischler, *A Mathematical History of Division in Extreme and Mean Ratio*, and "The Golden Number" (summary). See especially pp. 144-151 on the *Trattato d'abaco* and *De quinque corporibus regularibus* of Piero della Francesca and the *Divina proportione* of his student, Paccioli. For Piero's treatment see Daly Davis, *Piero della Francesca's Mathematical Treatises*, pp. 30-38. Reliable modern introductions to the geometry of the regular pentagon and decagon are Aaboe, *Episodes from the Early History of Mathematics*, pp. 54-72, and Lawlor, *Sacred Geometry*, pp. 36-37 and 48-64. For Euclid's treatment see Heath, *Euclid's Elements*, I, pp. 402-403 (II:11), and pp. 96-100 (IV:10).

Hermeticism. The literature is vast. See the articles in *Oxford Dictionary of Byzantium* and *Oxford Dictionary of the Renaissance.* A particularly useful source for its influence in magic, alchemy and science during the 15th-18th centuries is *Venice 2002 Exhibition Catalogue.* See also *Florence 1999-2000 Exhibition Catalogue* on the first translations in Italy.

Monograms in Greek and Latin. The best source is probably still Gardthausen, *Das alte Monogramm.* See also, for example, the article "Monogram" in *The Oxford Dictionary of Byzantium*, and the articles "Monogramm" and "AEIOU" in *Lexikon des Mittelalters*, also Ernst, *Carmen figuratum*, pp. 529-539. On the "silent alternative", that is, makers' and owners' marks without alphabetic appendages, see King, *The Ciphers of the Monks*, pp. 324-334, and the bibliography there cited.

Number notations. See Ifrah, *Histoire universelle des chiffres*, for useful introductions to the Roman and Hindu-Arabic notations, with some new information on medieval notations, including alphanumerical ones, in King, *The Ciphers of the Monks*, especially pp. 281-317. See also the fine article "Arabic Numerals" by Richard Lemay in *Dictionary of the Middle Ages.*

Organum Ptolemaei. See Zinner, *Astronomische Instrumente*, pp. 131-134, and King, *In Synchrony with the Heavens*, XIIb-15.

Bibliography

> "Books are full of the voices of the wise, full of lessons from antiquity, full of moral and legal wisdom, full of religion. Books live, they discourse and speak directly to us, they teach and instruct us, they bring us consolation. They show us things far remote from our times, and, as it were, place them before our eyes as if they were present today. So great is the power of books, so great their dignity, their grandeur, even their divinity, that without them we should all be rude and ignorant. Without books, we should have almost no memory of the past, no examples to follow, no knowledge of either human or divine affairs. Were it not for books, the same tombs that consume men's bodies would likewise bury their very names in oblivion." From Cardinal Bessarion's act of donation of his books in 1468 to the Doge and Senate of Venice, translated in Englander *et al.*, *Anthology of Medieval Sources on Culture and Belief*, p. 149.

Notes:~ Exhibition catalogues are usually listed by place, and festschrifts are listed by the name of the dedicatee, in both cases sometimes with a cross-reference by the name of the editor(s). Parts of surnames such as "de", "van" "von", *etc.*, are not observed in the alphabetical ordering. The Arabic article prefix al- is ignored in the ordering. Websites are listed at the end. If they are clearly associated with an individual, they are also listed in the bibliography under that name; otherwise, they may be listed under Anonymous.

A

Asger H. Aaboe, *Episodes from the Early History of Astronomy*, New York, Berlin, *etc.*: Springer, 2001.

— , *Episodes from the Early History of Mathematics*, (New Mathematical Library, 13), New York: Random House & The L. W. Singer Company, 1964.

Aaboe Festschrift: *From Ancient Omens to Statistical Mechanics – Essays on the Exact Sciences Presented to Asger Aaboe*, J. Lennart Berggren and Bernard R. Goldstein, eds., *Acta Historica Scientiarum Naturalium et Medicinalium* (Copenhagen) 39 (1987).

Mohammed Abu Zayed, David A. King and Petra G. Schmidl, "On astrolabes and poetry from Baghdad to al-Andalus: Heavenly Arabic verses and an enigmatic Judaeo-Arabic inscription", to appear in a memorial volume for Prof. Franz Rosenthal, edited by Jacob Lassner.

Silke Ackermann, "Dormant Treasures – The Zinner-Archive at Frankfurt University", *Nuncius – Annali di storia della scienza* (Florence) 16:2 (2001), pp. 711-722.

— , "Light on Byzantium – A Universal Sundial in the British Museum", in *Through a Glass Brightly – Studies in Byzantine and Medieval Art and Archaeology Presented to David Buckton*, Chris Entwistle, ed., Oxford: Oxbow Books, 2003, pp. 16-22.

Adler: see *Wolfenbüttel HAB 1987-1988 Exhibition Catalogue.*

Anna Akasoy, "Mehmed II as a Patron of Greek Philosophy and Science: Eastern and Western Perspectives", paper presented at the conference "The Renaissance and the Ottoman World", Warburg Institute and School of Oriental and African Studies, London, 26-27 April, 2006. [Not seen.]

Alberti, *De cifra*: *Leon Battista Alberti – A Treatise on Ciphers*, Niccolò Galimberti, ed., Turin: Galimberti Tipografia, 1997. [Italian version *Dello scrivere in cifra* published *ibid.*, 1994.]

— , *Della pittura*: *Leon Battista Alberti – Della Pittura – Über die Malkunst*, text, German translation and commentary by Oskar Bätschmann and Sandra Gianfreda, Darmstadt: Wissenschaftliche Buchgesellschaft, 2002.

Alcabitius: see al-Qabîsî, under Q.

Zofia Ameisenowa, *The Globe of Martin Bylika of Olkusz and Celestial Maps in the East and the West*, transl. by Andrzej Potocki, Warsaw, *etc.*: Akademi Nauk, 1959.

Amsterdam 1990 Exhibition Catalogue: Anthony J. Turner, ed., *Time*, The Hague: Tijd voor Tijd Foundation, 1990. [A catalogue of an exhibition held at the 'Nieuwe Kerk', Amsterdam, from Nov., 1990 to Jan., 1991.]

Amsterdam 2004 Conference Proceedings: see Oestmann *et al.*

Milton V. Anastos, "Pletho's Calendar and Liturgy", *Dumbarton Oaks Papers* 4 (1948), pp. 183-305.

Giovanni Anceschi, *Monogrammi e figure – Teorie e storie della progettazione di artefatti comunicativi*, Florence & Milan: La Casa Usher, 1981.

Kirsti Andersen, *The Geometry of an Art – The History of the Mathematical Theory of Perspective from Alberti to Monge*, (Sources and Studies in the History of Mathematics and Physical Sciences), New York, Berlin: Springer, 2006.

— , "Perspective and the Plan and Elevation Technique, in particular in the Work of Piero della Francesca", in *Wussing Festschrift*, pp. 1-23.

Anonymous 1, "Die Engelstellen im Triodion der byzantinischen Liturgie", available at www.himmelsboten.de/Engel/Gebete/Triodion.htm (2005).

Anonymous 2, "Urbino in the Age of Federico da Montefeltro", with sections 1: Urbino: The Historical and Geographical Context, 2: Educated Society in Urbino – Literature, Mathematics, Music, Painting, 3: The Legend of Urbino, available at www.uniurb.it/Uborse/worken.htm (2005).

Elmer H. Antonsen, "Linguistics and Politics in the 19th Century: The Case of the 15th Rune", *Michigan Germanic Studies* 6:1 (1980), pp. 1-16.

Mariano Apa, "Cultura figurativa e dibatto sull'astronomia ad Urbino nel secolo XV", in Cerboni Baiardi *et al.*, eds., *Federico di Montefeltro*, II, pp. 247-267.

Arezzo 1992 Exhibition Catalogue: *Con gli occhi di Piero – Abiti e gioielli nelle opere di Piero della Francesca*, Maria Grazia Ciardi Duprè and Guiliana Chesne Dauphiné Griffo, eds., Venice: Marsilio, 1992.

Arezzo-Sansepolcro 1992 Conference Proceedings: Marisa Dalai Emiliani and Valter Curzi, eds., *Piero della Francesca tra arte e scienza – Atti del Convegno Internazionale di Studi – Arezzo, 8-11 ottobre 1992 – Sansepolcro, 12 ottobre 1992*, Venice: Marsilio, 1996.

Aronberg Lavin: see Lavin.

L. R. N. Ashley, "Classical Pseudonyms in Europe at the Time of the Reformation", *Journal of the American Name Society* 14 (1966), pp. 193-196.

B

Jean Babelon, "Jean Paleologus et Ponce Pilate" *Gazette des Beaux-Arts*, 6e série, 4 (1930), pp. 365-375.

Pedro Bádenas de la Peña, "The Byzantine Intellectual Elites at the Court of Mehmet Fatih. Adaptation and Identity", available at www.filol.csic.es/bizantinos/Mehmet.html (September, 2005).

[Includes a discussion of George of Trebizond and George Amirutzes and their relationships with the Ottoman Sultan.]

Bätschmann & Gianfreda: see Alberti, *Della pittura*.

Remigius Bäumer and Leo Scheffczyk, eds., *Marienlexikon* (herausgegeben im Auftrag des Institutum Marianum Regensburg e.V.), 6 vols., St. Ottilien: EOS Verlag Erzabtei St. Ottilien, 1988-94.

Leo Bagrov, "The Maps of Regiomontanus", *Imago Mundi* 4 (1947), pp. 31-32.

A. Banck, *Byzantine Art in the Collections of the USSR*, Leningrad & Moscow: Sovietsky Knudozhnik, 1966.

James R. Banker, "Contributi alla cronologia della vita e delle opere di Piero della Francesca", *Arte cristiana* no. 823, vol. 92 (2004), pp. 248-258.

— , *The Culture of San Sepolcro during the Youth of Piero della Francesca*, Ann Arbor: University of Michigan Press, 2003.

— , "A Manuscript of the Works of Archimedes in the Hand of Piero della Francesca", *The Burlington Magazine* 147 (2005), pp. 165-169.

Barcelona 2004 Exhibition Catalogue: *Mediterraneum – Splendour of the Medieval Mediterranean – 13th-15th Centuries*, Elisenda Guedea, ed., Barcelona: Institut Europeu de la Mediterrània, 2004.

Ernst von Bassermann-Jordan, *The Book of Old Clocks and Watches*, transl. from the 4th edn. of *idem*, *Uhren*, by H. Alan Lloyd, London: George Allen & Unwin Ltd., 1964.

— , *Uhren – Ein Handbuch für Sammler und Liebhaber*, 9th edn. reworked by Hans von Bertele, Munich: Klinkhardt & Biermann, 1982.

Eugenio Battisti, *Piero della Francesca – Nuova edizione ...* , 2 vols., Milan: Electa, 1992.

T. D. Barnes, "Publilius Optatianus Porfyrius", *American Journal of Philology* 96 (1975), pp. 173-186.

Herbert Bauer and Georg Stolz, *Engelsgruß und Sakramentshaus in St. Lorenz zu Nürnberg*, Königstein im Taunus: Karl Robert Langewiesche Nachfolger, 1989 edn.

Silvio Bedini, "Falconi, Renaissance Astrologer and Astronomical Clock and Instrument Maker", *Nuncius – Annali di Storia della Scienza* (Florence) 14 (2004), pp. 56-60.

Befort: see Debauvais & Befort.

Pierre Béhar, *Les langues occultes de la Renaissance – Essai sur la crise intellectuelle de l'Europe au XVIe siècle*, Paris: Desjonquères, 1996.

Jim Bennett, *The Measurers – A Flemish Image of Mathematics in the Sixteenth Century* Oxford: Museum of the History of Science, 1995.

Richard Benz, *Die Legenda aurea des Jacobus de Voragine aus dem Lateinischen übersetzt*, first publ. 1955, Gütersloh: Gütersloher Verlagshaus, 2004.

Beretta *et al.*: see *Miniati Festschrift*.

Berggren & Goldstein: see *Aaboe Festschrift*.

Richard Bernheimer, *Wild Men of the Middle Ages – A Study in Art, Sentiment and Demonology*, Cambridge, Ma.: Harvard University Press, 1952, repr. New York: Octagon Books, 1970.

Carlo Bertelli, *Piero della Francesca – La forza divina della pittura*, Milan: Amilcare Pizzi, 1991.

Sergio Bettini, Italo Furlan, Lino Moretti *et al.*, eds., *Venezia e Bizanzio*, Venice: Electa, 1974.

Concetta Bianca, "L'Accademia del Bessarione tra Roma e Urbino", in Cerboni Baiardi *et al.*, eds., *Federico di Montefeltro*, III, pp. 61-79.

— , *Da Bisanzio a Roma – Studi sul Cardinale Bessarione*, Rome: Roma nel Rinascimento, 1999.

Bibliotheca sanctorum, 15 vols., Città del Vaticano: Istituto Giovanni XXIII della Pontificia Università Lateranense, 1961-1970.

Bibliotheca sanctorum orientalium – Enciclopedia dei santi – le chiese orientali, 2 vols., Rome: Città Nuova, 1998-1999.

Bielefeld Catalogue: *Wissenschaftliche Instrumente und Sonnenuhren – Kunstgewerbe-sammlung der Stadt Bielefeld / Stiftung Huelsmann*, Dirk Syndram, ed., Munich: Callwey, 1989.

Maria Teresa Binaghi Olivari, *La Pinacoteca di Brera*, Florence: Bonechi, 1983.

Biographical Encyclopedia of Astronomers, Thomas Hockey *et al.*, eds., Dordrecht: Kluwer Academic Publishers, to appear in 2007.

Ludwik Birkenmajer, *Marcin Bylica z Olkusza ... ...* , Cracow: Akademii Umiejetnosci, 1892. [See also the summary and review in German listed under Karlinski.]

Nancy Bisaha, *Creating East and West – Renaissance Humanists and the Ottoman Turks*, Philadelphia: University of Pennsylvania Press, 2004.

Bernhard Bischoff, *Paläographie des römischen Altertums und des abendländischen Mittelalters*, (first published 1956), Berlin, Bielefeld & Munich: Erich Schmidt, 1979, transl. as *Paléographie de l'antiquité romaine et du moyen âge occidental*, Paris: Picard, 1985, and as *Latin Palaeography – Antiquity and the Middle Ages*, Cambridge, *etc.*: Cambridge University Press, 1990.

— , "Übersicht über die nichtdiplomatischen Geheimschriften des Mittelalters", *Mitteilungen des Instituts für österreichische Geschichtsforschung* 62 (1957), pp. 1-27, repr. in *idem*, *Mittelalterliche Studien*, III, pp. 120-148, with tables III-IV.

— , *Mittelalterliche Studien – Ausgewählte Aufsätze zur Schriftkunde und Literatur-geschichte*, 3 vols., Stuttgart: Anton Hiersemann, 1966.

— , "Die sogenannten 'griechischen' und 'chaldäischen' Zahlzeichen des abendländischen Mittelalters", in *Scritti di paleografia e diplomatica in onore di Vincenzo Federici*, Florence, 1944, pp. 325-334, reprinted in *idem*, *Mittelalterliche Studien*, I, pp. 67-73, and pls. 6-7.

Jeremy Black, *Italy and the Grand Tour*, New Haven, Ct. & London: Yale University Press, 2003.

Alighiero Boetti: see Lauter.

Vojtech Bohác, Michael Kunzler, Oleksandr Petrynko, and Andreas A. Thiermeyer, eds., *Apostolos – Die biblischen Lesungen, Troparien und Kondakien – Antiphonen und alle anderen liturgischen Texte für die Feier der heiligen Liturgie und des Stundengebets in den heiligen Kirchen des byzantinischen Ritus nach den Tagen des Kirchenjahres*, Paderborn: Bonifatius, 2005.

Marinella Bonvini Mazzanti, *Battista Sforza Montefeltro – Una "principessa" nel Rinascimento italiano*, Urbino: QuattroVenti, 1993.

Bosch: see Chabàs & Bosch.

Bott: see *Nuremberg 1992-93 Exhibition Catalogue*.

Leicester Bradner, "The Neo-Latin Epigram in Italy in the Fifteenth Century", *Medievalia et Humanistica – An American Journal for the Middle Ages and Renaissance* 8 (1954), pp. 62-70.

J. Bridgeman, ""Belle considerazione": Dress in the Work of Piero della Francesca", *Apollo* 136 (1992), pp. 218-225.

— , ""*Troppo belli e troppo eccellenti*": Observations on Dress in the Work of Piero della Francesca", in Wood, ed., *The Cambridge Companion to Piero della Francesca*, pp. 76-90.

Attilio Brilli, *Alla ricerca di Piero della Francesca – Itinerari in Toscana, Umbria, Marche, Emilia Romagna*, Milan: Electa, 1990, 1991, new edn., 1994.

Brown: see Fortini Brown.

Francesca Romana Bruscolini, *Allo Alifante el cor l'Aquila morse – Federico da Montefeltro, la sua* librària *et la consultazione* on line, Urbino: Montefeltro, 2004.

Brussels 1982 Exhibition Catalogue: Jacqueline Lafontaine-Dosogne, ed., *Splendeur de Byzance*, Brussels: Musées royaux d'art et d'histoire, 1982.

Brussels SGB 1984 Exhibition Catalogue: Georges Baptiste *et al.*, *La mesure du temps dans les collections belges*, Brussels: Société Générale de Banque, 1984.

F. Buecheler and E. Lommatzsch, *Carmina Latina Epigraphica*, *pars posterior* of A. Riese, *Anthologia Latina*, fascs. I, II, III, Leipzig, 1895, 1897, 1926.

Danuta Burczyk-Marona, "Zegar sloneczny i astrolabium profesora Jana Brozka w zbiorach Muzeum Uniwersytetu Jagiellonskiego // The Sun-dial and Astrolabe of Professor Jan Brozek in the Jagellonian University Collection," *Opuscula Musealia (Zeszytu Naukowe Uniwersytetu Jagiellonskiego)* [Warsaw and Cracow], 3 (1988), in Polish with an English summary on pp. 113-115.

Charles Burnett, "King Ptolemy and Alchandreus the Philosopher: The Earliest Texts on the Astrolabe and Arabic Astrology at Fleury, Micy and Chartres", *Annals of Science* 55 (1998), pp. 329-368.

Burnett *et al.*: see *Pingree Festschrift*.

C

Marinella Calisi, *Storia e strumenti del Museo astronomico e Copernicano di Roma*, Rome: Osservatorio astronomico di Roma, 2000.

Maurizio Calvesi, "*La Flagellazione* nel quadro storico del convegno di Mantova e dei progetti di Mattia Corvino", in Lavin, ed., *Piero della Francesca and his Legacy*, pp. 115-125.

— , *Piero della Francesca*, New York: Rizzoli International Publications, 1998. [Contains a survey of earlier interpretations of Piero's "Flagellation" and some superb illustrations.]

Cambridge WMHS 1994 Exhibition Catalogue: Jim Bennett and Domenico Bertolini Meli, *Sphaera mundi – Astronomy Books in the Whipple Museum 1478-1600*, Cambridge: Whipple Museum of the History of Science, 1994.

Campbell: see *The Oxford Dictionary of the Renaissance*.

Charles Du Fresne Du Cange, *Glossarium ad scriptores mediae et infimae graecitatis*, Leiden, 1688 edn., repr. 2 pts. in 1 vol., Graz: Akademische Druck- und Verlagsanstalt, 1958.

— , *Glossarium mediae et infimae latinitatis*, first published Paris, 1678, 1883-87 edn., repr. 10 pts. in 5 vols., Graz: Akademische Druck- und Verlagsanstalt, 1954.

Adriano Cappelli, *Lexicon abbreviaturarum – Dizionario di abbreviature latine ed italiane*, Milan: Ulrico Hoepli, 1990 (reprint of the 6th edn., 1961).

Giacomo E. Carretto, "Bessarione e il Turco", in *Venice 1994 Exhibition Catalogue*, pp. 260-274.

David Carrier, "Piero della Francesca and his Interpreters: Is there Progress in Art History?", *History and Theory – Studies in the Philosophy of History* 26:2 (1987), pp. 150-165.

Carter: see Wittkower & Carter.

Brendan Cassidy: see *Princeton 1990 Colloquium Proceedings*.

Marina Cristina Castelli, "Piero della Francesca : *Flagellazione di Cristo*", entry in *Urbino 1992 Exhibition Catalogue*, pp. 118-121.

Casulleras & Samsó: see *Vernet Festschrift*.

Giuseppe Centauro and Enzo Settesoldi, *Piero della Francesca – committenza e pittura nella Chiesa di S. Francesco ad Arezzo (con nuovi documenti inediti)*, Poggibonsi: Lalli, 2000.

Giorgio Cerboni Baiardi, Giorgio Chittolini and Piero Floriani, eds., *Federico di Montefeltro*, 3 vols., I: *Lo stato*, II: *Le arti*, III: *La cultura*, Rome: Bulzoni, 1986.

Josep Chabàs and Daniel Bosch, *L'astrolabi pla – Guia per a la construcció i utilització*, Barcelona: Institut de Ciències de l'Educació de la Universitat Politècnica de Catalunya, 1987.

José Chabás and Bernard R. Goldstein, *The Alphonsine Tables of Toledo*, (Archimedes – New Studies in the History and Philosophy of Science and Technology, vol. 8), Dordrecht, Boston & London: Kluwer, 2003.

François Charette, *Mathematical Instrumentation in Fourteenth-Century Egypt and Syria – The Illustrated Treatise of Najm al-Dîn al-Misrî*, Leiden & Boston: Brill, 2003.

— and David A. King, *The Universal Astrolabe of Ibn al-Sarrâj – A History of a Medieval Mathematical Jewel*, to appear.

— and Petra G. Schmidl, "Al-Khwârizmî and Practical Astronomy in Ninth-Century Baghdad – The Earliest Extant Corpus of Texts in Arabic on the Astrolabe and Other Portable Instruments", *SCIAMVS – Sources and Commentaries in Exact Sciences* (Kyoto, Hokuto) 5 (2004) pp. 101-198.

— — , "A Universal Plate for Timekeeping by the Stars by Habash al-Hâsib: Text, Translation and Preliminary Commentary", *Suhayl – Journal for the History of the Exact and Natural Sciences in Islamic Civilisation* (Barcelona) 2 (2001), pp. 107-159.

Charette: see also van Cleempoel *et al.*, *Astrolabes at Greenwich*.

Luciano Cheles, *The Studiolo of Urbino – An Iconographic Investigation*, Wiesbaden: Ludwig Reichert, for The Pennsylvania State University Press, University Park, 1986.

Chesne Dauphiné Griffo: see *Arezzo 1992 Exhibition Catalogue*.

Bernard Chevalier *et al.*, *Le Couronnement de la Vierge par Enguerrand Quarton*, a special issue of *Études Vauclusiennes* (Avignon) nos. 24-25 (1980-1981), 56 pp.

Chittolini: see Cerboni Baiardi *et al.*

Chlench: see *Gmunden 2006 Conference Proceedings.*

Hyacinthe Chobaut, "Documents inédits sur les peintres et verriers d'Avignon, du Comtat et de la Provence occidentale de la fin du XIVe siècle au premier tiers du XVIe siècle", *Mémoires de l'Académie de Vaucluse* 4 (1939/1940).

Christie's 1988 Catalogue: Christie's Amsterdam, *Important Judaica – Thursday 15 December 1988*, Amsterdam: Christie's, 1988.

Ciardi Duprè: see *Arezzo 1992 Exhibition Catalogue.*

Cieri Via: see *Urbino 1992 Conference Proceedings.*

Marshall Clagett, *Archimedes in the Middle Ages*, 3 vols., I: *The Arabo-Latin Tradition*, Madison: University of Wisconsin Press, 1964.

Kenneth Clark, *Piero della Francesca*, London: Phaidon Press, 1951, 2nd edn., 1969.

Koenraad van Cleempoel, *A Catalogue Raisonné of Scientific Instruments from the Louvain School, 1530 to 1600*, (*De diversis artibus*: Collections of studies from the International Academy of the History of Science, vol 65 (N.S. 28)), Turnhout: Brepols, 2002.

—, "Representations of Astrolabes in Western Art", in *idem et al.*, *Astrolabes at Greenwich*, pp. 99-111.

— , with contributions from Silke Ackermann, François Charette, Elly Dekker, Paul Kunitzsch, Richard Lorch, David Proctor and A. J. Turner, *Astrolabes at Greenwich – A Catalogue of the Astrolabes in the National Maritime Museum, Greenwich*, Oxford: Oxford University Press, & Greenwich: National Maritime Museum, 2005.

van Cleempoel: see also *Madrid FCA 1997 Exhibition Catalogue.*

Cecil H. Clough, "Cardinal Bessarion and Greek at the Court of Urbino", *Manuscripta* 8 (1964), pp. 160-171, repr. in *idem*, *The Duchy of Urbino*, VII.

—, *The Duchy of Urbino in the Renaissance*, London: Variorum, 1981.

—, "Federigo da Montefeltro's Artistic Patronage", *Journal of the Royal Society of Arts* 126 (1978), pp. 1-17, repr. in *idem*, *The Duchy of Urbino*, IX.

—, "Federigo da Montefeltro's Patronage of the Arts, 1468-1482", in *Journal of the Warburg and Courtauld Institutes* 36 (1973), pp. 129-144, repr. in *idem*, *The Duchy of Urbino*, VIII.

— , "Piero della Francesca: some problems of his art and chronology", *Apollo – The Magazine of the Arts* (London) 91 (April, 1970), pp. 278-289.

David R. Coffin, *The Villa in the Life of Renaissance Rome*, Princeton, N.J.: Princeton University Press, 1979.

Bruce Cole, *Piero della Francesca – Tradition and Innovation in Renaissance Art*, New York: IconEditions / HarperCollins, 1991.

The Coptic Encyclopedia, Aziz S. Atiya, ed., 9 vols., New York: Macmillan, *etc.*, 1991.

Edward Courtney, "Greek and Latin Acrostichs", *Philologus – Zeitschrift für Klassische Philologie* 134 (1990), pp. 3-13.

Coyne: see *Vatican 1982 Conference Proceedings.*

Anne Crane, "Under the Protection of the Divine Bessarion Called Cardinal I Arise in Rome ... 1462", *Antiques Trade Gazette* (London), 28 October, 1989, p. 27.

Antonio Criminisi, *Accurate Visual Metrology from Single and Multiple Uncalibrated Images*, Distinguished Dissertation Series, London, Berlin, Heidelberg: Springer, 2001.

—, [Views of the three-dimensional scenario of the "Flagellation" from different angles] at http://research.microsoft.com/vision/cambridge/3d/svm_movies/flagella_camera.mpeg and www.robits.ox.ac/uk/~criminisi/- and "Getting into the Picture", at http://plus.maths.org/issue23/features/criminisi/index.html

Alfred W. Crosby, *The Measure of Reality – Quantification and Western Society, 1250-1600*, Cambridge: Cambridge University Press, 1997.

Maximilian Curtze, "Der Briefwechsel Regiomontan's mit Giovanni Bianchini, Jacob von Speier und Christian Roder", in *idem*, *Urkunden zur Geschichte der Mathematik im Mittelalter und der Renaissance*, Leipzig, 1902, repr. New York, London: Johnson Reprints, 1968, pp. 185-336.

Curzi: see *Arezzo-Sansepolcro 1992 Conference Proceedings*.

D

Dal Poggetto: see Dal Poggetto.

Dalai Emiliani: see *Arezzo-Sansepolcro 1992 Conference Proceedings*, and Piero della Francesca.

Ormonde Maddock Dalton, "The Byzantine Astrolabe at Brescia", *Proceedings of the British Academy* (London) 12 (1926), pp. 133-146 and 3 pls.

van Damme: see *Madrid FCA 1997 Exhibition Catalogue*.

Gregor Damschen, "The Ancient Acrostic – A Bibliography", at www.telemachos.hu-berlin.de/esterni/akrostichon.html (2003).

Gustav Davidson, *A Dictionary of Angels including the Fallen Angels*, first published 1967, New York: The Free Press & New York *etc.*: Maxwell Macmillan, 1971.

Margaret Daly Davis, *Piero della Francesca's Mathematical Treatises – The* Trattato d'abaco *and* Libellus de quinque corporibus regularibus, Ravenna: Longo, 1977.

—, "Piero's Treatises: The Mathematics of Form", in Wood, ed., *The Cambridge Companion to Piero della Francesca*, pp. 134-151.

Karl H. Dannenfeldt, "The Renaissance Humanists and the Knowledge of Arabic", *Studies in the Renaissance* 2 (1955), pp. 96-117.

Linda T. Darling, "The Renaissance and the Middle East", in Ruggiero, ed., *A Companion to the Worlds of the Renaissance*, pp. 55-69.

Davis: see Daly Davis.

de: see the next part of the name.

Francis Debeauvais and Paul-André Befort, *Cueillir les étoiles – Autour des astrolabes de Strasbourg*, Strasbourg: Amis des Instruments des Sciences et des Astrolabes, 2002.

Debeauvais: see also *Strasbourg 1987 Exhibition Catalogue*.

Hertha von Dechend, "Erinnerungen an die Frühzeit des Instituts", in von Gotstedter, ed., *Ad radices*, pp. 3-11.

von Dechend: see also de Santillana & von Dechend.

Jan C. Deiman, "Imitations amongst the Mensing Instruments", in *Leiden MB 1999 Conference Proceedings*, pp. 98-112.

Dekker: see Kunitzsch & Dekker.

Delpont *et al.*: see *Paris IMA 2000-01 Exhibition Catalogue*.

Marcel Destombes, "Un astrolabe carolingien et l'origine de nos chiffres arabes", *Archives internationales d'Histoire des sciences* 58-59 (1962), pp. 1-45, repr. in *M. Destombes (1905-1983) : Contributions sélectionnées à l'Histoire de la Cartographie et des Instruments Scientifiques*, G. Schilder *et al.*, eds., Utrecht: Hes, & Paris: A. G. Nizet, 1987, pp. 153-200.

— and E. S. Kennedy, "An Introduction to *Kitâb al-'Amal bi-'l-Asturlâb*", an English introduction to the Arabic text of al-Sûfî's treatise on the use of the astrolabe, Hyderabad, Deccan, 1966, repr. in Kennedy *et al.*, *Studies in the Islamic Exact Sciences*, pp. 405-447.

Dictionary of the Middle Ages, Joseph Reese Strayer, gen. ed., 13 vols., New York: Charles Scribner's Sons (for the American Council of Learned Societies), 1982-89.

Dictionary of Scientific Biography, 14 vols. and 2 supp. vols., New York: Charles Scribner's Sons, 1970-80.

Dictionnaire d'archéologie chrétienne et de liturgie, Fernand Cabrol and Henri Leclercq, eds., 15 pts. in 30 vols., Paris: Letouzey et Aîné, 1913-53.

Dictionnaire de biographie française, 19 vols. to date, Paris: Letouzey et Ainé, 1933 to present.

Dizionario biografico degli Italiani, 65 vols. to date, Rome: Istituto della Enciclopedia Italiana, 1960 onwards.

Franz Dornseiff, *Das Alphabet in Mystik und Magie*, 2nd edn., Leipzig, Berlin: B. G. Teubner 1925.

du: see the next part of the name.

Sven Dupré, "Galileo, Mathematical Instruments and Orthographic Projection", *Bulletin of the Scientific Instrument Society* no. 69 (2001), pp. 10-20.

E

Sabine Eiche, ed., *Ordine et officij de casa de lo illustrissimo Signor Duca de Urbino*, Urbino: Accademia Raffaello, 1999.

Eiche: see also Smith.

Eisenhardt: see Linhard & Eisenhardt.

Enciclopedia cattolica, 12 vols., Vatican City: Ente per l'Enciclopedia cattolica e per il libro cattolico, 1948-1954.

The Encyclopaedia of Islam, new edn., 12 vols., Leiden: Brill, 1960 to present.

Polychronis K. Enepekides, "Die Wiener Legation des Kardinals Bessarion in den Jahren 1460-61 unter Berücksichtigung der neuentdeckten urkundlichen Quellen in Wien", in *Miscellanea Marciana di studi bessarionei*, (*Medioevo e Umanesimo*, 24), Padua: Antenore, 1976, pp. 69-82.

David Englander, Diana Norman, Rosemary O'Day and W. R. Owens, *Culture and Belief in Europe 1450-1600 – An Anthology of Sources*, Oxford & Cambridge, Ma.: Blackwell, in association with the Open University, 1990, repr. 1993.

EPACT: Website "Epact: Scientific Instruments of Medieval and Renaissance Europe", located at

www.mhs.ox.ac.uk/epact,

including descriptions of individual instruments which can be accessed using various key-words.

Ulrich Ernst, *Carmen figuratum – Geschichte des Figurengedichtes von den antiken Ursprüngen bis zum Ausgang des Mittelalters*, Cologne, Weimar, Vienna: Böhlau, 1991. [A universal history of configured letters and acrostics.]

—, "The Figured Poem: Towards a Definition of Genre", *Visible Language* 20:1 (1986), pp. 8-27.

—, "Zahl und Mass in den Figurengedichten der Antike und des Frühmittelalters", *Miscellanea mediaevalia* (Thomas-Institut, Cologne) 16 (1982), pp. 310-332.

Ernst: see also *Wolfenbüttel HAB 1987-1988 Exhibition Catalogue*.

Helen Evans, Melanie Holcomb and Robert Hallman, *The Arts of Byzantium*, a special issue of *The Metropolitan Museum of Art Bulletin* 58:4 (2001), New York: The Metropolitan Museum of Art, 68 pp.

James Evans, *The History and Practice of Ancient Astronomy*, New York & Oxford: Oxford University Press, 1998.

—, "The Material Culture of Greek Astronomy", *Journal for the History of Astronomy* 30 (1999), pp. 237-307. [Overlooks the 1062 astrolabe.]

"Evelyn", *Piero della Francesca – Monarca della pittura ai suoi dì*, Città di Castello: S. Lapi, 1912. [Dedicated to her husband, Piero Franceschi Marini.]

F

G. Nicco Fasola, ed., *Piero della Francesca – De prospectiva pingendi*, Florence: Sansoni, 1942, repr. Florence: Casa Editrice Le Lettere, 1984. [The definitive text.]

M. D. Feld, "Constructed Letters and Illuminated Texts: Regiomontanus, Leon Battista Alberti, and the Origins of Roman Type", *Harvard Library Bulletin* 28 (1980), pp. 357-379.

Felice Feliciano, *Alphabetum Romanum, Vat. Lat. 6852 aus der Bibliotheca Apostolica Vaticana*, 2 vols., Zurich: Belser, 1985.

Angelo Ferretti Torricelli, "Catalogo degli strumenti scientifici delle collezioni dei Civici musei di storia e arte e dell'Ateneo di Brescia", in *Aspetti della società bresciana nel Settecento*, Roberto Navarrini, ed., Brescia: Grofo, 1981, pp. 199-208.

Edgarda Ferri, *Piero della Francesca – Storia e misteri del maestro della luce*, Milan: Arnaldo Mondadori, 2001.

Fiaccadori: see *Venice 1994 Exhibition Catalogue*.

J. V. Field, *The Invention of Infinity – Mathematics and Art in the Renaissance*, Oxford, New York, Tokyo: Oxford University Press, 1997.

—, *Piero della Francesca: A Mathematician's Art*, New Haven, Ct. & London: Yale University Press, 2005.

—, "Piero della Francesca's Mathematics", in Wood, ed., *The Cambridge Companion to Piero della Francesca*, pp. 152-170.

—, "Rediscovering the Archimedean Polyhedra: Piero della Francesca, Luca Pacioli, Leonardo da Vinci, Albrecht Dürer, Daniele Barbaro, and Johannes Kepler", *Archive for History of Exact Sciences* 50 (1997), pp. 241-289.

— and Michael T. Wright, *Early Gearing – Geared Mechanisms in the Ancient and Medieval World*, London: The Science Museum, 1985.

— & —, "Gears from the Byzantines: A Portable Sundial with Calendrical Gearing", *Annals of Science* 42 (1985), pp. 87-138, repr. in *eidem* and Donald R. Hill, *Byzantine and Arabic Mathematical Gearing*, London: The Science Museum, 1985.

Floriani: see Cerboni Baiardi *et al.*

Florence Bargello Catalogue: Mara Miniati, ed., *Strumenti scientifici della Collezione Carrand*, Florence: Museo Nazionale del Bargello, 1991.

Florence 1954 Catalogue: *Catalogo degli strumenti del Museo di Storia della Scienza*, Florence: Leo S. Olschki, 1954.

Florence 1987-88 Exhibition Catalogue: Mara Miniati, ed., *L'Età di Galileo – Il secolo d'oro della scienza in Toscana*, Florence: Istituto e Museo di Storia della Scienza, 1987.

Florence 1999-2000 Exhibition Catalogue: Sebastiano Gentile and Carlos Gilly, *Marsilio Ficino e il ritorno di Ermete Trismegisto / Marsilio Ficino and the Return of Hermes Trismegistus*, Florence: Centro Di, 1999.

Henri Focillon, *Piero della Francesca*, Paris: Armand Colin, 1952, with an Italian translation of the same title, Parma: Pratiche Editrice, 1992.

Menso Folkerts, *The Development of Mathematics in Medieval Europe – the Arabs, Euclid, Regiomontanus*, London: Ashgate-Variorum, 2006.

— , "Luca Pacioli and Euclid", in *Sansepolcro 1994 Conference Proceedings*, pp. 219-231, repr. in *idem*, *The Development of Mathematics in Medieval Europe*, XI.

— , "Die mathematischen Studien Regiomontans in seiner Wiener Zeit", in Hamann, ed., *Regiomontanus-Studien*, pp. 175-209, repr. in idem, *Development of Mathematics*, V.

— , "Piero della Francesco and Euclid", in *Arezzo-Sansepolcro 1992 Conference Proceedings*, pp. 293-312, repr. in *idem*, *Development of Mathematics*, X.

— , "Regiomontans Euklidhandschriften", *Sudhoffs Archiv – Zeitschrift für Wissenschafts-geschichte* 58 (1974), pp. 149-164.

— , "Regiomontanus' Role in the Transmission of Mathematical Problems", in *From China to Paris: 2000 Years Transmission of Mathematical Ideas*, Yvonne Dold-Samplonius, Joseph W. Dauben, Menso Folkerts and Benno van Dalen, eds., Stuttgart: Franz Steiner, 2002, pp. 411-428.

Folkerts Festschrift: a *Festschrift* for Prof. Menso Folkerts, Munich, edited by Richard P. Lorch and Paul Kunitzsch, in press.

Mario Fondelli, "La restituzione della pavimentazione della *Flagellazione di Cristo* di Piero della Francesca", in *Urbino 1992 Exhibition Catalogue*, pp. 502-503.

Patricia Fortini Brown, "*Sant'Agostino nello studio* di Carpaccio, un ritratto nel ritratto", in *Venice 1994 Exhibition Catalogue*, pp. 303-319.

Gino Franceschini, "La morte di Gentile Brancaleoni (1457) e di Buonconte da Montefeltro", *Archivio storico lombardo*, n.s., 2 (1937), pp. 489-500.

della Francesca: see Piero.

Karen Frank, "[On the arrival of the head of St Andrew in Italy and its political importance]", paper presented at the Third Annual University of California at Santa Barbara Medieval Studies Graduate Student Conference: Self, Community, and Artifact in the Middle Ages, 2004, summary available at medievalstudies.ucsb.edu/events/conference2004.html#frank (no title is given) .

Frankfurt 2006 Exhibition Catalogue: Jochen Sander *et al.*, *Cult Image – Altarpiece and Devotional Painting from Duccio to Perugino*, (text in German and English), Petersberg (D): Michel Imhof, 2006.

Du Fresne: see Du Cange.

Marcus Frings, *Mensch und Maß – Anthropomorphe Elemente in der Architekturtheorie des Quattrocento*, Weimar: Verlag und Datenbank für Geisteswissenschaften, 1998. [The book is printed in A5 format sideways (it is not clear why), so that it has the same shape as Piero's "Flagellation".]

G

Galimberti: see Alberti, *De cifra*.

Joseph S. Galland, *An Historical and Analytical Bibliography of the Literature of Cryptology*, (Northwestern University Studies in the Humanities, no. 10), Evanston, Ill.: Northwestern University, 1945, repr. New York: AMS Press, 1970.

Paolo Galluzzi, *Les ingénieurs de la Renaissance de Brunelleschi à Léonard de Vinci*, Florence: Museo di Storia della Scienza, & Paris: Cité des Sciences et de l'Industrie, 1995, and *Gli ingegnieri del Rinascimento da Brunelleschi a Leonardo da Vinci*, Florenze: Giunti, 1996.

Paolo Garbini, "Corti e comete: lo scritto di Nicolò Conti sulla cometa del 1456", in *Urbino 1992 Conference Proceedings*, pp. 435-447.

Salvador García Franco, *Catálogo crítico de los astrolabios existentes en España*, Madrid: Instituto Histórico de la Marina, 1945.

Viktor Gardthausen, *Das alte Monogramm*, Leipzig: Karl W. Hiersemann, 1924.

—, *Griechische Paläographie*, 2 vols., I: *Das Buchwesen im Altertum und im byzantinischen Mittelalter*, and II: *Die Schrift, Unterschriften und Chronologie im Altertum und im byzantinschen Mittelalter*, 2nd edn., Leipzig: Veit, 1911-1913, repr. Leipzig: Zentral-antiquariat der DDR, 1978.

Vittoria Garibaldi, *The National Gallery of Umbria (at Perugia)*, Milan: Mondadori Electa, 2002, repr., 2004.

Geneva MAH 1985 Exhibition Catalogue: *Treasures of Islam, published in association with the Musée d'art et d'histoire, Geneva*, London: Sotheby's & Philip Wilson, 1985.

Gentile: see *Florence 1999-2000 Exhibition Catalogue*.

Gianfreda & Bätschmann: see Alberti, *Della pittura*.

Sharon Gibbs with George Saliba, *Planispheric Astrolabes from the National Museum of American History*, Washington, D.C.: Smithsonian Institution Press, 1984.

Creighton Gilbert, "Piero della Francesca' *Flagellation*: The Figures in the Foreground", *Art Bulletin* 53 (1971), pp. 41-51.

—, "On Subject and Not-Subject in Italian Renaissance Pictures", *Art Bulletin* 34 (1952), pp. 202-216.

Gilly: see *Florence 1999-2000 Exhibition Catalogue* and *Venice 2002 Exhibition Catalogue*.

Owen Gingerich, "Reprint of Gunther's *Astrolabes*", *Journal for the History of Astronomy* 9:1 (1978), pp. 69-70.

Carlo Ginzburg, *The Enigma of Piero della Francesca, new edition with appendixes*, transl. by Martin Ryler and Kate Soper, London & New York: Verso, 2000, first edn. published in Italian as *Indagini su Piero*, 1981, second edition as *Indagini su Piero – Il Battesimo, Il ciclo di Arezzo, La flagellazione di Urbino*, Turin: Einaudi, 1994, repr. 2001.

F. K. Ginzel, *Handbuch der mathematischen und astronomischen Chronologie*, 3 vols., Leipzig: J. C. Hinrichs, 1906-1914, repr. Zwickau: Ullmann, 1958.

Alain Girard, *Le Couronnement de la Vierge d'Enguerrand Quarton du musée de Villeneuve-lès-Avignon – Un chef d'œuvre de l'Occident médiéval en terre gardoise*, [Nîmes]: Conseil Général du Gard & Musée de l'art sacré du Gard, 2005.

Giusti: see *Sansepolcro 1994 Conference Proceedings*.

E. Glowatzki and H. Göttsche, *Die Tafeln des Regiomontanus – Ein Jahrhundertwerk*, (*Algorismus – Studien zur Geschichte der Mathematik und der Naturwissenschaften*, Menso Folkerts, ed., Heft 2), Munich: Institut für Geschichte der Naturwissenschaften, 1990.

Gmunden 2006 Conference Proceedings: *Johannes von Gmunden (ca. 1384 - 1442) – Astronom und Mathematiker*, Rudolf Simek and Kathrin Chlench, eds., (*Studia Medievalia Septentrionalia*, 12), Vienna: Fassbaender, 2006.

Bernard R. Goldstein, "Levi ben Gerson: On Instrumental Errors and the Transversal Scale", *Journal for the History of Astronomy* 8 (1977), pp. 102-112.

— , *Theory and Observation in Ancient and Medieval Astronomy*, London: Variorum Reprints, 1985.

Goldstein: see *Aaboe Festschrift* and Chabás & Goldstein.

Goldstein Festschrift: *Astronomy and Astrology from the Babylonians to Kepler – Essays Presented to Bernard R. Goldstein on the Occasion of his 65th Birthday*, Peter Barker, Alan C. Bowen, José Chabás, Gad Freudenthal and Y. Tzvi Langermann, eds., 2 pts., *Centaurus* 45 (2003) and 46 (2004).

Ernst Hans Gombrich, "Introduction: Aims and Limits of Iconography," in *idem*, *Symbolic Images – Studies in the Art of the Renaissance*, London: Phaidon, 1972, 3rd edn. repr. in *Gombrich on the Renaissance*, vol. 2, London: Phaidon, 1985.

Anton von Gotstedter, ed., *Ad radices – Festband zum fünfzigjährigen Bestehen des Instituts für Geschichte der Naturwissenschaften Frankfurt am Main*, Stuttgart: Franz Steiner, 1994.

Penelope Gouk, *The Ivory Sundials of Nuremberg 1500-1700*, Cambridge: Whipple Museum of the History of Science, 1988.

Thalia Gouma-Peterson, "Piero della Francesca's Flagellation: an Historical Interpretation", *Storia dell'arte* (Rome and Florence) 28 (Sept.-Dec. 1976), pp. 217-233.

Grafton: see *Washington LC 1993 Exhibition Catalogue*.

Ivor Grattan-Guinness, ed., *Companion Encyclopedia of the History and Philosophy of the Mathematical Sciences*, 2 vols., London & New York: Routledge, 1994.

— , "Numerology and Gematria", in *idem*, ed., *Companion Encyclopedia* ... ... , II, pp. 1585-1592.

Grayson: see Piero della Francesca.

Graz 1989 Exhibition Catalogue: *Zeitmesser von der Sonnenuhr zum Räderwerk – Ausstellungskatalog*, Eva Marko and Lukas Stolberg, eds., Graz: Steiermärkisches Landesmuseum Johanneum, Abteilung für Kunstgewerbe, 1989.

Greenwich Astrolabe Booklet: *The Planispheric Astrolabe*, Greenwich: National Maritime Museum, 1976.

Greenwich 1999-2000 Exhibition Catalogue: Kristen Lippincott, with Umberto Eco, E. H. Gombrich and others, *The Story of Time*, London: Merrell Holberton in association with the National Maritime Museum, 1999.

Greppin Catalogue: *La Collection Greppin - Instruments scientifiques*, Zurich: Galerie Koller, 1975.

Johannes Gründler *et al.*, eds., *Matthias Corvinus und die Renaissance in Ungarn 1458-1541*, (Katalog des Niederösterreichischen Landesmuseums, N.F., Nr. 118), Vienna: Amt der Niederösterreichischen Landesregierung, 1982.

Gualdoni: see Simonini & Gualdoni.

Margherita Guarducci, *L'epigrafia greca dalle origini al tardo impero*, 4 vols., Rome: Istituto Poligrafico dello Stato, 1967-1978.

Guedea: see *Barcelona 2004 Exhibition Catalogue*.

Robert T. Gunther, *The Astrolabes of the World ...* , 2 vols., Oxford: The Clarendon Press, 1932, repr, in 1 vol., London: The Holland Press, 1976. [Long out of date, but still very useful.]

Dimitri Gutas, *Greek Thought, Arabic Culture – The Graeco-Arabic Translation Movement in Baghdad and Early 'Abbâsid Society (2nd-4th/8th-10th Centuries)*, London & New York: Routledge, 1998.

H

Stephanie Haarländer, *Rabanus Maurus zum Kennenlernen – Ein Lesebuch mit einer Einführung in sein Leben und Werk*, Mainz: Bistum Mainz, 2006.

Sergei Hackel, ed., *The Byzantine Saint – University of Birmingham Fourteenth Spring Symposium of Byzantine Studies*, San Bernardino, Ca.: R. Reginald – The Borgo Press, 1983.

Helmar Härtel, "Ludolfus Borchdorp de Brunswik", *Braunschweigisches Jahrbuch* 68 (1987), pp. 113-120.

Edwin Hall, "More about the Detroit van Eyck: The Astrolabe, the Congress of Arras and Cardinal Albergati", *The Art Quarterly* 34 (1971), pp. 180-201.

Paul Halsall, "Greek Abbreviations" available at www.fordham.edu/halsall/ikon/greekabb.html (January, 2006).

Günther Hamann, ed., *Regiomontan-Studien*, (*Veröffentlichungen der Kommission für Geschichte der Mathematik, Naturwissenschaften und Medizin*, Hefte 28, 29, 30 = *Sitzungsberichte der Österreichischen Akademie der Wissenschaften, Phil.-Hist. Klasse*, Band 364), Vienna, 1980.

Georg Hartmann: see Lamprey.

J. Hartmann, "Die astronomischen Instrumente des Kardinals Nikolaus Cusanus", *Abhandlungen der Königlichen Gesellschaft der Wissenschaften zu Göttingen, Math.-Phys. Klasse*, N. F. Band X, Nr. 6 (Berlin: Wiedmannsche Buchhandlung), 1919, 56 pp. and 12 plates.

Willy Hartner, "The Islamic Astronomical Background to Nicholas Copernicus", *Studia Copernicana* 13 (1975), pp. 7-16, repr. in *idem*, *Oriens-Occidens II*, pp. 316-325.

— , "The Mercury Horoscope of Marcantonio Michiel of Venice – A Study in the History of Renaissance Astrology and Astronomy", *Vistas in Astronomy* 1 (1955), pp. 84-138, repr. in *idem*, *Oriens-Occidens*, I, pp. 440-495.

— , *Oriens-Occidens – Ausgewählte Schriften zur Wissenschafts- und Kulturgeschichte – Festschrift zum 60. Geburtstag*, Hildesheim: Georg Olms, 1968 (I), and *Oriens-Occidens – Ausgewählte Schriften zur Wissenschafts- und Kulturgeschichte, Band II*, Yasukatsu Maeyama, ed., Hildesheim, *etc.*: Georg Olms, 1984 (II).

— , "The Principle and Use of the Astrolabe", a chapter in *A Survey of Persian Art from Prehistoric Times to the Present*, Arthur Upham Pope, ed., 9 vols., London & New York:

Oxford University Press, 1938-39, III, pp. 2530-2554, repr. with the original plates in *Astrolabica*, no. 1, Feb. 1978, Paris: Société Internationale de l'Astrolabe, and article "Asturlâb" in *The Encyclopaedia of Islam*, 2nd edn., both repr. in *idem*, *Oriens-Occidens*, I, pp. 287-311 and pp. 312-318.

—, "Ptolemäische Astronomie im Islam und zur Zeit des Regiomontanus", in Hamann, ed., *Regiomontan-Studien*, pp. 109-124.

—, "The Vaso Vescovali in the British Museum – A Study on Islamic Astrological Iconography", *Kunst des Orients* 9 (1975), pp. 99-130, repr. in *idem*, *Oriens-Occidens II*, pp. 214-245.

Hartner Festschrift: *ΠΡΙΣΜΑΤΑ – Naturwissenschaftsgeschichtliche Studien – Festschrift für Willy Hartner*, Yasukatsu Maeyama and Walter G. Saltzer, eds., Wiesbaden: Franz Steiner, 1977.

Frank Rutger Hausmann, "Untersuchungen zum neulateinischen Epigramm Italiens im Quattrocento", *Humanistica Lovaniensia – Journal of Neo-Latin Studies* 21 (1972), pp. 1-35.

Hans-Wilhelm Haussig, *The History of Byzantine Civilization*, translated by J. M. Hussey from *Kulturgeschichte von Byzanz*, 2nd edition of 1966, New York & Washington: Praeger, 1971. [Ignores the 1062 astrolabe.]

Gerald S. Hawkins and David A. King, "On the Orientation of the Ka'ba", *Journal for the History of Astronomy* 13 (1982), pp. 102-109, repr. in King, *Islamic Mathematical Astronomy*, XII.

Thomas L. Heath, *The Thirteen Books of Euclid's Elements*, 3 vols., 2nd edn., Cambridge: Cambridge University Press, 1925, repr. New York: Dover, 1956.

—, *The Works of Archimedes ... and "The Method of Archimedes"*, first published 1897 and 1912, repr. New York: Dover, 1953.

van Heertum: see *Venice 2002 Exhibition Catalogue*.

Tofigh Heiderzadeh, "Patronage, Networks and Migration: Turco-Persian Scholarly Exchanges in the 15th, 16th and 17th Centuries", *Archives internationales d'Histoire des sciences* 55 (2005), pp. 419-434.

Heinke, "Classical Cryptography", *The Classical Journal* (Lawrence, Kansas) 58 (1962), pp. 113-121.

Lothar Heiser, *Die Engel im Glauben der Orthodoxie*, (SOPHIA – Quellen östlicher Theologie, Bd. 13), Trier: Paulinus-Verlag, 1976.

Philip Hendy, *Piero della Francesca and the Early Renaissance*, London: Weidenfeld and Nicholson, 1968.

G. L. Hersey, *Pythagorean Palaces – Magic and Architecture in the Italian Renaissance*, Ithaca, N.Y., & London: Cornell University Press, 1976.

Roger Herz-Fischler, *A Mathematical History of Division in Extreme and Mean Ratio*, Waterloo, Ontario: Wilfred Laurier University Press, 1987.

—, "The Golden Number, and Division in Extreme and Mean Ratio", in Grattan-Guinness, ed., *Companion Encyclopedia of the History and Philosophy of the Mathematical Sciences*, II, pp. 1576-1584.

George Francis Hill, *The Development of Arabic Numerals in Europe Exhibited in Sixty-Four Tables*, Oxford: Clarendon Press, 1915.

James Hilton, *Chronograms*, 3 vols., London: Elliott Stock, 1882-1895. [Not available for cosultation: see White.]

Hockey: see *Biographical Encyclopedia of Astronomers*.

J. Hoffman, "Piero della Francesca's "Flagellation" – A Reading from Jewish History", *Zeitschrift für Kunstgeschichte* 44 (1981), pp. 340-357.

Vincent Foster Hopper, *Medieval Number Symbolism – Its Sources, Meaning, and Influence on Thought and Expression*, New York: Columbia University Press, 1938.

Hoskin: see *Vatican 1982 Conference Proceedings*.

Barnabas Hughes, O.F.M., *Regiomontanus on Triangles* – De triangulis omnimodis *by Johann Müller, otherwise known as Regiomontanus, translated with an introduction and notes*, Madison, Milwaukee & London: The University of Wisconsin Press, 1967.

James Hutton, *The Greek Anthology in Italy to the Year 1800*, Ithaca, N.Y.: Cornell University Press, 1935.

I

Georges Ifrah, *Histoire universelle des chiffres*, 2nd edn., 2 vols., Paris: Robert Laffont, 1994. [The first edition of 1981 is available in translation as *From One to Zero – A Universal History of Numbers*, New York, *etc*.: Penguin, 1988, and *Universalgeschichte der Zahlen*, Frankfurt, New York: Campus, 1986.]

Ekmeleddin Ihsanoglu, *Science, Technology and Learning in the Ottoman World*, Aldershot & Burlington, Vt.: Ashgate – Variorum, 2004.

— *et al.*, *Osmanlı astronomi literatürü tarihi – History of Astronomy Literature during the Ottoman Period*, 2 vols.; *Osmanlı matematik literatürü tarihi – History of Mathematics Literature during the Ottoman Period*, 2 vols.; and *Osmanlı cografya literatürü tarihi – History of Geographical Literature during the Ottoman Period*, 2 vols., Istanbul: Research Centre for Islamic History, Art and Culture (IRCICA), 1997, 1999, and 2000.

Ihsanoglu Festschrift: *Essays in honour of Ekmeleddin Ihsanoglu*, Mustafa Kacar and Zeynep Durukal, eds., 2 vols., Istanbul: Research Centre for Islamic History, Art and Culture (IRCICA), 2006.

Jozef IJsewijn, *Companion to Neo-Latin Studies*, Amsterdam, New York, Oxford: North-Holland, 1977.

— with Dirk Sacré, *Companion to Neo-Latin Studies*, Part II: *Linguistic, Literary, Philological and Editorial Questions*, second entirely rewritten edition, Amsterdam, New York, Oxford: North-Holland & Leuven: Leuven University Press, 1998.

Pierre Ioannou, "Un opuscule inédit du cardinal Bessarion. Le panégyrique de Saint Bessarion, anachorète égyptien", *Analecta Bollandiana* (Brussels) 65 (1947), pp. 107-138.

J

S. A. Jayawardene, "Towards a Biography of Luca Pacioli", *Sansepolcro 1994 Conference Proceedings*, pp. 19-28.

Joannou: see Ioannou.

Clemens Jöckle, *Encyclopedia of Saints* (translated from the 1997 German original), London: Parkgate, 1997.

Alexander Jones, *An Eleventh-Century Manual of Arabo-Byzantine Astronomy*, (*Corpus des astronomes byzantins*, III), Amsterdam: J. C. Gieben, 1987.

—, "David Pingree (1933-2005)", *Journal for the History of Astronomy* 37 (2006), pp. 229-230.

— , "Later Greek and Byzantine Astronomy", in C. Walker, ed., *Astronomy before the Telescope*, pp. 98-109.

K

David Kahn, *The Codebreakers – The Story of Secret Writing*, New York: Macmillan, 1967, 4th edn., New York: Scribner, 1996.

F. Karlinski, Summary and review in German of Birkenmajer, *Marcin Bylika z Olkusza* (in Polish), in *Anzeiger der Akademie der Wissenschaften in Krakau*, 1892, (Krakau: Universitätsdruckerei, 1893), pp. 98-110. [The author is named on p. 69.]

Wolfgang Kaunzner, "Regiomontanus als Mathematiker", in Hamann, ed., *Regiomontan-Studien*, pp. 125-145.

Kazhdan: see *The Oxford Dictionary of Byzantium*.

Martin J. Kemp, *Behind the Picture – Art and Evidence in the Italian Renaissance*, New Haven, Ct., & London: Yale University Press, 1997.

— , *The Science of Art: Optical Themes in Western Art from Brunelleschi to Seurat*, New Haven, Ct., & London: Yale University Press, 1990.

Edward Stewart Kennedy, *Astronomy and Astrology in the Medieval Islamic World*, Aldershot & Brookfield, Vt.: Ashgate–Variorum, 1998.

— , "Spherical Astronomy in al-Kâshî's *Khâqânî Zîj*", *Zeitschrift für Geschichte der arabisch-islamischen Wissenschaften* (Frankfurt am Main) 2 (1990), pp. 1-46, repr. in *idem*, *Astronomy and Astrology in the Medieval Islamic World*, VII.

— , Colleagues and Former Students, *Studies in the Islamic Exact Sciences*, David A. King and Mary Helen Kennedy, eds., Beirut: American University of Beirut, 1983.

— and David Pingree, *The Astrological History of Mâshâ'allâh*, Cambridge, Mass.: Harvard University Press, 1971.

Kennedy Festschrift: *From Deferent to Equant – Studies in the History of Science in the Ancient and Medieval Near East in Honor of E. S. Kennedy*, David A. King and George Saliba, eds., *Annals of the New York Academy of Sciences* 500 (1987).

Kennedy: see also Destombes & Kennedy.

Eckhard Keßler and Heinrich C. Kuhn, eds., *Germania latina – Latinitas teutonica – Politik, Wissenschaft, Humanismus – Kultur vom späten Mittelalter bis in unsere Zeit*, 2 vols., Munich: Wilhelm Fink, 2003.

Khalili Collection Instrument Catalogue: *The Nasser D. Khalili Collection of Islamic Art*, vol. XII: *Science, Tools and Magic*, by Francis Maddison and Emilie Savage-Smith, 2 pts., London: The Nour Foundation, in association with Azimuth Editions & Oxford University Press, 1997. [See also the review in King, "Cataloguing Islamic Astronomical Instruments".]

David A. King, "The Astrolabe Depicted in the Intarsia of the *Studiolo* of Archduke Federico in Urbino", in Vetrano, ed., *La scienza del Ducato di Urbino – The Science of the Dukedom of Urbino*, pp. 101-139.

— , "An Astrolabe from Einbeck *ca.* 1330", to appear in *Folkerts Festschrift*.

— , "Die Astrolabiensammlung des Germanischen Nationalmuseums (in Nürnberg)", in *Nuremberg 1992-93 Exhibition Catalogue*, I, pp. 101-114, and II, pp. 568-602. [Includes descriptions of the 1457 astrolabe associated with Peuerbach and the 1462 astrolabe of Regiomontanus, .]

—, "Astrolabis de la Catalunya medieval", in *La ciència en la història dels Països Catalans*, I: Dels àrabs al Renaixement, Barcelona: Institut d'Estudis Catalans, 2004, pp. 161-204.

—, "Astronomical Alignments in Medieval Islamic Religious Architecture", *Annals of the New York Academy of Sciences* 385 (1982), pp. 303-312, repr. in King, *Islamic Mathematical Astronomy*, XIII.

—, "Astronomical Instruments between East and West", in *Krems 1993 Conference Proceedings*, pp. 143-198. [On the use of instruments as historical sources; also presents a list of 40 instruments from 15th-century Vienna (some 45 are known now).]

—, "On the Astronomical Tables of the Islamic Middle Ages", *Studia Copernicana* 13 (1975), pp. 37-56, repr. in *idem*, *Islamic Mathematical Astronomy*, I. [Long outdated.]

—, *Astronomy in the Service of Islam*, Aldershot: Variorum, 1993.

—, "(Byzantine Astronomy)", a review of David Pingree, *The Astronomical Works of Gregory Chioniades* (1985-86) and Alexander Jones, *Byzantine Astronomical Manual* (1987), in *Isis* 82 (1991), pp. 116-118.

—, "Cataloguing Islamic Astronomical Instruments", *Bibliotheca Orientalis* 57 (2000), cols. 247-258. [An essay review of the *Khalili Collection Instrument Catalogue*.]

—, *The Ciphers of the Monks – A Forgotten Number Notation of the Middle Ages*, (Boethius – Texte und Abhandlungen zur Geschichte der Mathematik und der Naturwissenschaften, Band 44), Stuttgart: Franz Steiner, 2001. [Contains new information on medieval European number-notations, alphabetical codes, and astrolabes.]

—, "The Cult of St. Wilgefortis in Flanders, Holland, England and France", in *Am Kreuz – Eine Frau: Anfänge – Abhängigkeiten – Aktualisierungen*, Sigrid Glockzin-Bever and Martin Kraatz, eds., in *Ästhetik – Theologie – Liturgik* (Münster: LIT Verlag), 26 (2003), pp. 55-97.

—, "The Earliest Known European Astrolabe in the Light of Other Early Astrolabes", in Stevens *et al.*, eds., *The Oldest Latin Astrolabe*, pp. 359-404.

—, *Islamic Astronomical Instruments*, London: Variorum, 1987, repr. Aldershot: Variorum, 1995.

—, "Islamic Astronomy", in C. Walker, ed., *Astronomy before the Telescope*, pp. 143-174.

—, *Islamic Mathematical Astronomy*, London: Variorum, 1986, 2nd rev. edn., Aldershot: Variorum, 1993.

—, *Mathematical Astronomy in Medeval Yemen – A Bio-Bibliographical Survey*, Malibu, Ca.: Undena, 1983.

—, "Reflections on some new studies on applied science in Islamic societies (8th-19th centuries", *Islam & Science* 2:1 (2004), pp. 43-56. [Specifically addressed to Muslim readers.]

—, "A Remarkable Italian Astrolabe from *ca.* 1300 – Witness to an Ingenious [Islamic] Tradition of Non-Standard Astrolabes", in *Miniati Festschrift*, pp. 29-52. [A new updated version connecting the piece to the astrolabe in a painting by Jan van Eyck is in *idem*, *In Synchrony with the Heavens*, XIIId.]

—, "A Remarkable Mathematical and Astronomical Instrument from Baghdad *ca.* 900", to appear. [Based on *idem*, "A remarkable mathematical and astronomical instrument signed by Nastûlus ... " in the catalogue of Sotheby's of London, *Arts of the Islamic World, October 11, 2006*, pp. 76-83 (lot 87).

—, Review of G. Turner, *Elizabethan Instrument Makers* (2000), in *Journal of the History of Collections* (Oxford), 15:1 (2003), pp. 147-150.

— , "The Star-Names on Three 14th-Century Astrolabes from Spain, France and Italy", in *Kunitzsch Festschrift*, pp. 307-333.

— , *In Synchrony with the Heavens – Studies in Astronomical Timekeeping and Instrumentation in Medieval Islamic Civilization*, 2 vols., I: *The Call of the Muezzin (Studies I-IX)*, and II: *Instruments of Mass Calculation (Studies X-XVIII)*, Leiden: Brill, 2004-2005. [See especially XIIIa on Islamic and European astrolabes as historical sources.]

— , "Two Astrolabes for the Ottoman Sultan Bayezit II", in *Ihsanoglu Festschrift*, I, pp. 439-459. [Also in *idem*, *In Synchrony with the Heavens*, XIVe.]

— , *World-Maps for Finding the Direction and Distance to Mecca – Innovation and Tradition in Islamic Science*, Leiden: Brill, and London: Al-Furqan Islamic Heritage Foundation, 1999.

— and Kurt Maier, "The Medieval Catalan Astrolabe of the Society of Antiquaries, London", in *Vernet Festschrift*, II, pp. 673-718.

— and Julio Samsó, with a contribution by Bernard R. Goldstein, "Astronomical Handbooks and Tables from the Islamic World (750-1900): An Interim Report", *Suhayl – Journal for the History of the Exact and Natural Sciences in Islamic Civilisation* (Barcelona) 2 (2001), pp. 9-105. [A summary is in the article "Zîdj" in *The Encyclopaedia of Islam*.]

— and Gerard L'E. Turner, "The Astrolabe Presented by Regiomontanus to Cardinal Bessarion in 1462", *Nuncius – Annali di Storia della Scienza* (Florence) 9 (1994), pp. 165-206, repr. in Turner, *Renaissance Astrolabes and their Makers*, IV [A]; also "The Astrolabe Dedicated to Cardinal Bessarion by Regiomontanus in 1462", in *Venice 1994 Exhibition Catalogue*, pp. 340-367 [B]. [A description by Turner of the 1462 astrolabe supplemented with details of ten other astrolabes from the Vienna school by King. In B, some of the captions have been muddled.]

King: see also Abu Zayed, King & Schmidl.

Kilian Kirchhoff, *Die Ostkirche betet – Hymnen aus den Tagzeiten der byzantinischen Kirche – Die Vorfastenzeit*, Leipzig: Jakob Hegner, 1934.

Erich Kleineidam, *Universitas Studii Erffordensis – Überblick über die Geschichte der Universität Erfurt*, II: *Spätscholastik, Humanismus und Reformation 1461-1521*, 2nd edn., Leipzig: Benno, 1992.

Hans Gunther Klemm, *Georg Hartmann aus Eggolsheim (1489-1564) – Leben und Werk eines fränkischen Mathematikers und Ingenieurs*, Forchheim (D): Ehrenbürg-Gymnasium (*Wissenschaftliche und künstlerische Beiträge*, Heft 8), 1990, 104 pp.

Joel L. Kraemer, *Humanism in the Renaissance of Islam – The Cultural Revival during the Buyid Age*, Leiden: E. J. Brill, 1986, 2nd revd. edn., 1992.

Krems 1993 Conference Proceedings: Harry Kühnel, ed., *Kommunikation zwischen Orient und Okzident – Alltag und Sachkultur*, (*Sitzungsberichte der Phil.-Hist. Klasse der ÖAW*, Band 619), Vienna: Österreichische Akademie der Wissenschaften, 1994.

Karl Krumbacher, "Die Akrostichis in der griechischen Kirchenpoesie", *Sitzungs-Berichte der philosophisch-philologischen und historischen Klasse*, 1903, IV, Munich: Bayerische Akademie der Wissenschaften, 1904.

— , *Geschichte der byzantinischen Litteratur von Justinian bis zum Ende des Oströmischen Reiches (527-1453)*, 2nd edn., Munich, 1897, repr. New York: Burt Franklin, •date•.

Wilhelm Kühlmann, "CAMENA – Lateinische Texte der Frühen Neuzeit", available at www.uni-mannheim.de/mateo/camena.html (Jan., 2006).

See also Schibel.

Kerstin Kühnast, *Studien zum Studiolo zu Urbino*, (Inaugural-Dissertation zur Erlangung des Doktorgrades der Philosophischen Fakultät der Universität zu Köln), Cologne, 1987.

Kühnel: see Krems.

Kühnel Festschrift: Gertrud Blaschitz, Helmut Hundsbichler, Gerhard Jaritz and Elisabeth Vavra, eds., *Symbole des Alltags – Alltag der Symbole – Festchrift für Harry Kühnel zum 65. Geburtstag*, Graz: Akademische Druck- u. Verlagsanstalt, 1992.

Paul Kunitzsch, *The Arabs and the Stars*, Northampton: Variorum, 1989.

— , *Arabische Sternnamen in Europa*, Wiesbaden: Otto Harrassowitz, 1959.

— , "The Stars on the Astrolabe", in van Cleempoel *et al.*, *Astrolabes at Greenwich*, pp. 41-46.

—, "The Star Catalogue Commonly Appended to the Alphonsine Tables, *Journal for the History of Astronomy* 17 (1986), pp. 89-98, repr. in *idem*, *The Arabs and the Stars*, XXII.

— , *Typen von Sternverzeichnissen in astronomischen Handschriften des zehnten bis vierzehnten Jahrhunderts*, Wiesbaden: Otto Harrassowitz, 1966.

— and Elly Dekker, "The Stars on the Rete of the So-Called 'Carolingian Astrolabe'", in *Vernet Festschrift*, II, pp. 655-672.

Kunitzsch Festschrift: *Sic itur ad astra – Studien zur Geschichte der Mathematik und Naturwissenschaften – Festschrift für den Arabisten Paul Kunitzsch zum 70. Geburtstag*, Menso Folkerts and Richard P. Lorch, eds., Wiesbaden: Otto Harrassowitz, 2000.

Kunitzsch: see also *Folkerts Festschrift*.

L

Juan María Laboa, ed., *Atlante storico del monachesimo orientale e occidentale*, Milan: Jaca Book SpA, 2002, with a German version *Mönchtum in Ost und West – Historischer Atlas*, Regensburg: Schnell & Steiner, 2003.

Lotte Labowsky, article "Bessarione" in *Dizionario biografico degli Italiani*, vol. IX, 1967, pp. 686-696.

— , *Bessarion's Library and the Biblioteca Marciana – Six Early Inventories*, (*Sussidi Eruditi*, 31), Rome: Edizioni di Storia e Letteratura, 1979.

— , "Cardinal Bessarion's Precious Manuscript of Euclid", *Jahrbuch der österreichischen Byzantinistik* 22 (1979), pp. 129-139.

Lafontaine-Dosogne: see *Brussels 1982 Exhibition Catalogue*.

John P. Lamprey, *Hartmann's Practika – A Manual for Making Sundials and Astrolabes with the Compass and Rule, written from 1518 to 1528 by Georg Hartmann, translated and edited by John Lamprey 2002*, privately published (John Lamprey, P. O. Box 336, Bellvue, Colorado 80512, U.S.A.).

Birgit Laskowski, *Piero della Francesca 1416/17-1492*, Cologne: Könemann, 2002, translated from *Meister der italienischen Kunst – Piero della Francesca, ibid.*, 1998.

Rolf Lauter, ed., *Alighiero Boetti – Mettere al mondo il mondo*, Frankfurt am Main: Museum für moderne Kunst & Galerie Jahrhunderthalle Hoechst (Cantz), 1998.

Jan Lauts and Irmlind Luise Herzner, *Federico da Montefeltro – Herzog von Urbino – Kriegsherr, Friedensfürst und Förderer der Künste*, Munich & Berlin: Deutscher Kunstverlag, 2001.

Marilyn Aronberg Lavin, "*Monarca della Pittura*: Piero and his Legacy", in *eadem*, ed., *Piero della Francesca and his Legacy*, pp. 9-20.

—, *Piero della Francesca*, London & New York: Phaidon, 2002.

—, "Piero della Francesca's "Flagellation": The Triumph of Christian Glory", *Art Bulletin* 50 (1968), pp. 321-342.

—, "Piero della Francesca's Iconographic Innovations at Arezzo", in *Princeton 1990 Colloquium Proceedings*, pp. 139-155.

—, *Piero della Francesca: 'The Flagellation'*, London & New York: Viking, 1972, and Chicago: University of Chicago Press, 1990.

—, ed., *Piero della Francesca and his Legacy*, (Center for Advanced Study in the Visual Arts, Studies in the History of Art 48, Symposium Papers XXVIII), Washington, D.C.: National Gallery of Art, 1995.

Robert Lawlor, *Sacred Geometry – Philosophy and Practice*, New York: Crossroad, 1982.

Simona Lecchini Giovannoni and Marco Collareta, eds., *Disegni di Santi di Tito (1536-1603)*, Florence: Leo S. Olschki, 1985.

Legenda aurea: see Benz.

Ernst and Johanna Lehner, *Astrology and Astronomy – A Pictorial Archive of Signs and Symbols*, first published 1964, repr. New York: Dover, 2005.

Leiden MB 1999 Conference Proceedings: *Scientific Instruments: Originals and Imitations – The Mensing Collection – Proceedings of a Symposium held at the Museum Boerhaave, Leiden, 15-16 October 1999*, Peter de Clercq, ed., Leiden: Museum Boerhaave, 2000.

W. Levitan, "Dancing at the End of the Rope: Optatian Porfyry and the Field of Roman Verse", *Transactions of the American Philological Association* 115 (1985), pp. 245-269.

Charlton T. Lewis and Charles Short, *A Latin Dictionary founded on Andrew's edition of Freund's Latin dictionary, revised, enlarged, and in great part rewritten*, Oxford: The Clarendon Press, (1st edn., 1879), 1966 edn.

Lexikon der christlichen Ikonographie, Engelbert Kirschbaum S.J. *et al.*, eds., 16 vols. (Allgemeine Ikonographie / Ikonographie der Heiligen), Rome, Freiburg, Basle & Vienna: Herder, 1968-76, repr. 1990 and 1994.

Lexikon des Mittelalters, Robert Henri Bautier *et al.*, eds., Munich, *etc.*: Artemis & Winkler, 1980-99.

Henry George Liddell and Robert Scott, *A Greek-English Lexicon, revised and augmented ... with a supplement*, (1st edn., 1843), Oxford: The Clarendon Press, 1968 edn.

Ronald Lightbown, Article "Franceschi (Della Francesca), Piero (dei)" in *Dizionario biografico degli Italiani*, vol. 49 (1997), pp. 624-637.

—, *Piero della Francesca*, New York, London & Paris: Abbeville Press, 1992.

Frank Linhard and Peter Eisenhardt, eds., *Naturwissenschaftliche Raum- und Zeitbegriffe in den wissenschaftlichen Disziplinen des 17. und 18. Jahrhunderts*, to appear in *Zeitsprünge* (Zentrum zur Erforschung der Frühen Neuzeit, Frankfurt am Main), Heft 1/2 (2007).

Linton Catalogue: *Collection Leonard Linton et de divers amateurs*: *Instruments scientifiques, livres anciens*, [a catalogue of the Leonard Linton collection offered for sale by E. Libert & A. Castor and A. Brieux in 1980], Paris: Alain Brieux, 1980.

Linz 1990 Exhibition Catalogue: *Mensch und Kosmos – Oberösterreichische Landesausstellung 1990*, Wilfried Seipel, ed., 2 vols., Linz: Oberösterreichisches Landesmuseum, 1990.

Raimondo Loenertz, article "Bessarione" in *Enciclopedia Cattolica*, II, Vatican City: Ente per l'Enciclopedia Cattolica e per il libro cattolico, 1949, cols. 1492-1498.

Fabrizio Lollini, "Ancora la 'Flagellazione': addenda di bibliografia e di metodo", *Bollettino d'Arte* 67 (May-June 1991), pp. 149-150.

— , "Bessarione e le arti figurative", in *Venice 1994 Exhibition Catalogue*, pp. 148-170.

— , "L'iconografia di Bessarione: *Bessarion pictus*", in *Venice 1994 Exhibition Catalogue*, pp. 275-284.

— , "Una possibile connotazione antiebraica della "Flagellazione" di Piero della Francesca" *Bollettino d'arte* 65 (1991), pp. 1-28.

Enrico Ferdinando Londei, "Ipotesi sulla collocazione originaria della *Flagellazione* di Piero della Francesca – Indizi di percorso", *Art e dossier* (Florence), no. 71 (Sept., 1992), pp. 30-32.

— , "La scena della 'Flagellazione' di Piero della Francesca – la sua identificazione con un luogo di Urbino del Quattrocento, *Bollettino d'Arte* 65 (Jan.-Feb. 1991), pp. 29-66.

Roberto Longhi, *Piero della Francesca,* ••, 1927; *Piero della Francesca – translated from the Italian by Leonard Penlock*, London & New York: Frederick Warne, 1930; *Piero della Francesca*, translated by David Tabbat, ••: Sheep Meadow, 2003• – see the review cited as Penny, "Journey to Arezzo".

Richard P. Lorch, "The Astronomical Instruments of Jâbir ibn Aflah and the Torquetum", *Centaurus* 20 (1976), pp. 11-34, repr. in *idem*, *Arabic Mathematical Sciences – Instruments, Texts, Transmission*, Aldershot & Brookfield, Vt.: Variorum, 1995, XVI.

— , *Al-Farghânî – On the Astrolabe – Arabic Text with Translation and Commentary*, Stuttgart: Franz Steiner, 2005.

Lorch: see also *Folkerts Festschrift* and *Kunitzsch Festschrift*.

Piero Lotito, *La notte di Emil Vrána*, Cinisello Balsamo (Milano): San Paolo, 2000. [A modern novel around "The Flagellation".]

Walther Ludwig, *Miscella Neolatina*, Astrid Steiner-Weber, ed., 3 vols., Hildesheim, Zurich, New York: Georg Olms, 2005.

Gianfrancesco Lusini, *Bessarione – Orazione dogmatica sull'unione dei greci e dei latini*, Naples: Vivarium, 2001.

M

Maccagni: see Piero della Francesca.

Mack Smith: see Smith.

Ruth J. Macrides, "Saints and Sainthood in the Early Palaiologan Period", in Hackel, ed., *The Byzantine Saint*, pp. 67-87.

Francis R. Maddison, "Hugo Helt and the Rojas Astrolabe Projection", *Revista da Faculdade de Ciências* (Coimbra) 39 (1966), pp. 195-251 and 2 fold-outs.

— , "Medieval Scientific Instruments and the Development of Navigational Instruments in the XVth and XVIth Centuries", *Revista da Universidade de Coimbra* 24 (1969), printed separately, Agrupamento de Estudos de Cartografia Antiga, XXX, Secção de Coimbra, 1969.

[—], *A Supplement to a Catalogue of Scientific Instruments in the Collection of J. A. Billmeir, Esq., C. B. E.*, London: Frank Partridge & Sons, 1957.

Madrid FCA 1997 Exhibition Catalogue: Jacques van Damme, Koenraad van Cleempoel, Gerard L'E. Turner, *et al.*, *Instrumentos scientificos del siglo XVI – La corte española y la escuela de Lovaina*, Madrid: Fundación Carlos de Amberes, 1997, with an English text without illustrations published separately as *Scientific Instruments in the Sixteenth*

Century: The Spanish Court and the Louvain School. [Catalogue of an exhibition held at the Foundation during 26.11.1997-1.2.1998.]

Anna Maria Maetzke, *Piero della Francesca*, Milan: Silvana, 2003.

— & D. Gallopi Nappini, *Il Museo civico di Sansepolcro*, Florence: Cassa di Risparmio di Firenze, 1988.

Yasukatsu Maeyama, *Astronomy in Orient and Occident – Selected papers on its cultural and scientific history*, Hildesheim, Zurich, New York: Georg Olms, 2003.

Maeyama: see also *Hartner Festschrift*.

Henry Maguire, *The Icons of their Bodies: Saints and their Images in Byzantium*, Princeton, N.J.: Princeton University Press, 1996.

Maier: see King & Maier.

Mainz 1980 Exhibition Catalogue: Wilhelm Weber, ed., *Rabanus Maurus in seiner Zeit 780-1980*, Mainz: Mittelrheinisches Landesmuseum, 1980.

Mainz 2000 Exhibition Catalogue: Winfried Wilhelmy, *Dragon, Griffin and Courtly Love – Mainz Tapestries from the Late Gothic Period*, Mainz: Philipp von Zabern, 2000.

Mainz 2006 Exhibition Catalogue: Winfried Wilhelmy *et al*., *Rabanus Maurus – Auf den Spuren eines karolingischen Gelehrten*, Mainz: Philipp von Zabern, 2006.

Alexios von Maltzew, *Menologion der Orthodox-Katholischen Kirche des Morgenlandes*, Berlin: Karl Siegismund, 1901.

Alessandro Marchi, "Piero della Francesca : *La Flagellazione*", entry in *Urbino 2005 Exhibition Catalogue*, pp. 183-185.

Marchi: see also *Urbino 2005 Exhibition Catalogue*.

Jo Marchant, "Who's Who in Renaissance Italy – Hidden codes in Piero della Francesca's *Flagellation*" (title on cover) and "A Leap of Faith" (title of article), *Nature – The International Weekly Journal of Science*, vol. 446 (29 March 2007), pp. 488-491.

Marko & Stolberg: see *Graz 1989 Exhibition Catalogue*.

R. Massin, *Alphabets*, and Ségolène Le Men, *Abécédaires*, Paris: Musée-Galerie de la Seita, 1986. [Catalogue of two exhibitions in 1986-87, published in a single volume.]

William H. McNeill, *Venice: The Hinge of Europe, 1081-1797*, Chicago: University of Chicago Press, 1974, repr. Chicago: Midway Reprints, 1986. Italian version *Venezia – Il cardine d'Europa, 1081-1797*, Rome: Il Veltro, 1979.

Igor P. Medvedev, "Solar Cult in Plethon's Philosophy?" *Byzantina* (Thessalonica) 13 (1985), pp. 739-749.

Aloys Meister, *Die Anfänge der modernen diplomatischen Geheimschrift – Beiträge zur Geschichte der italienischen Kryptographie des XV. Jahrhunderts*, and *Die Geheimschrift im Dienste der päpstlichen Kurie von ihren Anfängen bis zum Ende des XVI. Jahrhunderts*, Paderborn: Ferdinand Schöningh, 1902 and 1906. [The standard works on cryptography in Italy from the 14th to the 16th century.]

Le Men: see Massin & Le Men.

Giovanni Mercati, *Per la cronologia della vita e degli scritti di Niccolò Perotti arcivescovo di Siponto*, (*Studi e testi*, XLIV), Rome: Biblioteca Apostolica Vaticana, 1925.

(R. P.) E. Mercenier, *La prière des églises de rite byzantin*, 2 vols. in 3, Chevetogne (B): Éditions (du Monastère) de Chevetogne, 1947-1953.

J. C. J. Metford, *Dictionary of Christian Lore and Legend*, London: Thames and Hudson, 1983.

Rudolf Mett, *Regiomontanus – Wegbereiter des neuen Weltbildes*, Stuttgart & Leipzig: B. G. Teubner, and Zurich: VDF Hochschulverlag AG an der ETH Zürich, 1996.

Erich Meuthen, "Zum Itinerar der deutschen Legation Bessarions (1460-61)", *Quellen und Forschungen aus italienischen Bibliotheken und Archiven* 37 (1957), pp. 328-333.

Heinz Meyer and Rudolf Suntrup, *Lexikon der mittelalterlichen Zahlenbedeutung*, (Münstersche Mittelalter-Schriften, Band 56), Munich: Wilhelm Fink, 1987.

Adam Mez, *The Renaissance of Islam*, transl. by Khuda Bakhsh, London, 1937, from the German original *Die Renaissance des Islams*, Heidelberg, 1922, repr. Hildesheim, *etc.*: Olms, 1968.

Henri Michel, Traité *de l'astrolabe*, Paris: Gauthier-Villars, 1947, repr. Paris: Librarie Alain Brieux, 1989. [See now Morrison.]

Luigi Michelini Tocci, "Federico da Montefeltro e Ottaviano Ubaldini dalla Carda", in Cerboni Baiardi *et al.*, eds., *Federico di Montefeltro*, I, pp. 297-344.

—, *Storia di un mago e di cento castelli*, Pesaro: Cassa di Risparmio di Pesaro, 1986.

—, "Ottaviano Ubaldini della Carda e una inedita testimonianza sulla battaglia di Varna (1444)", in *Mélanges Eugène Tisserant*, VII, Vatican City: Biblioteca Apostolica Vaticana, 1964 (*Studi e Testi* 237), pp. 97-130 and 4 pls.

Middeldorf Festschrift: *Festschrift U. Middeldorf*, A. Kosegarten and P. Tigler, eds., 2 vols. (text and illustrations), Berlin: Walter de Gruyter, 1968.

Miniati: see *Florence 1987-88 Exhibition Catalogue* and *Florence Bargello Catalogue*.

Miniati Festschrift: *MUSA MUSAEI – Studies on Scientific Instruments and Collections in Honour of Mara Miniati*, Marco Beretta, Paolo Galluzzi and Carlo Triarico, eds., Florence: Leo S. Olschki, 2003.

H. Minkoff, "As Simple as ABC: What Acrostics in the Bible Demonstrate", *Bible Review* 13 (1997), pp. 27-34 and 46-47.

Ludwig Mohler, *Kardinal Bessarion als Theologe, Humanist und Staatsmann*, 3 vols., Paderborn: Ferdinand Schöningh, 1923 and 1942, repr. Aalen: Scientia, 1967. [Still the standard work.]

John Monfasani, "Alexinus Celadenus and Ottaviano Ubaldini: an epilogue to Bessarion's relationship with the court of Urbino", *Bibliothèque d'Humanisme et Renaissance* (Geneva) 46 (1984), pp. 95-110, repr. in *idem*, *Byzantine Scholars in Renaissance Italy*, IX.

—, "The Averroism of John Argyropoulos ... ... ", *Essays in the Renaissance* 5 (Florence, 1993), pp. 157-208, repr. in *idem*, *Greeks and Latins in Renaissance Italy*, II.

—, *Byzantine Scholars in Renaissance Italy: Cardinal Bessarion and Other Emigrés: Selected Essays*, Aldershot: Variorum, 1995.

—, *Collectanea Trapezuntiana – Texts, Documents, and Bibliographies of George of Trebizond*, (Medieval & Renaissance Texts & Studies, 25; The Renaissance Society of America: Renaissance Texts Series, 8), Binghamton, N.Y., 1984.

—, *George of Trebizond: A Biography and a Study of his Rhetoric and Logic*, Columbia Studies in the Classical Tradition, 1, Leiden: Brill, 1976.

—, *Greeks and Latins in Renaissance Italy – Studies on Humanism and Philosophy in the 15th Century*, London: Ashgate-Variorum, 2004.

—, "Platina, Capranica, and Perotti: Bessarion's Latin eulogists and his date of birth", publ. Padua, 1986, repr. in *idem*, *Byzantine Scholars in Renaissance Italy*, VI. [Presents evidence that Bessarion was born in 1408 rather than the generally-accepted 1403.]

Marina Montesano, "Pagan gods and the religion of the 'infidels' in the Western imagination and iconography between the Middle Ages and the Renaissance", paper in Italian ("Gli dei pagani e la religione degli "infedeli" ... ") presented at the 20th International Congress of Historical Sciences, Sydney, 2005.

Jim Morrison, *The Astrolabe*, an updated translation of Michel, *Traité de l'astrolabe*, privately distributed, available at www.astrolabes.org .

John Mortimer, *Summer's Lease*, London: Penguin, 1988. [A very English novel about "the Piero trail".]

Muhammad Mrayati, Yahya Mir 'Alam and M. Hassan Tayyan, *New Contributions to the History of Cryptology and Origins of Arab Cryptography and Cryptanalysis*, I: Analysis and Editing of Three Arabic Manuscripts – al-Kindi, Ibn 'Adlan, Ibn ad-Durayhim, unpublished manuscript (*ca.* 1995?), now published as *Series on Arabic Origins of Cryptology*, translated by Said M. al-Asaid, revised by Mohammed I. Al-Suwaidel, Ibrahim H. Al-Kadi and Marwan al-Bawab, vol. 1: Al-Kindi's Treatise on Cryptanalysis, vol. 2: Ibn Adlan's Treatise: A Manual on Cryptanalysis written for King al-Ashraf, vol. 3: Ibn al-Durayhim's Treatise on Cryptanalysis, Riyadh: King Abdulaziz City for Science and Technology and King Faisal Center for Research and Islamic Studies, 2003-04.

Anscari M. Mundó, "Analyse paléographique de l'astrolabe 'carolingien'", in Stevens *et al.*, eds., *The Oldest Latin Astrolabe,* pp. 303-321 and 9 figs.

Eduard de Muralt, *Essai de chronographie byzantine*, 2 vols., St Petersburg, 1855, repr. Paris: Orient Éditions & Amsterdam: Adolf M. Hakkert, 1963.

Peter Murray, "A Note on the Iconography of Piero", in *Middeldorf Festschrift*, pp. 175-179 and pl. XCVI.

N

Der neue Pauly – Enzyklopädie der Antike: see Pauly.

Otto Neugebauer, *Astronomy and History – Selected Essays*, New York, Berlin, Heidelberg, Tokyo: Springer, 1983.

— , "Astronomical and Calendrical Data in the Très Riches Heures", App. C in Millard Meiss, *French Painting in the Time of Jean de Berry – The Limbourgs and their Contemporaries*, Paris: Institut de France, 1974, pp. 421-432 and 480-481, repr. in *idem*, *Astronomy and History*, pp. 507-520.

— , "The Early History of the Astrolabe", *Isis* 40 (1949), pp. 240-256, repr. in *idem*, *Astronomy and History – Selected Essays*, New York, *etc.*: Springer, 1983, pp. 278-294.

— , *Ethiopic Astronomy and Computus*, (Österreichische Akademie der Wissenschaften, Philosophisch-Historische Klasse, Sitzungsberichte, 347. Band, Veröffentlichungen der Kommission für Geschichte der Mathematik, Naturwissenschaften und Medizin, Heft 22), Vienna: Österreichische Akademie der Wissenschaften, 1979.

— , *The Exact Sciences in Antiquity*, 2nd edn., New York: Dover, 1969.

— . "From Assyriology to Renaissance Art", *Proceedings of the American Philosophical Society* 133 (1989), pp. 391-403.

— , *A History of Ancient Mathematical Astronomy*, 3 vols., New York, *etc.*: Springer, 1975.

— , "The Study of Wretched Subjects" *Isis* 42 (1951), p. 111, repr. in *idem*, *Astronomy and History*, p. 3.

— and H. B. van Hoesen, *Greek Horoscopes*, *Memoirs of the American Philosophical Society* 48 (1959).

Neugebauer: see also Swerdlow.

Neulateinisches Jahrbuch – Journal of Neo-Latin Language and Literature, Marc Laureys and Karl August Neuhausen, eds., Hildesheim, Zurich, New York: Georg Olms, 1 (1999) onwards.

J. F. Niermeyer, *Mediae latinitatis lexicon minus – A Medieval Latin – French/English Dictionary*, Leiden: Brill, 1984.

John D. North, *The Ambassadors' Secret – Holbein and the World of the Renaissance*, London & New York: Hambledon & London, 2002.

—, "The Astrolabe", *Scientific American* 230 (1974), pp. 96-106, repr. in *idem*, *Stars, Minds and Fate*, no. 14. [The best overview of the theory and use of the instrument.]

—, "The Astrolabe and the Imagination", *Bulletin of the Scientific Instrument Society* no. 64 (2000), pp. 3-6.

—, *Horoscopes and History*, London: The Warburg Institute, 1986.

—, "Opus quorundam rotarum mirabilium", *Physis* 8 (1966), pp. 337-372, repr. in *idem*, *Stars, Minds and Fate*, no. 11: pp. 135-169.

—, *Richard of Wallingford: An edition of his writings with introductions, English translation and commentary*, 3 vols., Oxford: The Clarendon Press, 1976.

—, "Scrutinizing Astrolabes", *Journal for the History of Astronomy* 35 (2004), pp. 485-487.

—, *Stars, Minds and Fate – Essays in Ancient and Medieval Cosmology*, London & Ronceverte, W.V.: Hambledon, 1989.

—, "The Western Calendar – "Intolerabilis, Horribilis, et Derisibilis". Four Centuries of Discontent", in *Vatican 1992 Conference Proceedings*, pp. 75-113.

Nuremberg 1992-93 Exhibition Catalogue: Gerhard Bott, ed., *Focus Behaim-Globus*, 2 vols., Nuremberg: Germanisches Nationalmuseum, 1992. [See especially II, pp. 512-514 on the 1461 ephemeris from Vienna (not Nuremberg, as stated) featuring angels like the one on the 1462 astrolabe, and pp. 581-589 on the 1462 astrolabe and a related one, possibly by Georg Peuerbach, dated 1457.]

O

Günther Oestmann, *Schicksalsdeutung und Astronomie – Der Himmelsglobus des Johannes Stoeffler von 1493*, Stuttgart: Württembergisches Landesmuseum, 1993.

—, H. Darrel Rutkin and Kocku von Stuckrad, eds., *Horoscopes and Public Spheres – Essays on the History of Astrology*, (Religion and Society, vol. 42), Berlin & New York: Walter de Gruyter, 2005.

C. T. Onions, *The Oxford Dictionary of English Etymology*, Oxford: Clarendon Press, 1966.

Harold Osbourne, *The Oxford Companion to the Decorative Arts*, Oxford & New York: Oxford University Press, 1975, 1991 repr.

The Oxford Classical Dictionary, 3rd revd. edn., Simon Hornblower and Anthony Spawforth, eds., Oxford: Oxford University Press, 2003.

The Oxford Dictionary of Byzantium, prepared at Dumbarton Oaks, Alexander P. Kazhdan, gen. ed., 3 vols., New York & Oxford: Oxford University Press, 1991.

The Oxford Dictionary of the Renaissance, Gordon Campbell, ed., Oxford: University Press, 2003.

P

Paderborn 2001 Exhibition Catalogue: Christoph Stiegemann, ed., *Byzanz – Das Licht des Ostens – Kult und Alltag im Byzantinischen Reich vom 4. bis 15. Jahrhundert*, Mainz: Philipp von Zabern, 2001.

Antonio Panaino, *I Magi evangelici – Storia e simbologia tra Oriente e Occidente*, (Quaderni di Flaminia, 7), Ravenna: Longo, 2004.

Antonio Paolucci, *Piero della Francesca*, Florence: Cantini, 1989.

Paris BN 1973 Exhibition Catalogue: Edmond Pognon and Mireille Pastoureau, eds., *Nicolas Copernic ou La révolution astronomique*, Paris: Bibliothèque Nationale, 1973.

Paris IMA 2000-01 Exhibition Catalogue: *Les Andalousies de Damas à Cordoue*, Éric Delpont *et al.*, eds., Paris: Institut du Monde Arabe, 2000.

Paris IMA 2005 Exhibition Catalogue: *L'âge d'or des sciences arabes*, Ahmed Djebbar, gen. ed., Paris: Institut du Monde Arabe & Actes Sud, 2005.

Paris ML 2004 Exhibition Catalogue: Dominique Thiébaut, with Philippe Lorentz and François-René Martin, *Primitifs français – Découvertes et redécouvertes*, (Musée du Louvre, du 27 février au 17 mai 2004), Paris: Réunion des Musées Nationaux, 2004.

Johann David Passavant, *Rafael von Urbino und sein Vater Giovanni Santi*, 3 vols., Leipzig: F. A. Brockhaus, 1839 (I-II) and 1858 (III); transl. as *Raphael d'Urbin et son père Giovanni Santi*, 2 vols., Paris: Vve Jules Renouard, 1860.

Pauly: *Der neue Pauly – Enzyklopädie der Antike*, 9 vols. to date, Stuttgart & Weimar: J. B. Metzler, 1996 to present.

Pauly-Wissova, *Realencyclopädie*: *Paulys Realencyclopädie der classischen Altertumswissenschaft – Neue Bearbeitung von Georg Wissowa*, numerous volumes and supplements, Stuttgart: Metzler & Druchenmüller, 1893-1974.

Richard Pearse, "The Coptic Pseudo Gospel of Judas (Iscariot)", at: http://www.tertullian.org/rpearse/manuscripts/gospel_of_judas/ (May, 2005).

Olaf Pedersen, "European Astronomy in the Middle Ages", in C. Walker, ed., *Astronomy before the Telescope*, pp. 175-186.

—, *A Survey of the Almagest*, Odense: Odense University Press, 1974.

Pedersen: see also *Vatican 1982 Conference Proceedings*.

Nicholas Penny, "Journey to Arezzo", *London Review of Books* 25:8 (17 April 2003). [An essay review of David Tabbat's 2003 translation of Roberto Longhi, *Piero della Francesca*, surveying the history of British interest in Piero.]

Chiara Pertusi, "La Flagellazione di Piero della Francesca e le fonti letterarie sulla caduta di Constantinopoli", *Quaderni della Rivista di Bizantinistica* (Bologna) 12 (1994), 33 pp.

H. Petz, "Urkundliche Nachrichten über den literarischen Nachlaß Regiomontans und B. Walthers 1478-1522", *Mitteilungen des Vereins für Geschichte der Stadt Nürnberg* 7 (1888), pp. 237-262.

Gerhard Pfohl, ed., *Das Epigramm – Zur Geschichte einer inschriftlichen und literarischen Gattung*, Darmstadt: Wissenschaftliche Buchgesellschaft, 1969.

Piero della Francesca – Libellus de quinque corporibus regularibus (Codice Vaticano Urbinate Latino 632), 3 vols., Cecil Grayson, Marisa Dalai Emiliani, Carlo Maccagni, eds., Florence: Giunti, 1995.

David Pingree, "The Greek Influence on Early Islamic Mathematical Astronomy", *Journal of the American Oriental Society* 93 (1973), pp. 32-43.

—, "A Greek List of Astrolabe Stars", in *Kunitzsch Festschrift*, pp. 474-477.

—, "Indian Influence on Sasanian and Early Islamic Astronomy and Astrology", *The Journal of Oriental Research* (Madras) 34-35 (1964-66/1973), pp. 118-126.

—, "A New Look at [Dürer's] *Melencolia I*", *Journal of the Warburg and Courtauld Institutes* (London) 43 (1980), pp. 257-258.

—, ed., *The Latin Picatrix*, London: The Warburg Institute, 1986.

Pingree Festschrift: *Studies in the History of the Exact Sciences in Honour of David Pingree*, Charles Burnett, Jan P. Hogendijk, Kim Plofker and Michio Yano, eds., Leiden & Boston: Brill, 2004.

Pingree: see also Kennedy & Pingree.

Donal Edgar Pitcher, *An Historical Geography of the Ottoman Empire from earliest times to the end of the sixteenth century*, Leiden: Brill, 1972.

Martin Plessner, "Hermes Trismegistus and Arabic Science", *Studia Islamica* 2 (1954), pp. 45-59.

Paolo dal Poggetto, *La Galleria Nazionale delle Marche e le altre Collezione nel Palazzo Ducale di Urbino*, Urbino: Novamusa del Montefeltro, and Rome: Istituto Poligrafico e Zecca dello Stato, 2003.

Dal Poggetto: see also *Urbino 1992 Exhibition Catalogue*.

Pognon & Pastoureau: see *Paris BN 1973 Exhibition Catalogue*.

Renato Polacco, "La stauroteca del cardinal Bessarione", in *Venice 1994 Exhibition Catalogue*, pp. 368-378.

Maria Louisa Polichetti, *Il Palazzo di Federico da Montefeltro – restauri e ricerci* (1) *rilievi* (2), 2 vols., Urbino: QuattroVenti, 1985.

John Pope-Hennessy, *The Piero della Francesca Trail*, London: Thames and Hudson, 1991.

—, "Whose Flagellation?", *Apollo – The Magazine of the Arts* (London) 124 (Sept. 1986), pp. 162-165.

Emmanuel Poulle, "L'astrolabe", *La Recherche* (Paris) 178 (juin 1986), pp. 756-765.

—, "La fabrication des astrolabes au moyen âge," *Techniques et Civilisations* no. 22 (vol. IV, no. 4) (1955), pp. 117-124.

—, *Les instruments astronomiques du Moyen Âge* [A/B], 1st edn. in *Le Ruban rouge*, no. 32 (mars 1967), pp. 18-29, repr. by the Museum of the History of Science, Oxford: Selected Off-prints, no. 7, 1969 [A]; 2nd edn. with different plates in *Astrolabica* 3, Paris: Société internationale de l'Astrolabe, 1983 [B].

Derek J. de Solla Price, "Fake Antique Scientific Instruments", in *Actes du VIIIe Congrès International d'Histoire des Sciences, Florence, 3-9 sept., 1956*, pp. 380-394 (separatum paginated 1-15).

—, "The First Scientific Instrument of the Renaissance", *Physis – Rivista internazionale di storia della scienza* (Florence) 1 (1959), pp. 26-30. [The first description of the 1462 astrolabe.]

—, "Gears from the Greeks – The Antikythera Mechanism: A Calendar Computer from *ca.* 80 B.C.", *Transactions of The American Philosophical Society*, N.S., 64:7 (1974), repr. New York: Science History Publications, 1975.

—, "Instruments in the British Museum", an essay review of Ward, *Catalogue of Instruments in the British Museum*, in *Journal for the History of Astronomy* 13 (1982), pp. 129-132.

—, "Instruments of Reason. Centerpieces of Renaissance Science have become objets d'art" *The Sciences* (New York Academy of Sciences), October 1981, pp. 15-17.

— , "Portable Sundials in Antiquity, including an account of a new example from Aphrodisias", *Centaurus* 14 (1969), pp. 242-266.

— , with Sharon L. Gibbs and Janice Henderson, *A Computerized Checklist of Astrolabes*, New Haven, Ct.: Yale University, Department of History of Science and Medicine, 1973. [Privately distributed. Still invaluable.]

Princeton 1990 Colloquium Proceedings: *Iconography at the Crossroads*, Brendan Cassidy, ed., Princeton, N.J.: Index of Christian Art, Department of Art and Archaeology, Princeton University, 1993.

Otto Prinz, ed., *Mittellateinisches Wörterbuch bis zum ausgehenden 13. Jahrhundert*, 3 vols. to date, Munich: C. H. Beck, 1967 to present.

Ptolemy, *Almagest*: see Toomer and also Pedersen.

— , *Tetrabiblos*, F. E. Robbins, ed. and transl., (The Loeb Classical Library Series), Cambridge, Ma.: Harvard University Press, & London: William Heinemann, 1940, repr. 1964.

Q

al-Qabîsî, *Introduction to Astrology*: Charles Burnett, Keiji Yamamoto and Michio Yano, *Al-Qabîsî (Alcabitius): The Introduction to Astrology*, (Warburg Institute Studies and Texts 2), London: The Warburg Institute, & Turin: Nino Aragno, 2004.

Bernard Quaritch catalogue no. 1300: *Cryptography – A collection on books and manuscripts*, London: Bernard Quaritch, 2002.

R

F. Jamil Ragep, "'Alî Qushjî and Regiomontanus: Eccentric Transformations and Copernican Revolutions", *Journal for History of Astronomy* 36 (2004), pp. 359-371.

— , "Freeing Astronomy from Philosophy – An Aspect of Islamic Influence on Science", *Osiris* 16 (2001), pp. 49-71.

Diether Roderich Reinsch, "Byzantinisches Herrscherlob für den türkischen Sultan. Ein bisher unbekanntes Gedicht des Georgios Amirutzes auf Mehmed den Eroberer", in *Cupido legum*, Ludwig Burgmann, Marie Theres Fögen, Andreas Schminck, eds., Frankfurt am Main: Löwenklau Gesellschaft, 1985, pp. 195-210.

Giulio Renzi, *Piero della Francesca – Storia, leggenda, profezia, teologia nelle pitture murali della Cappella Maggiore (Basilica di S. Francesco, Arezzo)*, Poggibonsi: Lalli, 1996. [A brief English version is in *idem*, *A Guided Visit of the Basilica of Saint Francis in Arezzo with Historical and Theological Explanations of Piero della Francesca's Frescoes*, Cortona: Colosci, 2004.]

Antonio Rigo, "Bessarione, Giovanni Regiomontano e i loro studi su Tolomeo a Venezia e Roma (1462-1464)", *Studi veneziani*, n.s., 21 (1991), pp. 49-110.

— , "Gli interessi astronomici del cardinal Bessarione", in *Venice 1994 Exhibition Catalogue*, pp. 104-117.

James Robinson, *The Secrets of Judas – The Story of the Misunderstood Disciple and his Lost Gospel*, [San Francisco]: HarperSanFrancisco / HarperCollins, 2006.

Bernd Roeck, *Mörder, Maler und Mäzene – Piero della Francesca's 'Geißelung' – Eine kumsthistorische Kriminalgeschichte*, Munich: C. H. Beck, 2006. [See also the review by Dirk Schümer.]

— and Andreas Tönnesmann, *Die Nase Italiens – Federico da Montefeltro, Herzog von Urbino*, Berlin: Klaus Wagenbach, 2005.

Alfred Rohde, *Die Geschichte der wissenschaftlichen Instrumente*, Leipzig: Klinkhardt & Biermann, 1923.

Silvia Ronchey, "Bessarione poeta e l'ultima corte di Bisanzio", in *Venice 1994 Exhibition Catalogue*, 1994, pp. 47-65.

— , *L'enigma di Piero – L'ultimo bizantino e la crociata fantasma nella rivelazione di un grande quadro*, Milan: Rizzoli, 2006. [The full set of bibiographical notes has been reduced by the publishers but, as indicated *ibid.*, p. 513, it is available at: www.rizzoli.rcslibri.it/enigmadipiero .]

— , "Malatesta/Paleologhi – un'alleanza dinastica per rifondare Bizanzo nel quindicesimo secolo", *Byzantinische Zeitschrift* 93:2 (2000), pp. 521-567.

— , "La 'Mummia' di Mistra. Bessarione, Cleopa Malaltesta e un abito di damasco veneziano", *Thesaurismata* 31 (2001), pp. 75-89.

Paul L. Rose, *The Italian Renaissance of Mathematics – Studies on Humanists and Mathematicians from Petrarch to Galileo*, Geneva: Droz, 1975. [Still a very important work.]

M. A. Rouse and R. H. Rouse, *Authentic Witnesses – Approaches to Medieval Texts and Manuscripts*, Notyre Dame, Ind.: University of Notre Dame Press, 1991.

Joseph Roy (pseudonym), "Sous le signe des étoiles – Du télescope de grand-papa à l'astrolabe du XVe siècle, des prix à décrocher la lune", *L'Exprès* (Paris), 13-19 October, 1989.

Guido Ruggiero, ed., *A Companion to the Worlds of the Renaissance*, Oxford: Blackwell, 2002.

Jean-Pierre de Rycke, ""*Convenerunt in unum*". La 'Flagellation' d'Urbino et l'union des deux églises", *Arte*, 3. serie, 63 (anno LVI – n. 667), sett. 2005, pp. 21-50 and 15 pls.

S

Abdelhamid I. Sabra, "The Appropriation and Subsequent Naturalization of Greek Science in Medieval Islam: A Preliminary Statement", *History of Science* 25 (1987), pp. 223-243, repr. in *idem*, *Optics, Astronomy and Logic – Studies in Arabic Science and Philosophy*, Aldershot (U.K.): Variorum, 1994, I.

George Saliba, *A History of Arabic Astronomy – Planetary Theories during the Golden Age of Islam*, (New York University Studies in Near Eastern Civilization, XIX), New York: New York University Press, 1994.

—, *Islamic Science and the Making of the European Renaissance*, Cambridge, Ma.: MIT Press, 2007.

Saliba: see also Gibbs & Saliba; and *Kennedy Festschrift*.

Walter G. Saltzer, ed., *Zur Einheit der Naturwissenschaften in Geschichte und Gegenwart*, Darmstadt: Wissenschaftliche Buchgesellschaft, 1990.

Saltzer Festschrift: *Der Weg der Wahrheit – Aufsätze zur Einheit der Wissenschaftsgeschichte – Festgabe zum 60. Geburtstag von Walter G. Saltzer*, Peter Eisenhardt, Frank Linhard and Kaisar Petanides, eds., Hildesheim: Georg Olms, 1999.

Friedrich Samhaber, *Der Kaiser und sein Astronom – Friedrich III. und Georg von Peuerbach*, Peuerbach: Stadtgemeinde, 1999.

— , *Die Zeitzither – Georg von Peuerbach und das helle Mittelalter*, Raab (Austria): Wambacher, 2000.

Julio Samsó, *Las ciencias de los antiguos en al-Andalus*, Madrid: MAPFRE, 1992.

— , *Islamic Astronomy and Medieval Spain*, Aldershot & Brookfield, Vt.: Variorum, 1994.

Samsó: see King & Samsó and *Vernet Festschrift*.

Sander: see *Frankfurt 2006 Exhibition Catalogue*.

Sansepolcro 1994 Conference Proceedings: *Luca Pacioli e la matematica del Rinascimento – atti del convegno internazionale di studi – Sansepolcro 13-16 aprile 1994*, Enrico Giusti, ed., Petruzzi, 1998.

Sansepolcro 1994 Exhibition Catalogue: *Luca Pacioli e la matematica del Rinascimento*, Enrico Giusti and Carlo Maccagni, eds., Florenze: Giunti, 1994.

Danièle Sansy, "Chapeau juif ou chapeau pointu? Esquisse d'un signe d'infamie", in *Kühnel Festschrift*, pp. 349-375.

Giorgio de Santillana and Hertha von Dechend, *Hamlet's Mill – An essay on myth and the frame of time*, first published 1969, fifth printing: Boston, Ma.: David R. Godine, 1999.

Sreeramula Rajeswara Sarma, "Astronomical Instruments in Moghal Miniatures", *Studien zur Indologie und Iranistik* 16-17 (1992), pp. 235-276.

Aydın Sayılı, *The Observatory in Islam*. Ankara, 1960, repr. New York: Arno, 1981.

Karl-Heinz Schaldach, "Das Ernst Zinner-Archiv", in von Gotstedter, ed., *Ad radices*, pp. 25-28.

Sara Schechner Genuth, "Astrolabes: A Cross-Cultural and Social Perspective", in Webster & Webster, *Western Astrolabes (in Chicago)*, pp. 2-25.

Hartmann Schedel, *Weltchronik – Faksimile-Druck nach dem Original von 1493 aus der ehemals Reichsstädtischen Bibliothek Lindau*, Lindau: Antiqua Verlag, & Starnberg: Proff, n. d.

Wolfgang Schibel, "CAMENA – Neulateinische Dichtung im World Wide Web", *Neolateinisches Jahrbuch* 3 (2001), pp. 211-219.

Schmidl: see Abu Zayed, King & Schmidl, and Charette & Schmidl.

Harry C. Schnur, *Lateinische Gedichte deutscher Humanisten: lateinisch und deutsch*, Stuttgart: Reclam, 1967.

V. Scholderer, "A Note on the Astrolabe of Regiomontanus", *The Library* (Transactions of the Bibliographical Society, publ. by Oxford University Press), 5th series, 15 (Sept., 1960), p. 209.

P. H. Scholfield, *The Theory of Proportion in Architecture*, Cambridge: Cambridge University Press, 1958.

Günther Schuhmann, "Kardinal Bessarion in Nürnberg", *Jahrbuch für fränkische Landesforschung* 34-35 (1975), pp. 447-465.

Dirk Schümer, review of Bernd Roeck, *Mörder, Maler und Mäzene*: "Das ist jetzt keine abendländisch-antitürkische Ruckrede – Ein kunstgeschichtlicher Kriminalfall gelöst: Bernd Roecks sensationelle Deutung von Piero della Francescas Bild "Geißelung"", *Frankfurter Allgemeine Zeitung*, Feuilleton, October 4, 2006.

Karl Schwarzinger, *Katalog der ortsfesten Sonnenuhren in Österreich*, Vienna: Österreichischer Astronomischer Verein, 1991, and later editions.

Schweinfurt 1993-1994 Exhibition Catalogue: Uwe Müller, ed., *450 Jahre Copernicus* De revolutionibus *– Astronomische und mathematische Bücher aus Schweinfurter Bibliotheken*, Schweinfurt: Stadtarchiv & Dr. Otto Schäfer-Stiftung, 1993, repr. 1998.

Alain P. Segonds, *Jean Philopon: Traité de l'astrolabe*, *Astrolabica* (Paris: Alain Brieux), 2 (1981).

Seipel: see *Linz 1990 Exhibition Catalogue.*

Maurizio Seracini, "Piero della Francesca: *Flagellazione di Cristo*", entry in *Urbino 1992 Exhibition Catalogue*, pp. 452-455. [Annotated images from radiographic analyses.]

Fuat Sezgin, *Geschichte des arabischen Schrifttums*, 12 vols. to date, Leiden: Brill, 1967-1984, thereafter Frankfurt: Institut für Geschichte der arabischen-islamischen Wissenschaften, especially Band III: Medizin, Pharmazie, Zoologie, Tierheilkunde bis *ca.* 430 Hijra (1970); IV: Alchimie, Chemie, Botanik, Agrikultur ... (1971); V: Mathematik ... (1974); VI: Astronomie ... (1978); VII: Astrologie, Meteorologie und Verwandtes ... (1979); and X-XII: Mathematische Geographie und Kartographie ... (2000). [Exhaustive bio-bibliographical surveys until *ca.* 1050.]

— *et al.*, *Klaudios Ptolemaios – Geography – Arabic Translation (1465 A.D.) – Reprint of the Faksimile Edition of the MS Ayasofya 2610*, Frankfurt: Institut für Geschichte der arabischen-islamischen Wissenschaften, 1987.

Michael Shank, "The Classical Tradition in Fifteenth-Century Vienna", in *Tradition, Transmission, Transformation – Proceedings of Two Conferences on Premodern Science Held at the University of Oklahoma,* F. Jamil Ragep and Sally P. Ragep, with Steven J. Livesey, eds., Leiden, New York & Cologne: Brill, 1996, pp. 115-136.

— , "The Cultural Context of Fifteenth-Century Astronomy in Piero della Francesca's 'Flagellation of Christ'", lecture delivered at The Institute of Medieval Studies, University of North Michigan, Kalamazoo, on 25th April, 2000 – see www.unm.edu/~medinst/programs/science%20lectures/optics00.html (Sept., 2005).

— , "Regiomontanus on Ptolemy, Physical Orbs, and Astronomical Fictionalism: Goldsteinian Themes in "Defence of Theon against George of Trebizond", *Perspectives on Science* (MIT Press) 10:2 (2002), pp. 179-207.

Simek: see *Gmunden 2006 Conference Proceedings.*

Kate Simon, *I Gonzaga – Storia e segreti ...* , Rome: Newton & Compton, 2001, repr. 2004, translated from the English original *A Renaissance Tapestry*, 1988.

Laura Simonini and Flaminio Gualdoni, *Carmi figurati greci e latini*, Pollenza (Macerata), 1979.

Robert de La Sizeranne, *Federico di Montefeltro – capitane, principe, mecenate (1422/11482)*, Urbino: Argalla, 1972, 2nd edn., 1979, translated from the original *Le vertueux condottière Federico de Montefeltro*, Paris : Hachette, 1927.

Denis Mack Smith, *Federigo da Montefeltro*, original 1961 English version with an annotated Italian translation and a preface by Sabine Eiche, Urbino: QuattroVenti, 2005.

Jack Spalding, *Santi di Tito*, New York & London: Garland, 1982.

Spitzer Collection Catalogue, A: *La Collection Spitzer – Antiquité, Moyen Âge, Renaissance*, Paris, 1893. [Not consulted. Summarized in B.]

— , B: *La Collection Spitzer – Antiquité, Moyen Âge, Renaissance*, vol. 5, Paris: Maison Quantin and Librarie Centrale, 1893, pp. 81-122 ("Instruments de Mathématiques"). [Summarized from A.]

Burkhard Stautz, "Ein Astrolab aus dem Jahr 1420", *NTM – Internationale Zeitschrift für Geschichte und Ethik der Naturwissenschaften, Technik und Medizin* (Basle: Birkhäuser), N.S., 5:3 (1997), pp. 142-159.

— , *Die Astrolabiensammlungen des Deutschen Museums und des Bayerischen Nationalmuseums*, Munich: Oldenbourg, 1999. [Includes a description of an astrolabe of the Regiomontanus type.]

— , "Die früheste bekannte Formgebung der Astrolabien", in von Gotstedter, ed., *Ad radices*, pp. 315-328. [On the design of the 1062 astrolabe and the earliest surviving Islamic astrolabe.]

— , *Untersuchungen von mathematisch-astronomischen Darstellungen auf mittelalterlichen Instrumenten islamischer und europäischer Herkunft*, Bassum (D): Verlag für Geschichte der Naturwissenschaften und der Technik, 1997. [Includes studies of the star-positions on the 1062 Byzantine astrolabe and the Regiomontanus-type astrolabes.]

Wesley M. Stevens, Guy Beaujouan and Anthony J. Turner, eds., *The Oldest Latin Astrolabe*, (*Physis* 32:2-3), Florence: Leo S. Olschki, 1995.

Stiegemann: see *Paderborn 2001 Exhibition Catalogue.*

[Georg Stolz *et al.*], *Der Englische Gruss des Veit Stoss zu St. Lorenz in Nürnberg*, Munich: Bayerisches Landesamt für Denkmalpflege (Arbeitsheft 16), 1983. [See also Bauer & Stolz.]

E. J. Stormon, "Bessarion before the Council of Florence – A Survey of his Early Writings (1423-1437)", *Byzantine Papers*, Proceedings of the First Australian Byzantine Studies Conference (17-19 May 1978: Australian Association for Byzantine Studies. ByzAustr, 1), E. and M. Jeffery and A. Moffatt, eds., Canberra, 1978, pp. 128-156.

Carole Stott, *Celestial Charts – Antique Maps of the Heavens*, New York: Crescent Books, 1991.

Strasbourg 1987 Exhibition Catalogue: Francis Debauvais, *Catalogue des livres et objets scientifiques anciens exposés à la Bibliothèque Nationale et Universitaire de Strasbourg*, Strasbourg, 1987.

D. J. Struik, "Paulus van Middelburg (1445-1533)", *Mededeelingen van het Nederlandsch Historisch Instituut te Rome* 5 (1925), pp. 79-118.

Kocku von Stuckrad, *Geschichte der Astrologie – Von den Anfängen bis zur Gegenwart*, Munich: C. H. Beck, 2003.

— , *Was ist Esoterik – Kleine Geschichte des geheimen Wissens*, Munich: C. H. Beck, 2004.

al-Sûfî, Abu'l-Husayn 'Abdu'r-Rahmân: see Destombes & Kennedy.

— , *Suwaru'l-kawâkib or Uranometry*, (Arabic text with English introduction), Hyderabad-Deccan: Osmania Oriental Publications Bureau, 1954.

Noel M. Swerdlow, "Astronomy in the Renaissance", in C. Walker., ed., *Astronomy before the Telescope*, pp. 187-230.

— , "The Derivation and First Draft of Copernicus's Planetary Theory – A Translation of the *Commentariolus* with Commentary", *Proceedings of the American Philosophical Society* 117 (1973), pp. 423-512.

— , "Otto E. Neugebauer", *Proceedings of the American Philosophical Society* 137:1 (March 1993), pp. 137-165, with a summary at
www.nap.edu/html/biomens/oneugebauer.html ,
as well as the bibliography compiled by Janet Sachs and Gerald Toomer in *Centaurus* 22 (1979), pp. 257-280.

— , "The Recovery of the Exact Sciences of Antiquity: Mathematics, Astronomy, Geography", in *Washington LC 1993 Exhibition Catalogue*, pp. 125-168. [For a more detailed study see the next entry.]

—, "Regiomontanus on the Critical Problems of Astronomy", in *Nature, Experiment, and the Sciences*, Trevor H. Levere and Wlliam R. Shea, eds., Dordrecht: Kluwer, 1990, pp. 165-195.

— , *The Renaissance of Astronomy in the Age of Humanism. Regiomontanus, Copernicus, Tycho, Kepler, Galileo*, to be published by MIT Press. [Contains a detailed account of the contents of the *Epitome* of the *Almagest* started by Peuerbach and completed by Regiomontanus, as well as a translation of the latter's inaugural lecture at Padua in 1464. Inspected prior to publication, thanks to the generosity of the author.]

Syndram: see *Bielefeld Catalogue*.

T

Antonio Secondo Tessari, "La <<Flagellazione>> di Piero della Francesca: ovvero l'instaurazione del Regno di Cristo", *Arte cristiana* no. 745, vol. 79 (1991), pp. 277-286.

Dominique Thiébaut: see *Paris ML 2004 Catalogue*.

Anne Tihon, *Études d'astronomie byzantine*, Aldershot: Variorum, 1994.

— , "Traités byzantins sur l'astrolabe", in Wesley M. Stevens, Guy Beaujouan and Anthony J. Turner, eds., *The Oldest Latin Astrolabe*, (*Physis – Rivista internazionale di storia della scienza* 32:2-3), Florence: Leo S. Olschki, 1995, pp. 323-357.

— and Raymond Mercier, *Georges Gémiste Pléthon – Manuel d'astronomie*, (Corpus des astronomes byzantins, IX), Louvain-la-Neuve: Academia Bruylant, 1998.

Tocci: see Michelini Tocci.

Tullio Tomba, "Gli astrolabi della Collezione Settala nella Pinacoteca Ambrosiana", *Atti della Fondazione Giorgio Ronchi* (Florence) 33:2 (1978), pp. 3-20, and 33:4 (1978), pp. 3-14.

— and Giuseppe Brusa, *Museo d'arti applicate – Strumenti scientifici - Orologi*, Milan: Electa, 1983.

Walter Tommasoli, *La vita di Federico da Montefeltro 1422/1482*, Urbino: Argalia, 1978. repr. 1995.

Gerald J. Toomer, *Ptolemy's Almagest*, London: Duckworth, 1984, and New York, *etc.*: Springer, 1984.

— , "Ptolemy and his Greek Predecessors", in C. Walker, ed., *Astronomy before the Telescope*, pp. 68-91.

Erich Trapp *et al.*, *Prosopographisches Lexikon der Palaiologenzeit*, 12 fascicles with addenda, Vienna: Österreichische Akademie der Wissenschaften, 1976-1995.

Paolo Trento, *L'astrolabio – storia, funzioni, costruzione*, Rome: Biblioteca del Vascello, 1989.

Angelo Turchini, "Un'ipotesi per la "Flagellazione" di Piero della Francesca", *Quaderni medievali* 14 (1982), pp. 61-93.

Anthony J. Turner, *Early Scientific Instruments – Europe 1400 - 1800*, London: Sotheby's Publications, 1987.

— , "Paris, Amsterdam, London: The Collecting, Trade and Display of Early Scientific Instruments, 1830-1930", in *Leiden MB 1999 Conference Proceedings*, pp. 23-39.

— , *Mathematical Instruments in Antiquity and the Middle Ages – An Introduction*, London: Vade-Mecum Press, 1994.

— , *The Time Museum – Catalogue of the Collection*, Bruce Chandler, gen. ed., vol. 1: *Time Measuring Instruments*, part 1: *Astrolabes – Astrolabe Related Instruments*, Rockford, Ill.: The Time Museum, 1985.

A. Turner: see also *Amsterdam 1990 Exhibition Catalogue*; and Stevens, Beaujouan & Turner.

Gerard L'E. Turner, "A Critique of the Use of the First Point of Aries in Dating Astrolabes", in *Kunitzsch Festschrift*, pp. 548-554.

—, "Derek John de Solla Price – 1922-1983", *Annals of Science* 41 (1984), pp. 105-107.

—, *Elizabethan Instrument Makers – The Origins of the London Trade in Precision Instrument Making*, Oxford: Oxford University Press, 2000.

—, "15th-Century German Astrolabe", in *Christie's South Kensington – Exceptional Scientific and Engineering Works of Art – 8 April 1998*, pp. 70-73 (lot 49). [Detailed description of Gunther, *Astrolabes*, no. 256, which bears an *organum Ptolemaei*.]

—, "The Regiomontanus Astrolabe Dated 1462", catalogue supplement to *Fine Scientific & Philosophical Instruments*, auction at Christie's South Kensington on September 28, 1989, 12 pp. [More detailed than Price (1959), but innocent of the ten related astrolabes from 15th-century Vienna described in King & Turner (1994).]

—, *Renaissance Astrolabes and their Makers*, Aldershot & Burlington, Vt.: Ashgate – Variorum, 2003. [Contains a reprint of King & Turner (1994), A.]

—, "Whitwell's Addition, c. 1595, to a Fourteenth-Century Quadrant", *Antiquities Journal* 85 (1993), pp. 454-455.

—, ed., *Gli strumenti*, vol. 1 of *Storia della scienza*, ed. Paolo Galluzzi, Turin: Giulio Einaudi, 1991.

G. Turner: see also King & Turner and *Madrid FCA 1997 Exhibition Catalogue*.

U

Gert Ueding, ed., *Historisches Wörterbuch der Rhetorik*, 8 vols. to date, Tübingen: Max Niemeyer, 1992 to present.

Urbino 1992 Conference Proceedings: *Città e Corte nell'Italia di Piero della Francesca – Atti del Convegno Internazionale di Studi di Urbino, 4-7 ottobre 1992*, Claudia Cieri Via, ed., Venice: Marsilio, 1996.

Urbino 1992 Exhibition Catalogue: *Piero e Urbino – Piero e le Corti rinascimentali*, Paolo Dal Poggetto, ed., Venice: Marsilio, 1992.

Urbino 2001 Commemorative Volume: Flavio Vetrano, ed., *La scienza del Ducato di Urbino – The Science of the Dukedom of Urbino*, Urbino: Accademia Raffaello, 2001.

Urbino 2005 Exhibition Catalogue: *Il Rinascimento a Urbino – Fra' Carnevale e gli artisti del Palazzo di Federico*, Alessandro Marchi and Maria Rosaria Valazzi, eds., Milan: Skira, 2005.

V

Valazzi: see *Urbino 2005 Exhibition Catalogue*.

van: see under the next part of the name.

Giorgio Vasari, *Le vite de' più eccellenti pittori scultori e architetti*, ? vols., Siena: Pazzini Carli, 1791, and 9 vols., Novara: Istituto Geografico de Agostini, 1967; also *Lives of the Artists*, 2 vols., selections transl. by George Bull and first published in 1965 and 1987, London: Penguin, 1987 edn. and 1987, respectively.

Vatican 1982 Conference Proceedings: *Gregorian Reform of the Calendar – Proceedings of the Vatican Conference to Commemorate its 400th Anniversary 1582-1982*, G. V. Coyne, M. A. Hoskin and O. Perdersen, eds., Vatican City: Pontificia Academia Scientiarum & Specolas Vaticana, 1983.

Venice 1994 Exhibition Catalogue: Gianfranco Fiaccadori, ed., *Bessarione e l'Umanesimo*, Naples: Vivarium, 1994. [Catalogue of an exhibition at the Biblioteca Nazionale Marciana in 1994: see also King & Turner above.]

Venice 1996 Bessarion Exhibition Catalogue: *I Greci in Occidente – La tradizione filosofica, scientifica, lettereria (dalle raccolte della Biblioteca Nazionale Marciana)*, Gianfranco Fiaccadori and Paolo Eleuteri, eds.,Venice: Il Cardo, 1996. [Catalogue of an exhibition at the Biblioteca Nazionale Marciana during 16.10.-07.12.1996.]

Venice 2002 Exhibition Catalogue: *Magia, Alchimia, Scienza dal '400 al '700 – L'influsso di Ermete Trismegisto / Magic, Alchemy, Science 15th-18th Centuries – The Influence of Hermes Trismegistus*, Carlos Gilly and Cis van Heertum, eds., 2 vols., Florence: Centro Di, 2002. [Catalogue of an exhibition at the Biblioteca Nazionale Marciana co-sponsored by the Bibliotheca Philosophica Hermetica, Amsterdam, during 30.5-15.9.2002.]

Vernet Festschrift: *From Baghdad to Barcelona – Studies in the Islamic Exact Sciences in Honour of Prof. Juan Vernet*, Josep Casulleras and Julio Samsó, eds., (Anuari de Filologia (Universitat de Barcelona) XX (1996) B-2), 2 vols., Barcelona: Instituto "Millás Vallicrosa" de Historia de la Ciencia Árabe, 1996.

Flavio Vetrano, *Il Gabinetto di Fisica dell'Università di Urbino: la sua Storia, il suo Museo / The old "Gabinetto di Fisica" of the Urbino University: its History, its Museum*, Rome: Istituto Poligrafico e Zecca dello Stato, 1996.

— , ed., *La scienza del Ducato di Urbino – The Science of the Dukedom of Urbino*, Urbino: Accademia Raffaello, 2001.

Robert Vincent, *Géometrie du nombre d'or*, Marseilles: Chalagam, 1999, 3rd edn, 2001, transl. as *Geometry of the Golden Section*, Marseilles: Chalagam, 2003.

Ernst Vogt, "Das Akrostichon in der griechischen Literatur", *Antike und Abendland – Beiträge zum Verständnis der Griechen und Römer und ihres Nachlebens* 13 (1967), pp. 80-95.

von: see under the next part of the name.

W

Jeanne van Waadenoijen, "La 'Flagellazione' di Piero della Francesca", *Arte Cristiana* 81, no. 756 (1993), pp. 183-198.

Christopher Walker, ed., *Astronomy before the Telescope*, London: British Museum Press, 1996.

D. P. Walker, "Esoteric Symbolism", in G. M. Kirkwood, ed., *Poetry and Poetics from Ancient Greece to the Renaissance – Studies in Honor of James Hutton*, Ithaca, N.Y.: Cornell University Press, 1975, pp. 218-232, repr. in *idem*, *Music, Spirit, and Language in the Renaissance*, XV.

—, *Music, Spirit, and Language in the Renaissance*, London: Variorum Reprints, 1985.

— , *Spiritual and Demonic Magic from Ficino to Campanella*, London: The Warburg Institute, 1958, repr. Nendeln (FL): Kraus, 1976, and again Philadelphia: University of Pennsylvania Press, 2000.

F. A. B. Ward, *A Catalogue of European Scientific Instruments in the Department of Medieval and Later Antiquities of the British Museum*, London: British Museum Publications, 1981. [See also Price, "Instruments in the British Museum".]

— , "The Earliest String-Gnomon Sundials", *Antiquarian Horology*, (June, 1976), pp. 800-801.

Washington LC 1993 Exhibition Catalogue: *Rome Reborn – The Vatican Library and Renaissance Culture*, Anthony Grafton, ed., Washington, D.C.: Library of Congress, 1993

Washington NGA 1992 Exhibition Catalogue: *Circa 1492 – Art in the Age of Exploration*, Jay A. Levenson, ed., Washington, D.C.: National Gallery of Art, and New Haven, Ct., & London: Yale University Press, 1991.

Diedrich Wattenberg, "Johannes Regiomontanus und die astronomischen Instrumente seiner Zeit", in Hamann, ed., *Regiomontan-Studien*, pp. 343-362.

Roderick and Marjorie Webster, *Western Astrolabes*, (vol. I of *Historical Scientific Instruments of the Adler Planetarium & Astronomy Museum*, Bruce Chandler, gen. ed., and Sara Schechner Genuth, ed.), Chicago, Ill.: Adler Planetarium & Astronomy Museum, 1998.

Kurt Weitzmann *et al.*, *The Icon*, New York: Alfred A. Knopf, 1982.

Klaus Wessel (and later also Marcell Restle), eds., *Reallexikon zur byzantinischen Kunst*, 6 vols. to date, Stuttgart: Anton Hiersemann, 1966 to present.

W. Weyh, "Die Akrostichis in der byzantinischen Kanonesdichtung", *Byzantinische Zeitschrift* 17 (1908), pp. 1-69.

W. H. White, "Chronograms", *The Library* (London), 4th series, vol. 4 (*Transactions of the Bibliographical Society*, 2nd series, vol. 4), 1924, pp. 59-74. [Based on the 3-volume 1882-95 work *Chronograms* by James Hilton.]

Peter Whitfield, *The Mapping of the Heavens*, San Francisco: Pomegranate Artbooks (in association with The British Library), 1995.

Joachim Wiesenbach, "Wilhelm von Hirsau. Astrolab und Astronomie im 11. Jahrhundert", in Klaus Schreiner, ed., *Hirsau St. Peter und Paul 1091-1991*, II: *Geschichte, Lebens- und Verfassungsformen eines Reformklosters*, Stuttgart: Konrad Theiss, 1991, pp. 109-156.

U. von Wilamowitz-Moellendorff, "Die griechischen Technopägnien", *Jahrbuch des Archäologischen Instituts des Deutschen Reiches* 14 (1899), pp. 51-59.

Wilhelmy: see *Mainz 2000 Exhibition Catalogue* and *Mainz 2006 Exhibition Catalogue*.

Emanuel Winternitz, "Quattrocento Science in the Gubbio Study", in *Bulletin of the Metropolitan Museum of Art*, Oct., 1942, pp. 104-116.

Rudolf Wittkower, "The Changing Concept of Proportion", *Dædalus – Journal of the American Academy of Arts and Sciences* (Winter 1960), pp. 199-215.

— and B. A. R. Carter, "The Perspective of Piero della Francesca's 'Flagellation'", *Journal of the Warburg and Courtauld Institutes* (London) 16 (1953), pp. 292-302.

Wolfenbüttel HAB 1987-1988 Exhibition Catalogue: Jeremy Adler and Ulrich Ernst, *Text als Figur – Visuelle Poesie von der Antike bis zur Moderne*, (Austellungskataloge der Herzog August Bibliothek, 56), 3rd edn., Weinheim: VCH, 1990.

Jeryldene M. Wood, ed., *The Cambridge Companion to Piero della Francesca*, Cambridge: Cambridge University Press, 2002.

Wussing Festschrift: *Amphora – Festschrift für Hans Wussing zum 65. Geburtstag*, Sergei S. Demidov, Menso Folkerts, David A. Rowe and Christoph J. Scriba, eds., Basel, Boston, Mass. & Berlin: Birkhäuser, 1992.

Wright: see Field & Wright.

Z

Maria Teresa Zanobini Leoni, "*La Flagellazione* di Urbino di Piero dei Franceschi, *Sound Sonda – Rivista di studi elettronici : Art processes and visual objects computer analysis* 3-4 (31 Mai 1978), pp. 92-96.

John W. Zarker, "Acrostic 'Carmina Latina Epigraphica'", *Orpheus* 13 (1966), pp. 125-151.

Ernst Zinner, *Alte Sonnenuhren an europäischen Gebäuden*, Wiesbaden: Franz Steiner, 1964.

— , *Deutsche und niederländische astronomische Instrumente des 11.-18. Jahrhunderts*, 2nd edn., Munich: C. H. Beck, 1967, repr. 1972.

— , "Einige Handschriften des Johann Regiomontanus", *Berichte des Historischen Vereins für die Pflege der Geschichte des ehemaligen Fürstbistums Bamberg* 100 (1964), pp. 315-323.

— , *Leben und Wirken des Joh. Müller von Königsberg genannt Regiomontanus*, 2nd edn., Osnabrück: Otto Zeller, 1968, transl. by Ezra Brown as *Regiomontanus: His Life and Work*, Amsterdam: North-Holland, 1990. [References are given to the German version.]

— , "Das mittelalterliche Lehrgerät für Sternkunde zu Regensburg und seine Beziehung zu Wilhelm von Hirsau", *Zeitschrift für Instrumentenkunde* 43 (1923), pp. 278-282.

Hans Zinsmeister, "Der Briefwechsel Regiomontans mit Giovanni Bianchini, Jacob von Speyer und Christian Roder", translated into German from the 1902 Latin edn. of M. Curtze, unpublished typescript, 53 pp. [Original in the Institute for the History of Science, Frankfurt, courtesy of the author.]

Marino Zorzi, "Cenni sulla vita e sulla figura di Bessarione", in *Venice 1994 Bessarion Exhibition Catalogue*, pp. 1-19.

— , *La Libreria di San Marco – Libri, lettori, società nella Venezia dei Dogi*, Milan: Arnaldo Mondadori, 1987.

Stefano Zuffi, *Il Quattrocento*, Milan: Electa Mondadori, 2004, with translations *L'art au XVe siècle*, Paris: Hazan, 2005, and *European Art in the Fifteenth Century*, Los Angeles: J. P. Getty Foundation, 2005.

— , *Il Rinascimento*, Milan: Electa Mondadori, 2002, with a French translation *La Renaissance*, Paris: Gründ, 2003.

Paul Zumthor, *Langue, Texte, Énigme*, Paris: Seuil, 1975.

Ronald Edward Zupko, *Italian Weights and Measures from the Middle Ages to the Nineteenth Century*, Philadelphia, Pa.: The American Philosophical Society, 1981.

Websites
(all addresses valid in May, 2005)

— www.bresciascienza.it/cityline/cult/zani/astrolabio.gif

on the 1062 astrolabe.

— www8.informatik.uni-erlangen.de/IMMD8/Services/Astrolab/html/geschichte.html

on the 1462 astrolabe.

— www.telemachos.hu-berlin.de/esterni/akrostichon.html

for a bibliography on Greek and Latin acrostics.

— www.himmelsboten.de/Engel/Gebete/Triodion.htm

on angels in the Byzantine liturgy.

— www.kcl.ac.uk/cch/pbe

for names on Byzantine seals and statues.

— www.fordham.edu/halsall/ikon/greekabb.html

for Halsall, "Greek Abbreviations".

— www.dartmouth.edu/~matc/math5.geometry/unit13/unit13.html
on the geometry of Piero's painting.

— www.artemis.jussieu.fr/hermes/hermes/actes/ac0102/flagellation.gif
on the basic geometry of Piero's painting.

— http://research.microsoft.com/vision/cambridge/3d/svm_movies/flagella_camera.mpeg and www.robits.ox.ac/uk/~criminisi/-
for animated views of the 3-dimensional scenario of Piero's painting.

— www.mcs.surrey.ac.uk/Personal/R.Knott/Fibonacci/fib.html
on the "golden number" 1.618... = 1 / 0.618... .

— ourworld.compuserve.com/homepages/mpaoli/
on Alberti.

— www.uniurb.it/Uborse/worken.htm
on Urbino in the 15th century.

— http://www.uniurb.it/PhysLab/Rinasci.html
on science in 15th-century Urbino.

— www.mhs.ox.ac.uk/epact
"Epact: Scientific Instruments of Medieval and Renaissance Europe", including descriptions of individual instruments in the major collections.

— www.uni-frankfurt.de/fb13/ign/instrument-catalogue.html
on astronomical instruments to *ca.* 1500.

— www.uni-frankfurt.de/fb13/ign/astronomy_in_baghdad/bibliography.html
on the Renaissance of astronomy in Baghdad in the 9th and 10th centuries.

— www.uni-frankfurt.de/fb13/ign/
on the Frankfurt Institute.

— http://web.uni-frankfurt.de/fb13/ign/Code.htm
for a 50-page introduction to the topic of this book and a silent lecture of over 200 pdf images and explanatory text.

Acknowledgements:~

I include here first of all a brief statement concerning the original discoveries in order to pay due credit to Berthold Holzschuh.

The fact that the Latin dedication on Regiomontanus' 1462 astrolabe is an acrostic with hidden references to the 1062 Byzantine astrolabe and the Greek poem on it was noted by Berthold Holzschuh in February 2005, in preparation for a presentation in the weekly Seminar on medieval astronomical instruments at Frankfurt. As luck would have it, he had already made a seminar presentation on the Greek inscription on the 1062 astrolabe in 2003.

Again, as luck would have it, in February 2005, I was involved in the final proofreading of a new book on astronomical instruments, mainly Islamic, but also medieval and Renaissance European. In this, I had presented new information on both the 1062 and 1462 astrolabes and on some newly-identified instruments from 15th-century Vienna (bringing the number now known to about 45). A brief and jumbled announcement of Berthold Holzschuh's discovery of the acrostic was inserted in that book.

The connection between the geometry of the 1462 epigram and that of Piero's painting was announced in the Seminar by Berthold Holzschuh in May 2005. He had found on the Internet a reference to Bessarion as a possible candidate for the bearded man. He knew the painting well since, many years before, he had had a copy of it hanging in his study. On seeing the painting again, he realized that it had two principal axes related to those of the epigram. He did not resist the temptation to lay an appropriately enlarged folio of the epigram over a copy of the painting.

The key to the solution to the enigma of Piero's painting had been found. Berthold Holzschuh made a presentation of his findings at the conference "Mandarini Bizantini – Il mondo intellettuale tra Oriente e Occidente (XIV-XV secoli)" held at the Fondazione Giorgio Cini in Venice in June, 2005. I don't think anybody believed him. Where to go from there?

The weekly seminars of the Summer Semester 2005 and Winter Semester 2005-06 were devoted to the manifold consequences of these two remarkable and fortuitous discoveries, and during this time the present text evolved. This text was written entirely by this author, who alone bears responsibility for its contents. Sections of the text, particularly the appendixes, were available from previous publications or earlier descriptions of astrolabes. Between October 2005 and May 2006 many different versions were circulated either in part or *in toto* within the framework of the Seminar, under the guise of the Preprint Series of the Institute.

My greatest debt is to Berthold Holzschuh. He discovered that the words of the epigram can be rearranged to convey a quite different meaning. He discovered that the epigram is an acrostic with several vertical axes, with hidden references to the 1062 Byzantine astrolabe. He then discovered that the geometry of the epigram and the geometry of the painting were related, and realized that, as a consequence, the two principal figures could be Bessarion and Regiomontanus.

Berthold Holzschuh's own initial findings – outlined in a 10-page set of notes prepared in September, 2005 – are alas not yet documented in publishable form. I wish that I could have persuaded him to write about the way in which he made these discoveries. Nevertheless, to Berthold Holzschuh goes *all* of the credit for the original inspiration for the present work and for many insights thereafter, but none of the responsibility for what I have written in it.

It was Robert Dietz, a 25-year-old Physics student and another member of my seminar, who recognized Sergios in Regiomontanus' epigram and who solved the mystery of the transposed labels on his solar and calendrical scales.

I extend my sincere gratitude to the other members of the Frankfurt seminar besides Berthold Holzschuh and Robert Dietz, namely, Dr. Helga Dittberner, Nadine Faerber, Siegfried Müller, Florian Roesch, Sven Ruhberg, Dr. Petra Schmidl, Christian Schwarz, and Dr. Alexander and Susanne Walland, for their scepticism, patience and encouragement as I attempted, showing them specific extracts from one version after another, to assess Berthold Holzschuh's discoveries and to come to terms with their manifold consequences. By the time I had finished, I think only a couple of them believed anything I was saying. In May 2006, we went back to talking about some other astrolabes.

*****

To my wife Patricia Cannavaro my heartfelt thanks for braving the exigencies of another new book 10 years after I thought I had put Regiomontanus and Bessarion behind us and so soon after the 2000 pages of *In Synchrony with the Heavens* had been published in June 2005. Patricia was there from the outset, and throughout, not always knowing where it all would lead, least of all, in June 2006, back to Urbino, where we had spent two months in the summer of 1980.

My close friend Professor James Bradburne guided me into the language, written and pictorial, and the intellectual games of the Renaissance, and followed the whole venture with unflagging enthusiasm and critical oversight. He introduced me to D. P. Walker and Martin Kemp, Giuseppe Centauro and Enzo Settesoldi, and many other authors, as well as to the subject of Piero's commissions, the earthquake of 1741, and much else:~ he deserves my very special thanks.

With another close friend, Kurt Maier, I have shared many experiences in Frankfurt for almost 20 years, especially the excitement of discoveries of one sort or another, many of which were made by him. With his critical eye he kindly read three versions of this essay, the last in March, 2007, identifying all manner of errors of orthography and style:~ he also merits my special thanks.

Several other people read parts of the manuscript in various stages or summaries thereof and also provided helpful comments. I shall not name them but they know who they are. Leon Battista Alberti at the end of his prologue to his *Della pittura* (1435-36)* had the perfect words for all these good people, who were aware of my own rather desperate need for feedback, positive or negative:

> "Piacciati adunque leggermi con diligenza, e se cosa vi ti par da emendarla, correggimi. Niuno scrittore mai fu sì dotto al quale non fussero utilissimi gli amici eruditi; e io in prima de te desidero essere emendato per non essere morso da' detrattori."

> "May it please you to read me with diligence; and if something appears to you that could be emended, correct me. No author was ever so learned that his erudite friends could not be of great use to him; and I would wish for myself above all to be emended by you, so that I am not bitten by detractors."

All of the remaining errors and misinterpretations are the responsibility of nobody other than myself. Specialists on Byzantium and Renaissance Italy and Quattrocento art will often be reminded that I have no expertise in these fields.

The information on various instruments, particularly those from the 15th-century Vienna school, was collected in the late 1980s and early 1990s. This was part of a larger long-term project to catalogue medieval instruments to *ca.* 1500. The support of the German Research Foundation is gratefully acknowledged (see further n. 410).

To our Institute's librarian, Ryszard Dyga, my thanks for help in obtaining copies of obscure publications. To our Institute's CEO, Wolf-Dieter Wagner, my thanks for keeping the Institute office going, sometimes single-handedly, whilst I was deeply involved with this new theme.

My thanks also to MacOrg of Frankfurt, who stepped in when both my Mac computers expired on the same day in January 2006, taking with them this text and many others and also paralyzing the backup disc. This all happened when I was trying to format the two squares of Alighiero Boetti's "Ordine e disordine" side by side. MacOrg duly rescued all the data and got me off to a fresh start with a new computer and new software.

To Dr Lorenza Mochi Onori and Dr Agnese Vastano of the Soprintendenza per il Patrimonio Storico, Artistico ed Etnoantropologico delle Marche in Urbino my thanks for providing superb digital images of the "Flagellation" and the

---

* Alberti, *Della pittura*, (Bätschmann & Gianfreda, eds.), pp. 64-65.

Byzantine angel, and for permission to publish these here. To Professor Roberto Mantovani, my thanks for images of the Urbino intarsia.

To Jeremy Collins, formerly of Christie's, London, my thanks for images of Regiomontanus' astrolabe for Bessarion. To its present owner, my thanks for promising to make the piece available for international exhibitions.

Other images were kindly provided by the Benaki Museum, in Athens, the former Museo dell'Età Cristiana in Brescia, the Museo di Storia della Scenza in Florence, the Germanisches Nationalmuseum in Nuremberg, and the Institut du Monde Arabe in Paris.

The announcement of the discoveries

As noted above, in June 2005, Berthold Holzschuh announced his two discoveries at the conference "Mandarini Bizantini – Il mondo intellettuale tra Oriente e Occidente (XIV-XV secoli)" held at the Fondazione Giorgio Cini in Venice. I was alas unable to attend, but I nevertheless extend my gratitude to Prof. Antonio Rigo for making it possible for Berthold Holzschuh to attend.

On May 29, 2006, I addressed a small group of colleagues in the history of science at a Memorial Seminar in honour of the late David Pingree held at the International Institute of Asian Studies at Leiden University. The Seminar was entitled "Empires and the Exact Sciences in Pre-Modern Eurasia". My paper was entitled "The End of the Byzantine Empire and the First Scientific Instrument of the Renaissance: from Regiomontanus' Acrostic to Piero della Francesca's "Flagellation of Christ"". My gratitude to Dr. Kim Plofker and Prof. Jan Hogendijk for making this possible.

In May 2006, Italy was still reeling over the final resolution of the enigma of Piero's painting in Silvia Ronchey's new book *L'Enigma di Piero*, which was widely reported in the Italian press.

On June 14, 2006, I addressed a small audience at the University of Urbino, in a lecture sponsored by the Department of Physics of the University and the Accademia Raffaello. My title was "The Epigram that inspired Piero's *Flagellation*". The seeds were sown, and I was ready to complete a brief introduction to the subject on the Internet page:

http://web.uni-frankfurt.de/fb13/ign/Code.htm .

My lecture in Urbino was organized by Prof. Roberto Mantovani of the Department of Physics at the University of Urbino and Dr. Guiliano Donnini of the Accademia Raffaelo. Duccio Alessandri and Barti Roopanand guided us back to the Renaissance in Rimini, Monterchi and Arezzo. Dr. Lia Navarra Baldesi, President of the Fondazione Piero della Francesca in San Sepolcro, kindly welcomed us and graciously suffered a 20-minute lecture in broken Italian on the discoveries. There was a full page coverage *in Il Resto del Carlino*, June 10 ("Il mistero della Flagellazione raccontato in un astrolabio") and a shorter article in *Corriere Adriatico*, June 15 ("*Flagellazione*, svelati i segreti").

On October 10, 2006, I presented a paper entitled "The Geometry of Piero's *Flagellation* and the Geometry of the Acrostic of Regiomontanus that Inspired it", at the conference "L'arte della matematica nello studio della prospettiva" in Rome and Urbino. I am grateful to Prof. Gian Italo Bischi of the University of Urbino for arranging this.

A few days earlier, the German-speaking world was alerted to a new book by Bernd Roeck. A review by Dirk Schümer in *Frankfurter Allgemeine Zeitung*, 4th October 2006) hailed the author's findings as "a sensation", adding that "the most puzzling, the most controversial, the previously unexplained Jack the Ripper case of art history is solved".

On November 16 I presented a paper "From Regiomontanus' Acrostic to Piero della Francesca's "Flagellation of Christ" at the Humboldt University in Berlin. This was on the occasion of a conference entitled "What Counts. The presence and medial function of numbers in the Middle Ages", and it was organized by Dr. Moritz Wedell. My thanks to him for this invitation.

From April 2007 onwards, matters were no longer in our hands. An article appeared in *Nature* (London) on March 29th. It was written by the news editor, Jo Marchant, and it focussed on the controversial aspects of the discoveries rather than the discoveries themselves. It gave undue coverage to one out of some 40 previous hypotheses, first proposed in 1968, of which the proponent claimed that it was the "most generally accepted" in spite of the two dozen different hypotheses that have been suggested during the past 40 years. Various authorities on this and that were cited and ventured cautious comments one way or the other. Some gave the impression of being totally unaware of the most basic information about my hypotheses. Two scholars concurred: "it seems arbitrary to divide the painting vertically at the eyes of Christ and of the bearded figure, and to split the poem at the B and the I". Marchant correctly observed: "King ... believes that some art historians will dismiss his work because they can't understand it," and she also put on record that: "some art historians have dismissed the idea entirely, describing it to *Nature* as "nonsensical", "embarrassing" and "utter rubbish"". The same kind of language was used in England in 1989-1990 to counter claims that an item auctioned at Christie's in 1989 was an astrolabe made by Regiomontanus for Cardinal Bessarion.

Notes on Berthold Holzschuh and David King

Berthold Holzschuh, born in a small village in Franconia in 1945, studied Latin and Greek at the Humanistisches Gymnasium in Eichstätt in Bavaria. He manages his family wood-construction company in Aschaffenburg, and he is passionately interested in old books on astronomy. Since 1995 he has been a regular participant in the Seminar on Medieval Astronomical Instruments in Frankfurt. In 2003 he prepared a seminar presentation on the Greek inscription on the 1062 astrolabe, and in February 2005, he prepared one on the Latin inscription on the 1462 astrolabe and discovered the acrostic. Having established the connection between the 1462 and 1062 astrolabes, he discovered in May 2005, that the geometry of Regiomontanus' epigram was intimately related to that of Piero's "Flagellation of Christ". He announced these discoveries at the conference "Mandarini Bizantini – Il mondo intellettuale tra Oriente e Occidente (XIV-XV secoli)" held at the Fondazione Giorgio Cini in Venice in June 2005.

David King, born in England in 1941, studied Mathematics at Cambridge University, Education at Oxford, and Near Eastern Languages and Literatures at Yale. During 1972-79 he directed a project in the history of Islamic astronomy at the American Research Center in Egypt, and catalogued the 2,500 medieval Arabic astronomical and mathematical manuscripts in the Egyptian National Library. During 1979-85 he was Professor of Arabic at New York University. From 1985 until his retirement in 2007 he was Professor of the History of Science at the Johann Wolfgang Goethe University in Frankfurt. He is a specialist in the history of astronomy within the context of Islamic civilization, especially the application of astronomy and mathematical geography to the needs of religious ritual. He has also published on medieval European astronomical instruments; a monastic cipher-notation that was used both to denote numbers and as a letter code; and on the cult of a very remarkable medieval saint who is much misunderstood in the modern literature. In the early 1990s he published Regiomontanus' 1462 astrolabe in the context of over 40 other instruments from the 15th-century Vienna workshop and its offshoots, including 10 other astrolabes with the same distinctive design. In 2001 he published the first analysis of the astrolabe depicted in the intarsia of Federigo's *studiolo* in Urbino, comparing it with an astrolabe made in Urbino in 1462.

INDEX

Notes:~ References are to pages, footnotes (number of page and number of note separated by a slash) and illustrations (as in text). No attempt has been made to index the 45-odd instruments from 15th-century Vienna listed in Appendix D; only those most relevant to this study are to be found below. Medieval and Renaissance names are included in the form in which they are best known: thus, for example, Piero della Francesca and Jakob von Speyer but Alberti, Leon Battista, and Berselius, Paschasius.

> "If (King's ideas are) accepted even as a plausible hypothesis, then it says that scientific instruments can legitimately be treated as historical documents, in the same way as paintings themselves have been. Scientific objects can be considered as part of the puzzle." James Bradburne, quoted in Jo Marchant, "A Leap of Faith" (2007), p. 491.

A

B

N

O

P